COMPUTATIONAL MODELING OF HOMOGENEOUS CATALYSIS

Catalysis by Metal Complexes

Volume 25

Editors:

Brian James, *University of British Columbia, Vancouver, Canada*
Piet W. N. M. van Leeuwen, *University of Amsterdam, The Netherlands*

Advisory Board:

The titles published in this series are listed at the end of this volume.

COMPUTATIONAL MODELING OF HOMOGENEOUS CATALYSIS

edited by

Feliu Maseras
Unitat de Química Física,
Universitat Autònoma de Barcelona,
Bellaterra, Catalonia, Spain

and

Agustí Lledós
Unitat de Química Física,
Universitat Autònoma de Barcelona,
Bellaterra, Catalonia, Spain

KLUWER ACADEMIC PUBLISHERS
DORDRECHT / BOSTON / LONDON

A C.I.P. Catalogue record for this book is available from the Library of Congress.

ISBN 1-4020-0933-X

Published by Kluwer Academic Publishers,
P.O. Box 17, 3300 AA Dordrecht, The Netherlands.

Sold and distributed in North, Central and South America
by Kluwer Academic Publishers,
101 Philip Drive, Norwell, MA 02061, U.S.A.

In all other countries, sold and distributed
by Kluwer Academic Publishers,
P.O. Box 322, 3300 AH Dordrecht, The Netherlands.

Printed on acid-free paper

Contents

Contributors

Harold Basch. Department of Chemistry, Bar-Ilan University, Ramat-Gan, 52900, Israel.

Carles Bo. Departament de Química Física i Inorgànica, Universitat Rovira i Virgili, Pl.Imperial Tarraco, 1, 43005 Tarragona, Spain.

Luigi Cavallo. Dipartimento di Chimica, Università di Salerno, Via Salvador Allende, I-84081, Baronissi (SA) Italy.

Jorge J. Carbó. Unitat de Química Física, Edifi C.n, Universitat Autònoma de Barcelona, 08193 Bellaterra, Barcelona, Spain.

Eric Clot. LSDSMS (UMR 5636), Case courrier 14, Université Montpellier II, Place Eugène Bataillon, 34095 Montpellier Cedex 5, France.

Miquel Duran. Institut de Química Computacional and Departament de Química, Universitat de Girona, E-17071 Girona, Catalonia, Spain.

Odile Eisenstein. LSDSMS (UMR 5636), Case courrier 14, Université Montpellier II, Place Eugène Bataillon, 34095 Montpellier Cedex 5, France.

Steven Feldgus. Department of Chemistry, University of Wisconsin-Madison, 1101 University Avenue, Madison, WI 53706, USA

Xin Huang. Deparment of Chemistry, The Hong Kong University of Science and Technology, Clear Water Bay, Kowloon, Hong Kong, People's Republic of China.

Clark R. Landis. Department of Chemistry, University of Wisconsin-Madison, 1101 University Avenue, Madison, WI 53706, USA

Zhengyang Lin. Deparment of Chemistry, The Hong Kong University of Science and Technology, Clear Water Bay, Kowloon, Hong Kong, People's Republic of China.

Agustí Lledós. Unitat de Química Física, Edifici Cn, Universitat Autònoma de Barcelona, 08193 Bellaterra, Catalonia, Spain.

Alessandra Magistrato. Laboratory of Inorganic Chemistry, Swiss Federal Institute of Technology, ETH Zentrum, CH-8092 Zürich, Switzerland.

Feliu Maseras. Unitat de Química Física, Edifici Cn, Universitat Autònoma de Barcelona, 08193 Bellaterra, Catalonia, Spain.

Artur Michalak. Department of Theoretical Chemistry, Faculty of Chemistry, Jagiellonian University, R. Ingardena 3, 30-060 Cracow, Poland.

Keiji Morokuma. Cherry L. Emerson Center for Scientific Computation, and Department of Chemistry,Emory University, Atlanta, Georgia, 30322, USA.

Djamaladdin G. Musaev. Cherry L. Emerson Center for Scientific Computation, and Department of Chemistry,Emory University, Atlanta, Georgia, 30322, USA.

Yasuo Musashi. Information Processing Center, Kumamoto University, Kurokami 2-39-1, Kumamoto 860-8555 Japan.

Notker Rösch. Institut für Physikalische und Theoretische Chemie, Technische Universität München, 85747 Garching, Germany.

Ursula Röthlisberger. Laboratory of Inorganic Chemistry, Swiss Federal Institute of Technology, ETH Zentrum, CH-8092 Zürich, Switzerland.

Shigeyoshi Sakaki. Department of Molecular Engineering, Graduate School of Engineering, Kyoto University, Sakyo-ku, Kyoto 606-8501, Japan.

Miquel Solà. Institut de Química Computacional and Departament de Química, Universitat de Girona, E-17071 Girona, Catalonia, Spain.

Thomas Strassner. Institut für Anorganische Chemie, Technische Universität München, Lichtenbergstrasse 4, D-85747 Garching, Germany.

Antonio Togni. Laboratory of Inorganic Chemistry, Swiss Federal Institute of Technology, ETH Zentrum, CH-8092 Zürich, Switzerland.

Maricel Torrent. Medicinal Chemistry Dept., Merck Research Laboratories, Merck & Co., West Point PA, USA.

Cristiana Di Valentin. Dipartimento di Scienza dei Materiali, Università degli Studi di Milano-Bicocca, via Cozzi 53, 20125 Milano, Italy.

Tom K. Woo. Department of Chemistry, The University of Western Ontario, London, Ontario, Canada, N6A 5B7.

Ilya V. Yudanov Boreskov Institute of Catalysis, Siberian Branch of the Russian Academy of Sciences, 630090 Novosibirsk, Russia

Tom Ziegler. Department of Chemistry, University of Calgary, University Drive 2500, Calgary, Alberta, Canada T2N 1N4.

Acknowledgements

We must first thank all the contributors for their positive response to our call, their dedicated effort and the timely submission of their chapters. We must also thank the people at Kluwer Academic Publishers, especially Jan Willem Wijnen and Emma Roberts for their support and patience with our not always fully justified delays.

We want to finally thank all the graduate students in our group during these last years: Gregori Ujaque, Lourdes Cucurull-Sánchez, Guada Barea, Jorge J. Carbó, Jaume Tomàs, Jean-Didier Maréchal, Nicole Dölker, Maria Besora, Galí Drudis and David Balcells. For creating the environment where the research, even the compilation of edited volumes, is a more enjoyable task.

Preface

This book presents an updated account on the status of the computational modeling of homogeneous catalysis at the beginning of the 21st century. The development of new methods and the increase of computer power have opened up enormously the reliability of the calculations in this field, and a number of research groups from around the world have seized this opportunity to expand enormously the range of applications. This text collects a good part of their work.

The volume is organized in thirteen chapters. The first of them makes a brief overview of the computational methods available for this field of chemistry, and each of the other twelve chapters reviews the application of computational modeling to a particular catalytic process. Their authors are leading researchers in the field, and because of this, they give the reader a first hand knowledge on the state of the art.

Much of the material has been certainly published before in scientific journals, but we think it has been never put together in a volume of this type. We sincerely find the result impressive, both in terms of quality and quantity. Without the restrictions of space and content imposed in scientific journals, the contributors have been able to provide a complete account of their struggles, and in most cases successes, in the tackling of the problems of homogeneous catlysis. The contributors were asked to emphasize the didactic and divulgative aspects in their corresponding chapters, and we think they have done a very good work in this concern.

The reader will be able to use the book as a reference to what has been already done, as a how-to guide to what he can do, or as an indicator of what he can expect to be done by others in the near future. Because of this, it should be of interest both to established researchers and to interested graduate students. As for the particular audience, it should be of interest both for experimental chemists interested in knowing more about the application of computational chemistry to their field, and for computational chemists interested in applications to homogeneous catalysis.

We think this book provides a faithful snapshot on what is the status of the field at this point in time, and we hope that it gives significant clues with respect to its evolution in the future. There are still important processes that have not been treated theoretically, and others that escape the current capabilities, either in terms of computer power or methodogolical development. We believe nevertheless that the remaining problems will be solved in due time, and that the future of the computational modeling of homogeneous catalysis will be a brilliant one, but this, only time will tell.

Bellaterra

Feliu Maseras, Agustí Lledós

Chapter 1

Computational Methods for Homogeneous Catalysis

Feliu Maseras* and Agustí Lledós*
Unitat de Química Física, Edifici Cn, Universitat Autònoma de Barcelona, 08193 Bellaterra, Catalonia, Spain

Abstract: The methods commonly used for the computational modeling of homogeneous catalysis are briefly reviewed, with emphasis on their accuracy and range of applicability. Special mention is made to extended Hückel, Hartree-Fock and derived methods, density functional theory, molecular mechanics and hybrid quantum mechanics/molecular mechanics methods.

Key words: Hartree-Fock, Density functional theory, Extended Hückel, Molecular mechanics, Quantum mechanics/molecular mechanics

1. INTRODUCTION

Strictly speaking, is there such a thing as "computational methods for homogeneous catalysis"? Probably not. But there are a number of the molecules involved in homogeneous catalysis thak make them definitely different from, for instance, polypeptides in solution or products of gas-phase mass spectrometry. And these peculiarities are better described by some computational approaches than by others. In this chapter we present a brief overview of these approaches. The goal is not to make a systematic description of the methods, which can be found elsewhere [1-5], but to present them in the context of homogeneous catalysis, and to provide the reader not familiar with theory with a sufficient background for the correct understanding and interpretation of the results presented in the following chapters.

Homogeneous catalysis involves transition metal complexes. The presence of transition metal atoms, with their valence d shells, introduces serious demands in the type of quantum chemical methods that can be applied. While methods like Hartree-Fock (HF) and local density functional

1

F. Maseras and A. Lledós (eds.), Computational Modeling of Homogeneous Catalysis, 1–21.

theory (DFT) usually give quantitative results for organic molecules, they provide only a qualitative description in transition metal systems. More computer demanding methods introducing electronic correlation (within the HF formalism), or using non local corrections (in the DFT formalism) are a must if one desires any quantitatively reliable estimation. Homogeneous catalysts are often bulky systems. Furthermore, it is usually the bulk of certain regions of the catalyst that is responsible for the most chemically appealing features of the systems, like enantioselectivy or regioselectivity. Consideration of large systems poses certainly a serious strain in computer resources, because the cost scales usually as some power of the number of electrons. The major computational alternative to quantum mechanics, molecular mechanics, which is certainly less computer demanding, has the problem of the scarceness of reliable parametrizations for transition metal complexes.

Both the requirement of accurate methods and the large size of the systems make the theoretical study of homogeneous catalysis quite demanding in terms of computer effort. Only the dramatic increase in computer power in the last decades has made the quantitative study of these problems affordable. The expected progress in the near future anounces nevertheless a bright future for this field.

Homogeneous catalysis is an area of chemistry where computational modeling can have a substantial impact [6-9]. Reaction cycles are usually multistep complicated processes, and difficult to characterize experimentally [10-12]. An efficient catalytic process should proceed fastly and smoothly and, precisely because of this, the involved intermediates are difficult to characterize, when possible at all. Computational chemistry can be the only way to access to a detailed knowledge of the reaction mechanism, which can be a fundamental piece of information in the optimization and design of new processes and catalysts.

2.   QUALITATIVE CALCULATIONS ON MODEL SYSTEMS

2.1   Extended Hückel and other semiempirical methods

The extended Hückel method [13] is an extension of the traditional Hückel method [14] expanding its range of applicability beyond planar conjugate systems. From a mathematical point of view, it consists simply in solving the matricial equation 1, where H is the hamiltonian matrix, C are the molecular orbital coefficients, S the overlap matrix and ε a diagonal matrix containing the orbital energies.

$$HC = SC\varepsilon \qquad\qquad (1)$$

In the usual formulation of the extended Hückel method, the elements of the hamiltonian matrix are computed according to a simple set of arithmetic rules, and do not depend on the molecular orbitals. In this way, there is no need for the iterations required by more sophisticated methods, and in practice the results may be obtained nowadays in a question of seconds for any reasonably sized complex.

Figure 1. σ donation interaction between metal and hydride (left), and π backdonation interaction metal and carbonyl (right) as computed by the extended Hückel method.

Despite its simplicity, this method occupies a prominent position in the history of theoretical transition metal chemistry. It was certainly the first method to be applied to this type of systems, with the first works in the 1960's, and its use led to many of the ideas that constitute nowadays central concepts of organometallic chemistry [15]. For instance, those of σ donation and π backdonation shown in Figure 1. Extended Hückel studies were able to provide a simple explanation to the preferred position of different ligands in 5-coordinate complexes [16], and were also instrumental in outlining the essential differences between C-H and H-H activation [17] and in explaining the nature of dihydrogen complexes [18]. A modification of this method was also the tool used for the identification of the role of relativistic effects on a variety of problems [19]. The extended Hückel method is in fact the base of the CACAO program [20] where, complemented with a user-friendly graphical interface, continues to be applied nowadays for qualitative studies of transition metal chemistry.

The utility of the extended Hückel method for qualitative analysis must nevertheless not hide its limitations when quantitative results are desired. Although it can be of some utility in predicting bond and dihedral angles, it is unappropriate for the prediction of bond distances or bond energies. As a result, it cannot be applied to any reaction where bonds are made or broken, making its application to catalytic cycles very limited. It is useful in giving a qualitative picture of bonding interactions in a particular structure, but its range of applicability stops there. EH calculations are generally used on model systems, where only the metal center and its immediate environment are introduced. A further approach to real systems by introducing, for instance, bulky ligands, would be inefficient because of the inability of the method to reproduce properly steric repulsions.

The continued success of the extended Hückel method in transition metal chemistry, where it was the method of choice until the mid 1980's is surely related to the problems of other semiempirical methods in this area of chemistry. While methods like MOPAC [21] or AM1 [22] have been extremely productive in the field of organic chemistry, they have found little success in transition metal chemistry. These methods are based in equation 2, similar to 1, but with the very significant difference that the Fock matrix F is computed from the molecular orbitals, in an iterative way, though through an approximate formula.

$$FC = SC\varepsilon \tag{2}$$

These semiempirical methods are significantly more economical than the more accurate Hartree-Fock method, but they require a parametrization which is not trivial in the case of transition metal atoms. Some attempts have been made to introduce d orbitals in the traditional semiempirical methods, and among these, one can cite MNDO/d [23], ZINDO [24] and PM3(tm) [25]. The strenghts and weaknesses of these methods in their application to transition metal complexes are well exemplified in a recent systematic study on the performance of PM3(tm) on a variety of systems including products of cyclometallation, molecular dihydrogen complexes and H-BR$_2$ σ complexes of titanium [26]. The performance of PM3(tm) in the study of these systems is found to range from excellent in the case of dihydrogen complexes to very poor in the case of H-BR$_2$ complexes. Moreover, the quality of the results for a particular system is very difficult to predict a priori. As a result, the application of these methods in the field of homogeneous catalysis appears as quite limited, although they cannot be neglected as a potential useful tool in specific topics like the evaluation of relative stability of conformations.

2.2   Hartree-Fock and local density functional theory

The Hartree-Fock (HF) and local density functional theory (local DFT) methods provide a first level of accurate quantitative approach to a number of problems in chemistry. Unfortunately, this is seldom the case in homogeneous catalysis, where they are more likely to provide only a qualitative picture.

$$H\psi = E\psi \tag{3}$$

The Hartree-Fock approach derives from the application of a series of well defined approaches to the time independent Schrödinger equation (equation 3), which derives from the postulates of quantum mechanics [27]. The result of these approaches is the iterative resolution of equation 2, presented in the previous subsection, which in this case is solved in an exact way, without the approximations of semiempirical methods. Although this involves a significant increase in computational cost, it has the advantage of not requiring any additional parametrization, and because of this the HF method can be directly applied to transition metal systems. The lack of electron correlation associated to this method, and its importance in transition metal systems, limits however the validity of the numerical results.

The Hartree-Fock method was in any case the method of choice for the first quantitative calculations related to homogeneous catalysis. It was the method, for instance, on a study of the bonding between manganese and hydride in Mn-H, published in 1973 [28]. The first studies on single steps of catalytic cycles in the early 1980's used the HF method [29]. And it was also the method applied in the first calculation of a full catalytic cycle, which was the hydrogenation of olefins with the Wilkinson catalyst in 1987 [30]. The limitations of the method were nevertheless soon noticed, and already in the late 1980's, the importance of electron correlation was being recognized [31]. These approaches will be discussed in detail in the next section.

In any case, these first HF attempts to a quantitative study left an approach which is still currently in use, namely the use of model systems. The experimental system is not introduced as such in the calculation because of its large size, but it is replaced by a smaller system which, hopefully, has the same electronic properties. The reaction of the Wilkinson catalyst mentioned above [30, 31] can serve as example. The accepted active species is $Rh(PPh_3)_2Cl$, but the calculations were carried out on $Rh(PH_3)_2Cl$, as shown in Figure 2. The replacement of the phenyl substituents of the phosphine ligands by hydrogen atoms is certainly a simplification from the experimental system, but it represents an enormous saving in terms of computational time.

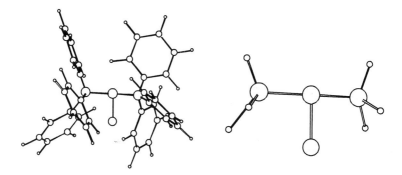

Figure 2. Real Rh(PPh$_3$)$_2$Cl active species (left) and the Rh(PH$_3$)$_2$Cl model used (right) in the calculations of the hydrogenation of ethylene by the Wilkinson catalyst [30, 31].

The density functional theory (DFT) [32] represents the major alternative to methods based on the Hartree-Fock formalism. In DFT, the focus is not in the wavefunction, but in the electron density. The total energy of an n-electron system can in all generality be expressed as a summation of four terms (equation 4). The first three terms, making reference to the non-interacting kinetic energy, the electron-nucleus Coulomb attraction and the electron-electron Coulomb repulsion, can be computed in a straightforward way. The practical problem of this method is the calculation of the fourth term E_{xc}, the exchange-correlation term, for which the exact expression is not known.

$$E_{DFT}[\rho] = T_s[\rho] + E_{ne}[\rho] + E_J[\rho] + E_{xc}[\rho] \tag{4}$$

A variety of expressions have been proposed for the exchange-correlation functional. Early works were based on the so called local density approximation (LDA), where it is assumed that the density can be locally treated as a uniform electron gas, or equivalently that the density is a slowly varying function. One of the most commonly used functionals within the LDA approach is the VWN correlation functional [33], used usually together with the exact (Slater) exchange functional (SVWN functional). The LDA approach is quite simple and computationally affordable. In fact, the LDA approach was used in the first applications of DFT to steps in homogeneous catalysis [34]. In this concern one can cite the study in 1989 of the migration of a methyl group from the metal to a carbonyl in CH$_3$Co(CO)$_4$ [35], a process relevant to hydroformylation. This approach was in any case soon discarded for transition metal chemistry because it underestimates the exchange energy and it seriously overestimates the correlation energy, with the result of leading to too large bond energies. As a result, the overall quality of the results is often not much better than that of the HF method [3].

Their use in homogeneous catalysis has nowadays a mostly qualitatively value, being replaced by the more accurate generalized gradient approximation (GGA), wich will be discussed in the next section.

This survey of theoretical methods for a qualitative description of homogeneous catalysis would not be complete without a mention to the Hartree-Fock-Slater, or Xα, method [36]. This approach, which can be formulated as a variation of the LDA DFT, was well known before the formal development of density functional theory, and was used as the more accurate alternative to extended Hückel in the early days of computational transition metal chemistry.

The methods described in this section were instrumental in the early computational modeling of homogeneous catalysis [34, 37], and are in a number cases the base of more accurate methods described later in this chapter. In any case, the qualitative accuracy they provide makes them of little application in present day research, with the only possible exception of the extended Hückel approach.

3.   QUANTITATIVE CALCULATIONS ON MODEL SYSTEMS

3.1   Hartree-Fock based methods

One of the more radical approximations introduced in the deduction of the Hartree-Fock equations 2 from the Schrödinger equation 3 is the assumption that the wavefunction can be expressed as a single Slater determinant, an antisymmetrized product of molecular orbitals. This is not exact, because the correct wavefunction is in fact a linear combination of Slater determinants, as shown in equation 5, where D_i are Slater determinants and c_i are the coefficients indicating their relative weight in the wavefunction.

$$\psi = \sum c_i D_i \tag{5}$$

The difference between the energy obtained with a single determinant (HF method) and that obtained by using the combination of all the possible determinants (full configuration interaction, full CI method) is called the correlation energy. It so happens that correlation energy is usually very important in transition metal compounds, and because of this, its introduction is often mandatory in computational transition metal chemistry.

There are a variety of methods for improving HF results with the introduction of electron correlation [27], and the discussion of their technical details is outside the scope of this chapter. They are in general based in some truncation of the full expansion defined by equation 5. There is a methodological distinction between dynamic and non-dynamic correlation that is however of some relevance for practical applications. Dynamic correlation can be defined as the energy lowering due to correlating the motion of the electrons, and is usually well described by methods giving a large weight to the most stable (HF) Slater determinant, and introducing the effect of the excited states as a minor perturbation. Non dynamic correlation, on the other hand, is the energy lowering obtained when adding flexibility to the wavefunction to account for near-degeneracy effects, and it usually requires a more ellaborate method [38].

A popular approach for introducing dynamic electron correlation in transition metal calculations is the use of methods based in the Møller-Plesset perturbational scheme [39]. In particular, the second level approach, the so called MP2 method provides a reasonable quality/price ratio, and has been extensively used in the modeling of homogeneous catalysis. In fact, it was the method of choice in the early 1990's [37] before the popularization of the DFT methods that will be described in next subsection. The MP2 method has been applied to the calculation of a number of catalytic cycles, like olefin hydrogenation [31] and hydroformylation [40]. The MP2 method is nowadays still a good alternative for calculations on model systems, and in favorable cases, it provides geometries within hundredths of Å of the correct values, and energies within few kcal/mol. Other methods for introducing dynamic correlation, like truncated configuration interaction (truncated CI) or generalized valence bond (GVB) [41] are less commonly used.

Another standard for quality/price is defined within the coupled cluster approach. In particular, the CCSD(T) method [42] is nowadays generally accepted as the most accurate method which can be applied systematically for systems of a reasonable size. One must nevertheless be aware of the high computational cost of the method, which is often used only for energy calculations on geometries optimized with other computational methods.

The treatment of systems where non-dynamic correlation is critical is quite more complicated from a methodological point of view. As mentioned above, non-dynamic correlation is associated to the presence of near-degeneracies in the electronic ground state of the system, which means that there are Slater determinants with a weight similar to that of the HF solution in equation 4. The problem of non-dynamic correlation is usually treated successfully by the CASSCF method [43] for organic systems. This method introduces with high accuracy the correlation in the orbitals involved in the near degeneracy, which constitute the so called active space. The problem in transition metal chemistry is that dynamic correlation is almost always necessary, and CASSCF neglects it. As a result, more sophisticated approaches must be used, among the which one can mention CASPT2 [44]. The use of this type of methods presents two important peculiarities. The first of them is its strong demand in terms of computational effort. The second inconvenient is the complexity of the setup of the calculation itself. The choice of the active space has to be made previous to the calculation, and it is seldom trivial. As a result, catalysts where non-dynamic correlation is important are still in the limits of what can be nowadays treated.

In summary, HF-based methods for the introduction of electron correlation constitute one of the two major alternatives for the quantitative calculation of homogeneous catalysis on model systems. Dynamic correlation can be well treated, and there is a quite well established hierarchy of the methods, with the MP2 method being the most used for geometry optimizations, and CCSD(T) being the main choice for highly accurate calculations. The treatment of non-dynamic correlation, luckily not always necessary, is more challenging, though significant progress can be made with the CASPT2 method.

3.2   Non local density functional theory

If the main limitations of HF theory are overcome by the introduction of electron correlation, those of density functional theory are expanded by the use of more accurate functionals. These functionals, that improve the uniform gas description of the LDA approach, are labeled as non-local or Generalize Gradient Approximation (GGA).

The GGA functionals are based in the same expression presented above for LDA functionals in equation 4, and they modify only the form of the exchange-correlation functional $E_{xc}[\rho]$. The GGA functionals are usually divided in two parts, namely exchange and correlation, and different expressions have been proposed for each of them. The exchange functionals which are more used nowadays are probably those labeled as B (or B88) [45] and Becke3 (or B3) [46], the latter containing a term introducing part of the HF exact exchange. As for correlation functionals, one should mention those by Lee, Yang and Parr [47] (labeled as LYP), Perdew [48] (known as P86), and Perdew and Wang [49] (PW91 or P91). Since the correlation and exchange functionals are in principle independent, different combinations of them can be used. For instance, it is common to find BLYP (B exchange, LYP correlation), BP86 (B exchange, P86 correlation) or B3LYP (B3 exchange, LYP correlation).

GGA DFT theory is extremely succesful in the calculation of medium size transition metal complexes [4, 34, 50]. As a result, it has been from the mid 1990's the method of choice for quantitative calculations of model systems of complexes involved homogeneous catalysis, complemented in some cases with single point more accurate CCSD(T) calculations on DFT optimized geometries. There are many examples of the success of non-local DFT theory in this field and, in fact, most of the chapters in this volume constitute good proof of this. GGA DFT competes well in terms of accuracy with the HF based MP2 method, with results usually close to experiment within hundredths of Å for geometries, and within a few kcal/mol for energies. Furthermore, DFT is much less demanding in terms of disk space, and scales better with respect to the size of the system as far as computer time is concerned.

DFT theory even seems to improve the performance of MP2 in cases where there is some small contribution of non dynamic correlation. This is seemingly the case in the BP86 computed first dissociation energies of a variety of metal carbonyls [51]. For instance, in the case of $Cr(CO)6$, the BP86 value is 192 kJ/mol, in exact (probably fortuitous) agreement with the (computationally most accurate) CCSD(T) value of 192 kJ/mol, but also reasonably close to the experimental value of 154±8 kJ/mol. In this case, the GGA DFT result improves clearly the local DFT SVWN value of 260 kJ/mol, and the MP2 result, wich is 243 kJ/mol. Comparable results can be found for the optimization of the Os-O distance in OsO_4 [52], which is relevant concerning olefin dihydroxylation.

On the other hand, it is found that DFT functionals currently available usually describe more poorly than MP2 the weak interactions due to dispersion, the so called van der Waals type interactions [53].

In any case, we consider that nowadays the only remaining disadvantage of DFT with respect to HF based methods has a conceptual origin. Within the Hartree-Fock framework there is a quite well defined hierarchy of methods, which define in a reliable way what one should do to improve a given result, whether if it is by including dynamic correlation or non-dynamic correlation. Within the DFT theory, one has only a list of functionals, the relative qualities of the which are mostly known from experience, and whose relative performance in front of a particular problem is sometimes difficult to predict. As a result, DFT offers an optimal quality/price ratio for systems that behave "properly", but it struggles badly when trying either to obtain highly accurate results or to deal with electronically complicated cases. In this latter cases, one has to resort to the usually more computationally demanding HF based methods. Because of this, we consider that these two general approaches to quantitative calculation of model systems in homogeneous catalysis are nowadays complementary, and will continue to coexist as useful alternatives at least for some time.

4.   CALCULATIONS ON REAL SYSTEMS

4.1   Molecular mechanics

The previous section has described how one can compute accurately a system of about 30 atoms including one transition metal. The problem is, as mentioned above, that these are usually not the real catalysts, but model systems where the bulky substituents have been replaced by hydrogen atoms. Calculations on model systems are usually at least indicative of the nature and the energy barriers of the steps involved in a catalytic cycle, but they are often unable to provide information on some of the most interesting features, namely enantioselectivity and regioselectivity. The reason for this failure is simply that selectivity is often associated to the presence of the bulky substituents which are deleted when defining the model system.

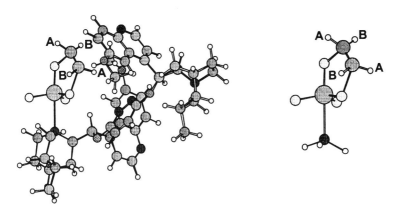

Figure 3. Schematic representation of the transition states for dihydroxylation of olefins catalyzed by the experimentally used OsO$_4$(DHQDZ) complex (left) and its simplified computational model OsO$_4$(NH$_3$) (right).

The correlation between bulky substituents and stereoselectivity is graphically shown in Figure 3, depicting the possible transition states in the dihydroxylation of a monosubstituted olefin by osmium tetroxide derivatives. This reaction is known to be selective [54], and the selectivity depends on whether the olefin substituent takes a position of type **A** or **B** in the transition state. The problem with calculations on a model system where the bulky base is replaced by NH$_3$ is that the positions **A** and **B** are completely symmetrical, and thus, they yield the same energy. In other words, the reaction would not be selective with this model system.

One of the main answers that computational chemistry has for the introduction of hundreds of atoms in a calculation is the use of molecular mechanics (MM) [55, 56]. Molecular mechanics is a simple "ball-and-spring" model for molecular structure. Atoms (balls) are connected by springs (bonds), which can be stretched or compressed with a certain energy cost. The energy expression for molecular mechanics has a form of the type shown in equation 6, where each term is a summation extended to all atoms involved.

$$E = E_{stretch} + E_{bend} + E_{tors} + E_{vdW} + E_{electrostatic} + \text{other terms} \qquad (6)$$

The size of the atoms and the rigidity of the bonds, bond angles, torsions, etc. are determined empirically, that is, they are chosen to reproduce experimental data. Electrons are not part of the MM description, and as a result, several key chemical phenomena cannot be reproduced by this method. Nevertheless, MM methods are orders of magnitude cheaper from a computational point of view than quantum mechanical (QM) methods, and because of this, they have found a preferential position in a number of areas of computational chemistry, like conformational analysis of organic compounds or molecular dynamics.

The application of molecular mechanics to transition metal systems is not as straightforward as in the case of organic systems because of the much larger variety of elements and atom types and the relative scarcity of experimental data to which parameters can be adjusted. Significant progress has been made in recent years [56-61], though its application to reaction mechanisms remains seriously complicated by the difficulty in describing changes in the coordination environment of the metal, and in locating transition states.

Despite their limitations, MM methods have been applied to homogeneous catalysis, and a recent review collects more than 80 publications on the topic [61]. The fact is that, before the relatively recent appearance of quantum mechanics / molecular mechanics (QM/MM) methods, described in the next subsection, MM methods were the only available tool for the introduction the real ligands in the calculations. The two reactions that have probably received more attention from pure MM calculations are enantioselective olefin hydrogenation [62-64] and homogeneous Ziegler-Natta olefin polymerization [65-67], the first studies dating from the late 1980's. The problem of the structure around the transition metal was solved by taking the results of QM calculations on model systems or by choosing structures from X-ray diffraction.

We consider that the application of standard force fields to homogeneous catalysis continues nowadays to be useful for preliminary qualitative descriptions, but that for a more quantitative description within the MM method one must necessarily go to the design of a specific force field for a given catalytic process, the so called "QM-guided molecular mechanics" (Q2MM) [68]. In this approach, a specific force field is defined for a particular reaction, using all available data from experiment and data obtained from accurate QM calculations of minima and transition states. The definition and testing of the parameters is usually quite time consuming, but once the process is done for a particular reaction, the effect of small variations in catalyst or substrate can be computed with a very low computational cost. This quite recent method has been applied succesfully to processes of asymmetric synthesis [69] and catalysis, in the case of dihydroxylation [70].

4.2 Quantum mechanics / molecular mechanics

The description of pure quantum mechanics (QM) methods presented in Section 3 has shown how in most cases they provide an accurate description of the electronic subtleties involved at the transition metal center of a catalytic process, but that they are unable to introduce the whole bulk of the catalyst substituents, which can be critical for selectivity issues. The description of pure molecular mechanics (MM) methods presented in subsection 4.1 has shown how these methods can easily introduce the steric bulk of the substituents, and accurately describe their steric interactions, but that they struggle badly when trying to describe properly the transition metal center and its immediate environment. The logical solution to this complementary limitations is to divide the chemical system in two regions, and to use a different description for each of them, QM for the metal and its environment, MM for the rest of the system. This is precisely the basic idea of hybrid quantum mechanics / molecular mechanics (QM/MM) methods.

$$E_{tot}(QM,MM) = E_{QM}(QM) + E_{MM}(MM) + E_{interaction}(QM/MM) \qquad (7)$$

In QM/MM methods, the total energy of a chemical system can be expressed as shown in equation 7, where the labels in parentheses make reference to the region, and those in subscript to the computational method. The total energy is thus the addition of three terms, the first one describing at a QM level the interactions within the QM region, the second one describing at an MM level the interactions within the MM region and the third one describing the interactions between the QM and the MM regions. The detailed form of this third term is not defined a priori, although it will be in principle composed of a QM and an MM description, as shown in equation 8, where the $E_{QM}(QM/MM)$ term can be roughly related to electronic effects and the $E_{MM}(QM/MM)$ term to steric effects [71]. The particular definition of each of the two terms in equation 8 gives rise to the variety of QM/MM methods available.

$$E_{interaction}(QM/MM) = E_{QM}(QM/MM) + E_{MM}(QM/MM) \qquad (8)$$

QM/MM methods have been around for some time in computational chemistry [72-74], with the first proposal of this approach being made already in the 1970's. The initial applications throughout the 1980's and the early 1990's were mostly concerned with the introduction of solvent effects, with a special focus on biochemical problems. The solvent is usually in the MM region, and its electronic effect on the QM region is often described by placing point charges. A lot of effort has been invested in the development of methods with a proper definition of these charges and their placement. The handling of covalent bonds across the QM/MM partition within this approach is not trivial [75], but these methods have reached a state of maturity which makes them quite widespread nowadays in computational biochemistry [76, 77].

The situation for transition metal chemistry has been somehow different, because, while the description of the electronic effects from the MM region may not be that important, the handling of covalent connections between the QM and MM regions is critical. The methods most succesful in handling this situation have been the closely related IMOMM [78] and ONIOM [79], the latter being actually a modification of the former. These schemes provide a computationally economical and methodologically robust method to introduce the steric effects of the MM region in the calculation, allowing the straightforward geometry optimization of both minima and transition states. The total cost of the calculation is only slightly larger than the corresponding QM calculation for the QM region. The downside of these methods is that in principle they neglect the electronic effects of the MM region on the QM atoms. Even this limitation can nevertheless be used advantageously to facilitate the analysis of the results [80].

Despite their relatively recent appearance in 1995, the IMOMM and ONIOM methods have been extremely productive in the computational modeling of homogeneous catalysis [81]. Not surprisingly, the have found the most succesful applications in cases where regioselectivity and enantioselectivity are critical. Among the reactions succesfully studied with the IMOMM and ONIOM methods, one can mention olefin dihydroxylation [82], homogeneous olefin polymerization by early [83] and late [84, 85] transition metal complexes, olefin hydrogenation [86], addition of diethylzinc to aldehydes [87] and hydroformylation [88]. A number of these applications are presented in detail in the remaining chapters of this book, and because of this they will not be discussed here.

Figure 4 presents the partition between the QM and MM regions in a typical QM/MM calculation, this one in particular from a theoretical study on hydroformylation [88]. The QM region is described by $HRh(CO)(PH_3)_2(CH_2=CHCH_3)$, including capping hydrogen atoms for the connections with the MM region. The rest of the system, consisting of most of the bulky benzoxantphos ligand, constitutes the MM region. The accurate description of the metal and its attached atoms, and the steric effects of the chelating phosphine are in this way introduced in the calculation, which predicts correctly the experimentally observed regioselectivity leading to the linear product. An interesting additional conclusion of this result is that the electronic effects of the phosphine substituents are not critical for selectivity.

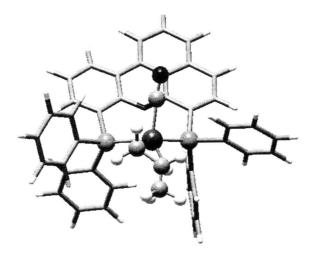

Figure 4. QM/MM partition used in the IMOMM calculation of one of the possible transition states of the insertion of propene into the Rh-H bond of HRh(CO)(benzoxantphos) [88].The QM atoms are shown as "balls and sticks", and the MM atoms are shown as "tube".

Hybrid quantum mechanics / molecular mechanics methods are nowadays a well established method for the computational modeling of homogeneous catalysis, being a very efficient option for the introduction of both the accurate calculation of electronic effects at the metal center and the steric effects associated to the presence of bulky ligands. The recent application of the methods in the field leaves still some margin for methodological improvement. Furthermore, QM/MM methods will also take advantage of any methodological progress in both pure QM and MM approaches. The application of these methods in homogeneous catalysis should therefore increase in the near future.

5.   CHALLENGES AHEAD

The content of this chapter so far has emphasized the successes of computational chemistry in the study of homogeneous catalysis. This is not by accident, because we intended to give the reader an idea of what can currently be done. From its reading, one could however get the wrong impression that everything is already solved from a methodological point of view, and that the only work left is the application of these methods to practical experimental problems. This is certainly not the case. This section collects some of the problems that still remain a challenge for computational chemistry, with special attention to what progress is currently being made.

First we will discuss the aspects related to the electronic description of the metal center and its immediate environment. The first problem which must be mentioned in this concern, and that has been already discussed above, is that of non dynamic correlation [38]. Non dynamic correlation appears associated to near degeneracy, that is, the presence of excited electronic state states very close in energy to the ground state. The treatment of this situations is still a challenge for theoretical methods, with the best choice being probably now the computationally demanding HF-based CASPT2 method [44]. The presence of important non dynamic correlation is not a rarity in homogeneous catalysis, and it is usually present whenever a metal-metal bond exists in a polynuclear complex.

A second problem in the electronic description of the reaction center is the presence of different spin states in the catalytic cycle, with the system changing between them in what is labeled as spin crossing [89]. In order to properly reproduce the crossing of two states one needs to start from a fairly accurate knowledge on the potential hypersurface of both of them, which in turn requires a good description of excited states, which is by no means easy. Research is currently active in the development of methods which can accomplish a good description of excited states in medium size transition metal complexes, the best current alternatives being probably CASPT2 [44] and time-dependent DFT (TDDFT) [90].   After one has a reasonable description of the two surfaces, its crossing must be explored both from the point of view of the distribution of nuclei and of the spin-orbit coupling of the electronic wavefunctions [91]. Spin crossing has not yet an easy solution from a computational point of view, but it has been explored theoretically for some catalytic processes [92, 93].

Apart from the solving of particularly complicated problems, methological development is also involving intense efforts in the direction of improving the performance of the available computational methods. This has an obvious impact in computational chemistry by allowing the study of larger and larger systems. Two particularly fruitful approaches in this concern are the use of pseudospectral methods [94] and of planar waves [95] within the DFT formalism.

The proper treatment of the electronic subtleties at the metal center is not the only challenge for computational modeling of homogeneous catalysis. So far in this chapter we have focused exclusively in the energy variation of the catalyst/substrate complex throughout the catalytic cycle. This would be an exact model of reality if reactions were carried out in gas phase and at 0 K. Since this is conspicously not the common case, there is a whole area of improvement consisting in introducing environment and temperature effects.

Reaction rates are macroscopic averages of the number of microscopical molecules that pass from the reactant to the product valley in the potential hypersurface. An estimation of this rate can be obtained from the energy of the highest point in the reaction path, the transition state. This approach will however fail when the reaction proceeds without an enthalpic barrier or when there are many low frequency modes. The study of these cases will require the analysis of the trajectory of the molecule on the potential hypersurface. This idea constitutes the basis of molecular dynamics (MD) [96]. Molecular dynamics were traditionally too computationally demanding for transition metal complexes, but things seem now to be changing with the use of the Car-Parrinello (CP) method [97]. This approach has in fact been already succesfully applied to the study of the catalyzed polymerization of olefins [98].

Homogeneous catalysis takes place usually in solution, and the nature of the solvent can seriously affect its outcome. There are two main approaches to the introduction of solvent effects in computational chemistry: continuum models and explicit solvent models [99]. Continuum models consider the solute inside a cavity within a polarizable continuum. They are quite succesful in capturing the essential qualitative aspects of solvation, but they have the big disavantage of neglecting any specific intermolecular interaction. The most popular continuum model is probably the polarizable continuum model (PCM) [100]. Explicit solvent models are formally simpler, because they consist in introducing the solvent molecules together with those of the solute. The problem is that because of the size of the system they have to use at the same time both QM/MM and MD, and the methods are not yet completely fit, though significant progress is being made [101].

There are therefore quite a few methodological challenges remaining for the computational modeling of homogeneous catalysis. This must prompt theoreticians to sharpen their tools and interested experimentalists to keep an eye on the development of new methods.

6.   CONCLUSIONS AND PERSPECTIVES

Recent progress in computer power and methodological algorithms has taken computational modeling to a level where it can make a valuable contribution to the understanding and improvement of the mechanisms operating in homogeneous catalysis. The standard and widespread non-local density functionals allow the calculation of all intermediates and transition states for most catalytic cycles with errors in energy barriers in the range of few kcal/mol. Higher accuracy can be obtained through single point calculations with the CCSD(T) method. The steric effect of bulky ligands is introduced in the calculation through hybrid quantum mechanics / molecular mechanics methods, leading to reliable quantitative predictions on the regioselectivity and enantioselectivity of the processes. Progress is also been made in the development of new methods to tackle reactions where these approaches fail, like those involving spin crossover or those without enthalpic barrier.

The tools are thus available for the computational elucidation of the mechanism of catalytic cycles. The only remaining question is whether this mechanistic knowledge is still necessary. Certainly, the highly efficient automation of tests provided by combinatorial chemistry [102] allows catalyst optimization without such mechanistic information. We believe however that the detailed knowledge of reaction mechanisms will continue to be, at least in selected cases, a valuable tool for the design of new and more efficient catalysts, and that computational modeling has become an extremely powerful tool to gain this knowledge.

ACKNOWLEDGMENT

Financial support from the Spanish DGES (Project No. PB98-0916-CO2-01) and the Catalan DURSI (Project No. 1999 SGR 00089). FM thanks also DURSI for special funding within the program for young distinguished researchers.

REFERENCES

1   *The Encylopaedia of Computational Chemistry*, Schleyer, P.v. R.; Allinger, N. L.; Clark, T.; Gasteiger, J.; Kollman, P. A.; Schaefer, H. F.; Schreiner, P. R., Eds.; Wiley: New York, 1998.
2   *Organometallic Bonding and Reactivity. Fundamental Studies*, Brown, J. M.; Hofmann, P., Eds., *Top. Organomet. Chem.* Vol. 4, Springer-Verlag: Heidelberg, 1999.

3   *Introduction to Computational Chemistry*, Jensen F., Wiley: New York, 1999.
4   Davidson, E. R. *Chem. Rev.* **2000**, *100*, 351.
5   *Computational Organometallic Chemistry*, Cundari, T., Ed., Marcel Dekker: New York, 2001.
6   *Theoretical Aspects of Homogeneous Catalysis*, van Leeuwen, P. W. N. M., Morokuma, K., van Lenthe, J. H., Eds., Kluwer: Dordrecht, 1995.
7   Musaev, D. G.; Morokuma, K. *Advances in Chemical Physics*, Rice, S. A.; Prigogine, I., Eds., Wiley: New York 1996; Vol. 45, p 61.
8   Torrent, M.; Solà, M.; Frenking, G. *Chem. Rev.* **2000**, *100*, 439.
9   Ziegler, T. *J. Chem. Soc., Dalton Trans.* **2002**, 642.
10 *Applied Homogeneous Catalysis with Organometallic Compounds*, Cornils, B.; Herrmann, W. A.; VCH-Wiley: New York, 1996.
11 *Comprehensive Asymmetric Catalysis*, Jacobsen, E. N.; Pfaltz, A.; Yamamoto, H., Eds.; Springer: Berlin, 1999.
12 *Inorganic and Organometallic Reaction Mechanisms*, Atwood, J. D.; Wiley: New York, 1997.
13 Hoffmann R. *J. Chem. Phys.* **1963**, *39*, 1397.
14 Hückel, E. *Z. Phys.* **1931**, *70*, 204.
15 *Orbital Interactions in Chemistry*, Albright, T. A.; Burdett, J. K.; Whangbo, M. H.; Wiley: New York, 1985.
16 Rossi, A. R.; Hoffmann R. *Inorg. Chem.* **1975**, *14*, 365.
17 Saillard, J. Y.; Hoffmann, R. *J. Am. Chem. Soc.* **1984**, *106*, 2006.
18 Jean, Y.; Eisenstein, O.; Volatron, F.; Maouche, B.; Sefta, F. *J. Am. Chem. Soc.* **1986**, *108*, 6587.
19 Pyykkö, P. *Chem. Rev.* **1988**, *88*, 563.
20 Mealli, C.; Proserpio, D. M. *J. Chem. Educ.* **1990**, *67*, 399.
21 Dewar, M. J. S.; Zoebisch, E. G.; Healy, E. F.; Stewart, J. J. P. *J. Am. Chem. Soc.* **1985**, *107*, 3902.
22 Stewart, J. J. P. *J. Comp. Chem.* **1989**, *10*, 209.
23 Thiel, W.; Voityuk, A. *Theor. Chim. Acta* **1992**, *31*, 391.
24 Anderson, W. P.; Edwards, W. D.; Zerner, M. C. *Inorg. Chem.* **1986**, *25*, 2728.
25 *A Guide to Molecular Mechanics and Molecular Orbitals Calculations in Spartan*, Hehre, W. J.; Yu, J.; Klunziguer, P. E, Wavefunction Inc: Irvine, CA, 1997.
26 Bosque, R.; Maseras, F. *J. Comput. Chem.* **2000**, *21*, 562.
27 *Modern Quantum Chemistry*, Szabo, A.; Ostlund, N. S.; 1st edition revised, McGraw-Hill: New York, 1989.
28 Bagus, P. S.; Schaefer, H. F., III *J. Chem. Phys.* **1973**, 58, 1844.
29 Dedieu, A. *Top. Phys. Organomet. Chem.* **1985**, *1*, 1.
30 Koga, N.; Daniel, C.; Han, J.; Fu, X. Y.; Morokuma K. *J. Am. Chem. Soc.* **1987**, *109*, 3455.
31 Daniel, C.; Koga, N.; Han, J.; Fu, X. Y.; Morokuma K. *J. Am. Chem. Soc.* **1988**, *110*, 3773.
32 *A Chemist's Guide to Density Functional Theory*, Koch, W.; Holthausen, M. C.; Wiley-VCH: Weinheim, 2000.
33 Vosko, S. J.; Wilk, L.; Nusair, M. *Can J. Phys.* **1980**, *58*, 1200.
34 Ziegler, T. *Chem. Rev.* **1991**, *91*, 651.
35 Versluis, L.; Ziegler, T. *J. Am. Chem. Soc.* **1989**, *111*, 2018.
36 (a) Schwarz, K. *Phys. Rev.* **1972**, *B5*, 2466. (b) Gopinathan, M. S.; Whitehead, M. A. *Phys. Rev.* **1976**, *A14*, 1.

37   Koga, N.; Morokuma, K. *Chem. Rev.* **1991**, *91*, 823.
38   Pierloot, K. *Computational Organometallic Chemistry*, T. Cundari, Ed.; Marcel Dekker: New York, 2001, page 123.
39   Møller, C; Plesset, M. S. *Phys. Rev.* **1934**, *46*, 618.
40   Matsubara, T.; Koga, N.; Ding, Y.; Musaev, D. G.; Morokuma, K. *Organometallics* *1997*, **16**, 1065.
41   Goddard, W. A.; Dunning, T. H.; Junt, W. J.; Hay, P. J. *Acc. Chem. Res.* **1973**, *6*, 368.
42   Purvis, G. D.; Bartlett, R. J. *J. Chem. Phys.* **1982**, *76*, 1910.
43   Roos, B. O. *Adv. Chem. Phys.* **1987**, *69*, 399.
44   Andersson, K.; Malmqvist, P. Å.; Roos, B. O. *J. Chem. Phys.* **1992**, *96*, 1218.
45   Becke, A. D. *Phys. Rev. B* **1988**, *38*, 3098.
46   Becke, A. D. *J. Chem. Phys.* **1996**, *104*, 1040.
47   Lee, C.; Yang, W.; Parr, R. G. *Phys. Rev. B* **1988**, *37*, 785.
48   Perdew, J. P. *Phys. Rev. B* **1986**, *34*, 7406.
49   Perdew, J. P.; Chevary, J. A.; Vosko, S. H.; Jackson, K. A.; Pederson, M. R.; Singh, D. J.; Filhais, C. *Phys. Rev. B* **1992**, *46*, 6671.
50   Görling, A.; Trickey, S. B.; Gisdakis, P.; Rösch, N. *Top. Organomet. Chem.* **1999**, *4*, 109.
51   Li, J.; Schreckenbach, G.; Ziegler, T. *J. Am. Chem. Soc.* **1995**, *117*, 486.
52   Ujaque, G.; Maseras, F.; Lledós, A. *Int. J. Quant. Chem.* **2000**, *77*, 544.
53   Meijer, E. J.; Sprik, M. *J. Chem. Phys.* **1996**, *105*, 8684.
54   Kolb, H. C.; VanNieuwenhze, M. S.; Sharpless, K. B. *Chem. Rev.* **1994**, *94*, 2483.
55   *Molecular Mechanics*, Burkert, U.; Allinger, N. L., ACS: Washington, 1982.
56   *Molecular Mechanics Across Chemistry*, Rappé, A K.; Casewit, C. J., University Science Books: Sausalito, CA, 1997.
57   Landis, C. R.; Root, D. M.; Cleveland, T. *Rev. Comput. Chem.* **1995**, *6*, 73.
58   Cundari, T. R. *J. Chem. Soc. Dalton Trans.* **1998**, 2771.
59   Comba, P. *Coord. Chem. Rev.* **1999**, *182*, 343.
60   White, D. P. *Computational Organometallic Chemistry*, T. Cundari, Ed.; Marcel Dekker: New York, 2001, page 39.
61   White, D. P.; Douglass, W. *Computational Organometallic Chemistry*, T. Cundari, Ed.; Marcel Dekker: New York, 2001, page 237.
62   Brown, J. M.; Evans, P. L. *Tetrahedron* **1988**, *44*, 4905.
63   Bogdan, P. L.; Irwin, J. J.; Bosnich, B. *Organometallics* **1989**, *8*, 1450.
64   Giovannetti, J. S.; Kelly, C. M.; Laandis, C. R. *J. Am. Chem. Soc.* **1993**, *115*, 5889.
65   Castonguay, L. A.; Rappé, A. K. *J. Am. Chem. Soc.* **1992**, *114*, 5832.
66   Kawamura-Kuribayashi, H.; Koga, N.; Morokuma, K. *J. Am. Chem. Soc.* **1992**, *114*, 8687.
67   Guerra, G.; Cavallo, L.; Moscardi, G..; Vacatello, M.; Corradini, P. *J. Am. Chem. Soc.* **1994**, *116*, 2988.
68   Norrby, P.-O. *Computational Organometallic Chemistry*, T. Cundari, Ed.; Marcel Dekker: New York, 2001, page 7.
69   Norrby, P.-O.; Brandt, P.; Rein, T. *J. Org. Chem.* **1999**, *64*, 5845.
70   Norrby, P.-O.; Rasmussen, T.; Haller, J.; Strassner, T.; Houk, K. N. *J. Am. Chem. Soc.* **1999**, *121*, 10186.
71   Maseras, F. *Top. Organomet. Chem.* **1999**, *4*, 165.
72   Warshel, A.; Levitt, M. *J. Mol. Biol.* **1976**, *103*, 227.
73   Singh, U. C.; Kollman, P. A. *J. Comput. Chem.* **1986**, *7*, 718.
74   Field, M. H.; Bash, P. A.; Karplus, M. *J. Comput. Chem.* **1990**, *11*, 700.

75 Bakowies, D.; Thiel, W. *J. Phys. Chem.* **1996**, *100*, 10580.
76 Gao, J. *Acc. Chem. Res.* **1996**, *29*, 298.
77 Monard, G.; Merz, K. M. *Acc. Chem. Res.* **1999**, *32*, 904.
78 Maseras, F.; Morokuma, K. *J. Comput. Chem.* **1995**, *16*, 1170
79 Dapprich, S.; Komáromi, I.; Byun, K. S.; Morokuma, K., Frisch, M. J. *J. Mol. Struct. (THEOCHEM)* **1999**, *461-462*, 1.
80 Maseras, F. *Computational Organometallic Chemistry*, T. Cundari, Ed.; Marcel Dekker: New York, 2001, page 159.
81 Maseras F. *Chem. Commun.* **2000**, 1821.
82 Ujaque, G.; Maseras, F.; Lledós, A. *J. Am. Chem. Soc.* **1999**, *121*, 1317.
83 Guerra, G.; Longo, P.; Corradini, P.; Cavallo, L. *J. Am. Chem. Soc.* **1999**, *121*, 8651.
84 Froese, R. D. J.; Musaev, D. G.; Morokuma, K. *J. Am. Chem. Soc.* **1998**, *120*, 1581.
85 Deng, L.; Margl, P.; Ziegler, T. *J. Am. Chem. Soc.* **1999**, *121*, 6479.
86 Feldgus, S.; Landis, C. R. *J. Am. Chem. Soc.* **2000**, *122*, 12714.
87 Vázquez, J.; Pericàs, M. A.; Maseras, F.; Lledós, A. *J. Org. Chem.* **2000**, *65*, 7303.
88 Carbó, J. J.; Maseras, F.; Bo, C.; van Leeuwen, P. W. N. M. *J. Am. Chem. Soc.* **2001**, *123*, 7630.
89 Harvey, J. N. *Computational Organometallic Chemistry*, T. Cundari, Ed.; Marcel Dekker: New York, 2001, page 291.
90 Petersilka, M.; Grossmann, U. J.; Gross, E. K. U. *Phys. Rev. Lett.* **1996**, *76*, 12.
91 Harvey, J. N.; Aschi, M. *Phys. Chem. Chem. Phys.* **1999**, *1*, 5555.
92 Linde, C.; Koliai, N.; Norrby, P.-O.; Akermark, B. *Chem. Eur. J.* **2002**, *8*, 2568.
93 Ogliaro, G.; Cohen, S.; de Visser, S. P.; Shaik, S. *J. Am. Chem. Soc.* **2000**, *122*, 12892.
94 Friesner, R. A. *J. Chem. Phys.* **1987**, *86*, 3522.
95 Blöchl, P. E. *Phys. Rev. B* **1994**, *50*, 17953.
96 Truhlar, D. G. *Faraday Discuss.* **1998**, *110*, 521.
97 Car, R.; Parrinello, M. *Phys. Rev. Lett.* **1985**, *55*, 2471.
98 Woo, T. K.; Blöchl, P. E.; Ziegler, T. *J. Phys. Chem. A* **2000**, *104*, 121.
99 *Solvent Effects and Chemical Reactivity*, Tapia, O.; Bertrán, J., Eds., Kluwer: Dordrecth, 1996.
100 Tomasi, J. *Chem. Rev.* **1994**, *94*, 2027.
101 Gao, J. L.; Truhlar, D. G. *Annu. Rev. Phys. Chem.* **2002**, *53*, 467.
102 Crabtree, R. H. *Chem. Commun.* **1999**, 1611.

Chapter 2

Olefin Polymerization by Early Transition Metal Catalysts

Luigi Cavallo

Dipartimento di Chimica, Università di Salerno, Via Salvador Allende, I-84081, Baronissi (SA) Italy

Abstract:　In this contribution we report about the computer modeling of the elementary steps of the propagation reaction for the polymerization of olefins with homogeneous catalysts based on early transition metals. Particular attention will be devoted to biscyclopentadienyl and monocyclopentadienylamido-based catalysts. Beside the coverage of literature data, the performances of various pure and hybrid density functional theory (DFT) approaches, and of classical *ab initio* Hartree-Fock (HF) and Møller-Plesset perturbative theory up to the second order (MP2) will be discussed through a systematic study of ethene insertion into the Zr–CH_3 σ-bond of the $H_2Si(Cp)_2ZrCH_3^+$ system. A comparison with singly and doubly excited coupled clusters single point calculations with a perturbative inclusion of triple excitations [CCSD(T)] will be also presented. The effects of the basis set on the insertion barrier will be discussed with a series of single point MP2 calculations. In the final sections we report about the origin of the regio- and stereoselectivity in the propene insertion with biscyclopentadienyl-based catalysts.

Key words:　Olefin polymerization, Ziegler-Natta catalysis

1.　INTRODUCTION

Many discoveries changed human life in the last fifty years, and the polymerization of olefins catalyzed by transition metals certainly can be considered among them. At first glance, the job requested to polymerization catalysts seems rather trivial. It consists in the enchainment of monomeric units by insertion of olefins into Mt–P (P = Polymeric chain) bonds. Despite its "simplicity" and the amount of work by many research groups both in the

F. Maseras and A. Lledós (eds.), Computational Modeling of Homogeneous Catalysis, 23–56.

academy and in the industry, this class of reaction is one of the "hot" topics in current chemistry.

The two seminal events in the field are the results of the ingenious work of the Nobel laureates Karl Ziegler and Giulio Natta. Ziegler discovered that $TiCl_3$ activated by AlR_3 (R = C_2H_5 or another alkyl group) was an effective catalysts for ethene polymerization in 1953 [1], while Natta and co-workers discovered the synthesis of stereoregular polymers in 1954, by using similar catalytic systems [2]. Attempts to provide soluble and chemically more defined and hence understandable models of the $TiCl_3$-based heterogeneous polymerization catalysts immediately followed. However, the early catalysts based on $Cp_2MtX_2/AlRCl_2$ or AlR_3 (Mt = Ti, Zr; X = Cl or alkyl group; Cp = cyclopentadienyl) met with limited success [3, 4].

For almost twenty years this field was substantially limited to a technological development of the heterogeneous catalysts until the serendipitous discovery of the activating effect of small amounts of water on the $Cp_2MtX_2/AlMe_3$ system [5] and the subsequent controlled synthesis of methylalumoxane (MAO) [6], which provided a potent cocatalyst able to activate group 4 metallocenes (and a large number of other transition metal complexes, too) towards the polymerization of ethene and virtually any 1-olefins. The immediate introduction of chiral and stereorigid metallocene-based catalysts allowed the synthesis of stereoregular polymers [7, 8]. After thirty years, the efforts of the scientific community succeeded in the homogeneous "replica" of the heterogeneous catalysts.

The so called "metallocene revolution" paved the route to an impressive and detailed understanding of, and control over, the mechanistic details of olefin insertion, chain growth and chain termination processes. The knowledge at atomic level of many mechanistic details has allowed for a fine tailoring of the catalysts, and the homogenous catalysts proved to be more flexible than the heterogeneous ones. It has been possible to tune the structure of these catalysts to obtain a series of new stereoregular polymers, in particular of a series of new crystalline syndiotactic polymers, to obtain a better control of the molecular mass distribution as well as, for copolymers, a better control of the comonomer composition and distribution, and to synthesize a new family of low-density polyethylene with long chain branches.

A main feature of the new homogeneous catalysts is that they can be "single site", that is they can include all identical catalytic sites. This can be a great advantage with respect to the heterogeneous catalytic systems, for which several sites with different characteristics are present. Several aspects relative to the catalytic behavior of these "single site" stereospecific catalysts have been described in some recent reviews [9-14].

Although the computer modeling of Ziegler-Natta polymerization reactions started with the heterogeneous catalytic systems [15-23], the discovery of the homogeneous catalysts gave to the theoretical community well-defined systems to work with. Most importantly, many systems are composed by roughly 20 atoms, and this allowed for studies on "real size" systems, and not on oversimplified models which are too far from the real catalysts (as often computational chemists are obliged, to make the systems treatable).. This stimulated cultural exchanges between experimentalists and theoreticians and, even more importantly, to establish strong connections between theoreticians and the industry. To date, from a theoretical point of view, probably this is the most investigated organometallic reaction.

This chapter covers the elementary steps which are relevant to the polymerization of olefins with group 4 catalysts, and special emphasis is dedicated to systems with a substituted biscyclopentadienyl-based ligand, or with a monocyclopentadienylamido-based ligand (the so-called constrained geometry catalysts, CGC) of Figure 1, since these are the most investigated (the mono-Cp systems to a less extent) and the ones of possible industrial relevance.

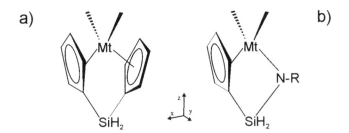

Figure 1. General structure of group 4 polymerization catalysts. Generic bent bis-Cp metallocene, part a), generic mono-Cp or constrained geometry catalyst (CGC), part b).

In particular, we will focus on the elementary steps which compose the propagation reaction. Beside an extensive coverage of literature data, the performances of various pure and hybrid density functional theory (DFT) approaches, and of classical *ab initio* Hartree-Fock (HF) and Møller-Plesset perturbative theory up to the second order (MP2) will be discussed through a systematic study of ethene insertion into the Zr–CH$_3$ σ-bond of the H$_2$Si(Cp)$_2$ZrCH$_3^+$ system. A comparison with singly and doubly excited coupled clusters single point calculations with a perturbative inclusion of triple excitations [CCSD(T)] will be also presented. The effects of the basis set on the insertion barrier will be discussed with a series of single point MP2 calculations.

Aspects concerning the regio- and stereochemical behavior of these catalysts in the stereospecific polymerization of propene (or 1-olefins, in general) will be not discussed in details since these topics are at the center of several reviews recently published [11, 12, 14, 24, 25]. Nevertheless, in the final sections we will briefly report about these points.

2.   GENERAL CONCEPTS

The most investigated group 4 bis-Cp and mono-Cp based catalysts are both pseudotetrahedral d^0 organometallic compounds in which the transition metal atom, beside the bis-Cp or mono-Cp ligand, bears two σ-ligands (usually Cl⁻ or CH_3^-). The bis-Cp or mono-Cp ligand remains attached to the metal during polymerization (for this reason they are also referred to as "ancillary" or "spectator" ligands) and actually defines the catalyst performances   (activity,   molecular   weights,   stereoselectivity, regioselectivity). It is well established that the active polymerization species is an alkyl cation (where the alkyl group is the polymeric growing chain). Therefore, one or both of the two σ-ligands are removed when the active catalyst is formed.

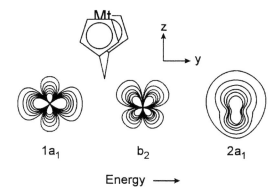

Figure 2. Contour diagram in the *yz* plane of the three most important molecular orbitals of the generic d^0 bent metallocene Cp_2Mt of C_{2v} symmetry. Solid and dashed lines correspond to positive and negative contour of the wave function.

All chemical transformations relevant to metal/olefin reactions occur at the three orbitals in the plane between the two Cp rings or between the Cp and the N atom in the case of the CGC catalysts. This plane is usually referred to as the "wedge" or belt of the catalyst. The first detailed analysis of the electronic structure of group 4 metallocenes, and the implications on their chemistry, was performed by Lauher and Hoffmann with simple

Extended Hückel calculations [26]. The metallocene equatorial belt is in the *yz* plane, and the C_2 axis is along the *z* axis.

Of the five frontier orbitals, the most important to the following discussion are the three low-lying $1a_1$, b_2 and $2a_1$ orbitals reported in Figure 2. All three orbitals have significant extent in the *yz* plane, which corresponds to the plane defining the equatorial belt of the metallocene. The b_2 orbital is chiefly d_{yz} in character, while the two a_1 orbitals in addition to contribution from the d_{x2-y2} and d_{z2} orbitals contain s and p_z contributions. The $1a_1$ orbital resembles a d_{y2} orbital and is directed along the *y* axis, while the $2a_1$ orbital is the highest in energy between the three orbitals, and points along the *z* axis. BP86 calculations we performed on the $H_2Si(Cp)_2Zr^{2+}$ fragment confirm this framework.

The mechanism generally accepted for olefin polymerization catalyzed by group 3 and 4 transition metals is reported in Figure 3, ands it is named after Cossee [27-29]. It substantially occurs in two steps; i) olefin coordination to a vacant site; ii) alkyl migration of the σ-coordinated growing chain to the π-coordinated olefin. Green, Rooney and Brookhart [30, 31] slightly modified this mechanism with the introduction of a stabilizing α-agostic interaction which would facilitate the insertion reaction. The key features of the insertion mechanism are that the active metal center must have an available coordination site for the incoming monomer, and that insertion occurs via chain migration to the closest carbon of the olefin double bond, which undergoes *cis* opening with formation of the new metal-carbon and carbon-carbon bonds. Consequently, at the end of the reaction the new Mt–chain σ-bond is on the site previously occupied by the coordinated monomer molecule.

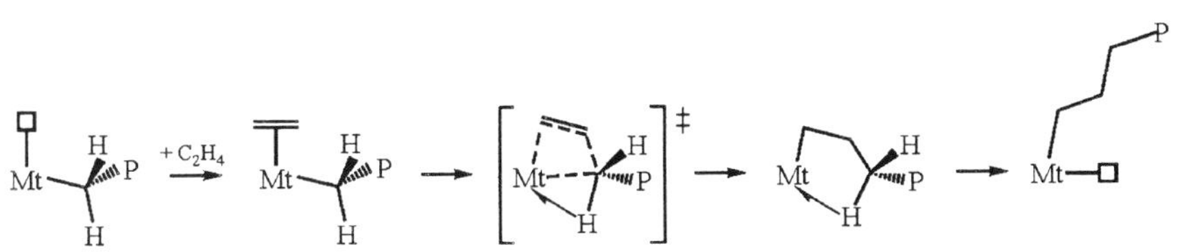

Figure 3. Modified Cossee mechanism for the polymerization of olefins with early transition metals. Green, Rooney and Brookhart introduced the presence of the adjuvant α-agostic interaction in the transition state.

3.   THE PROPAGATION REACTION

In the following sections we will focus on geometric and energetic aspects of the various species which compose the mechanism of Figure 3. We will start with the species prior olefin coordination (Section 3.1), we will then move to the olefin coordination step (Section 3.2), and we will terminate with the olefin insertion step (Section 3.3).

3.1   Olefin-free species

The position of the Zr–C(growing chain) σ bond in the absence of a further ligand (e.g. counterion, solvent, monomer) is of relevance for this class of reactions, and several authors investigated it theoretically. As for the simple model systems of the type $Cl_2TiCH_3^+$ and $H_2TiCH_3^+$, calculations based on classical *ab initio* [32-36], GVB [37], DFT [34-36] and CCSD(T) methods [36] suggested an off-axis geometry (see Figure 4), in agreement with the pioneering EHT analysis of Hoffmann and Lauher [26]. A systematic study by Ziegler and co-workers on various d^0 model systems of the type $L_2MtCH_3^{n+}$ (n = 0, 1), where Mt is a group 3 or 4 metal atom, and L is equal to CH_3^-, NH_2^- or OH^- [38], suggested an increased preference for the off-axis conformation as one moves down within a triad. This result was explained by a reduced steric pressure of the L ligands (which favors the on-axis geometry) bonded to a big metal at the bottom of the triad. Moreover, the energy of the $2a_1$ orbital which is responsible for on-axis bonding increases along the triad, and therefore the preference for the off-axis geometry is enhanced [38].

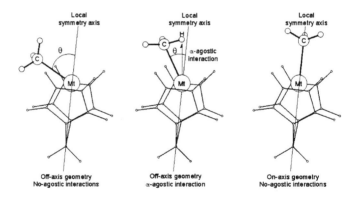

Figure 4. Minimum energy geometries of a generic bis-Cp olefin-free system.

When the models include the more representative Cp rings, the results obtained with different methods are contradictory. In fact, the HF and MP2 calculations on $H_2Si(Cp)_2MtCH_3^+$ and $Cp_2MtCH_3^+$ (Mt = Ti, Zr, Hf) [33, 35, 39, 40], as well as Car-Parrinello molecular dynamics symulations [41] suggested that the CH_3^- group is oriented along the symmetry axis, although in the crystalline structure of $[1,2-(CH_3)_2C_5H_3)]_2Zr(CH_3)^+\cdot CH_3B(C_6F_5)_3^-$ the methyl group is clearly off-axis [42]. Morokuma and co-workers suggested that the off-axis orientation of the methyl group in the crystalline structure could be due to the presence of the negative counterion. With the methyl group off-axis, a better electrostatic interaction between the two charged ions could be obtained.

On the contrary, the MP2 and DFT calculations of Ahlrichs on the $Cp_2TiCH_3^+$ system [35], and the DFT calculations of Ziegler on the $Cp_2TiCH_3^+$ and $H_2Si(Cp)_2ZrCH_3^+$ systems [43] suggested that the CH_3^- group is off-axis oriented. The recent analysis of Lanza, Fragalà and Marks indicated that already at the MP2 level, correlation effects favor the presence of a α-agostic interaction that pushes the Zr–C bond away from the local symmetry axis. The value of the Cp–Mt–Cp bending angle α is another key factor that influences the relative stability of the on- and off-axis geometries [37]. The GVB calculations of Goddard and co-workers showed that the on-axis geometry is favored by larger α values, due to an increased steric pressure of the Cp rings on the CH_3^- group, which clearly favors the on-axis geometry.

Since a systematic study of this point is still missing, we investigated the performances of different computational approaches through geometry optimizations of the $H_2Si(Cp)_2ZrCH_3^+$ species with different pure and hybrid DFT functionals, and at the HF and MP2 level of theory. The main geometrical parameters are reported in Table 1.

Table 1. Geometries and energies of the $H_2Si(Cp)_2ZrCH_3^+$ species. The angle θ and the α-agostic interaction are defined in Figure 4. A geometry with a α-agostic interaction was not found at the HF level.

QM Level	α-agostic interaction			No agostic interaction		
	θ (deg)	Zr–H (Å)	E (kcal/mol)	θ (deg)	Zr–H (Å)	E (kcal/mol)
BP86	64.6	2.29	0	67.0	3.02	1.8
BPW91	64.4	2.31	0	66.7	3.02	1.7
BLYP	61.9	2.32	0	62.7	3.01	1.7
B3LYP	28.9	2.31	0	60.4	2.97	1.2
B1LYP	27.5	2.31	0	58.0	2.95	1.3
HF	–	–	–	0.2		0
MP2	30.5	2.27	0	74.1	3.03	1.0

With the exception of the HF structure, in all cases the Zr–CH$_3$ bond is bent away from the local symmetry axis. When this deviation is larger, no agostic interactions were found, whereas to smaller values of the angle θ (see Figure 4) a α-agostic interaction is associated. We always found these two minimum energy situations. Differently, at the HF level we only found one structure of minimum energy, with the Zr–CH$_3$ bond perfectly aligned to the symmetry axis, and with no signs of agostic interactions. The pure BP86,

BPW91 and BLYP functionals predict that the angle θ is larger than 60°, independently of the presence or not of the agostic interaction, with slightly smaller values when the α-agostic interaction is present. Differently, the hybrid B1LYP and B3LYP functionals, and the MP2 method predict substantially smaller values of the angle θ (roughly 30°) for the α-agostic geometry. In all cases, the preferred geometry corresponds, by roughly 1.5 kcal/mol, to the structure with the α-agostic interaction. Finally, single point CCSD(T) calculations on the B3LYP geometries predict that the two structures are of substantially the same energy, since the α-agostic geometry is preferred by 0.2 kcal/mol only. These conclusions are in agreement with all the precedent studies which indicated that whatever geometry is favored, the potential energy surface for this swing motion of the Zr–C bond in the equatorial belt of the metallocene is relatively flat.

In the CGC catalysts, where the steric pressure of one of the Cp ligand is missing, off-axis geometries are more favored. In this case, the HF calculations of Lanza, Fragalà and Marks on the $H_2Si(Cp)(^tBuN)Mt(CH_3)^+$ (Mt = Ti, Zr) system indicated that for Ti the preference for the on-axis geometry is reduced by more than 5 kcal/mol relative to the off-axis geometry, whereas in case of Zr the on-axis geometry is not a minimum on the potential energy surface, but a transition state which connects two symmetrically related off-axis minimum energy situations. This detailed analysis substantially confirms the previous BP86 studies of Ziegler [44].

As for d^0 group III metallocenes, Goddard [37], Ziegler [38, 45] and co-workers found on-axis geometries for the Cp_2ScH [37], Cp_2ScCH_3 [45] and L_2MtCH_3 [38] (L = CH_3^-, NH_2^-, OH^-) species. The preferential on-axis geometry for all the neutral Sc species was ascribed to the higher s orbital contribution to bonding for group 3 metals with respect to group 4 metals.

As an alkyl group longer than a simple methyl group is σ bonded to the metal atom, the situation is different due to the possible formation of β- and γ-agostic bonds. With group 4 metallocenes, all authors substantially found a slightly off-axis geometry in the presence of a β-agostic interaction. The systematic study of systems of the type $L_2MtC_2H_5^{n+}$ (n = 0, 1), where Mt is a group 3 or 4 metal atom, and L is equal to CH_3, NH_2, OH, performed by Ziegler and co-workers, showed that the β-agostic bond only weakly perturbs the potential energy surface, which substantially remains similar to those present in the systems $L_2MtCH_3^{n+}$ [38].

As for the $H_2Si(Cp)_2ZrCH_3^+$ species, to test the performance of different computational approaches we performed geometry optimizations with different pure and hybrid DFT functionals, and at the HF and MP2 level of theory of the $H_2Si(Cp)_2ZrC_2H_5^+$ species. The main geometrical parameters are reported in Table 2.

Firstly, the HF and all the DFT methods, either pure or hybrid, predict that the Zr–C bond almost lies on the local symmetry axis. Differently, in the MP2 geometry the Zr–C bond is considerably bent away from the local symmetry axis, since the θ angle is close to 60°. This result is consistent with the calculations reported in Table 1 for the $H_2Si(Cp)_2ZrCH_3^+$ system, where the MP2 geometry resulted in the largest values of the θ angle. However, all the methodologies concord on the presence of a rather strong β-agostic interaction, since the Zr–H2 distance is in the range 2.10-2.15 Å. The BP86 functional results in the stronger agostic interaction (shortest Zr–H2 distance).

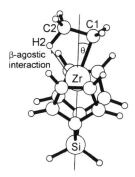

Figure 5. Minimum energy geometry of the $H_2Si(Cp)_2ZrC_2H_5^+$ species.

Table 2. Geometries of the β-agostic structure assumed by the $H_2Si(Cp)_2ZrC_2H_5^+$ species. The angle θ and the atoms numbering are defined in Figure 5.

QM Level	θ	Zr–C1–C2	C1–C2–H2	Zr–H2	C2–H2
	(deg)	(deg)	(deg)	(Å)	(Å)
BP86	1.8	84.8	113.3	2.10	1.18
BPW91	1.2	84.9	113.4	2.11	1.18
BLYP	1.9	85.4	112.9	2.13	1.18
B3LYP	4.3	85.5	113.2	2.13	1.16
B1LYP	5.2	85.4	113.2	2.13	1.16
HF	16.9	86.8	113.4	2.19	1.15
MP2	61.7	84.6	114.9	2.15	1.15

To check for the flatness of the potential energy surface relative to the swing motion of the Zr–C bond in the equatorial belt of the metallocene, we performed a geometry optimization at the B3LYP level, with the angle θ fixed at 61.7°, which is the value it assumes at the MP2 level. The optimized structure resulted to be only 0.8 kcal/mol higher in energy relative to the unconstrained B3LYP geometry, which suggests that the Zr-C bond is almost free to oscillate in the equatorial belt of the catalytic species.

Finally, it is clear that the presence of a γ- or a δ-agostic bond favor off-axis geometries, since the on-axis geometry would push the C atom which participates in the agostic interaction towards the Cp rings. In this respect, Ziegler calculated that for the $Cp_2Zr(n\text{-butyl})^+$ species the angle θ increases from 21° to 48° and finally to 56°, as the agostic interaction switches from β to γ and finally to δ [46].

3.2   Olefin coordination.

The electronics behind olefin coordination to group 4 cationic L_2MtR^+ species was studied in details by Marynick, Morokuma and co-workers [33, 47]. Their analysis indicated that olefin coordination is due to in-phase interactions between the olefin π orbital with metal orbitals corresponding to the $2a_1$, mainly, and to one lobe of the b_2 orbitals of Figure 2. A good overlap between the olefin π orbital and these metal orbitals is obtained also when the olefin is rotated by 90°, to assume a geometry in which the C=C double bond is perpendicular to the equatorial belt of the metallocene. Since group 4 cations contain d^0 metals, no back-bonding from the metal to the olefin $π^*$ orbital is present. These observations imply a small electronic barrier to olefin rotation, in agreement with the static *ab initio* calculations of Morokuma and co-workers, which calculated a barrier lower than 1 kcal/mol for olefin rotation in the system $Cl_2TiCH_3(C_2H_4)^+$ [33], and with the first principles molecular dynamics simulation of the $Cp_2ZrC_2H_5(C_2H_4)^+$ system by Ziegler and co-workers [48].

Several authors calculated the olefin uptake energy to olefin-free group 4 bis-Cp and mono-Cp polymerization catalysts. When the alkyl group is the simple methyl group, olefin coordination usually occurs in a barrierless fashion, and uptake energies in the range 15–30 kcal/mol (depending on the particular computational approach and/or metallocene considered) have been calculated [35, 39, 43, 49]. For neutral d^0 scandocenes, the interaction between the olefin and the metallocene is reduced due to the absence of the favorable electrostatic cation-olefin interaction [43, 50]. As a consequence, ethene uptake energies have been calculated to be roughly 20 kcal/mol lower than the corresponding uptake energies for the analogous cationic group 4 system.

Ethene uptake energies in the range 5–10 kcal/mol have been calculated for group 4 catalysts when alkyl groups longer than methyl are bonded to the metal atom, and β or γ-agostic interactions are present [43, 46, 49]. Also in these cases the olefin uptake is a barrierless process, unless bulky ligands as Me_5Cp rings are considered [49]. The substantially lower uptake energy values calculated in the presence of alkyl groups longer than methyl are ascribed to the presence of a β or γ-agostic interaction that stabilizes the olefin-free metallocene.

To examine systematically the performances of different computational approaches in evaluating geometry and energetics related to the coordination of olefins to group 4 catalysts, we performed geometry optimizations with different pure and hybrid DFT functionals, and at the HF and MP2 level of theory, of the $H_2Si(Cp)_2ZrCH_3(C_2H_4)^+$ complex. The main geometrical parameters are reported in Table 3. Since two different geometries were found for the $H_2Si(Cp)_2ZrCH_3^+$ species, for each methodology we considered the approach of ethene to the metal atom to both the $H_2Si(Cp)_2ZrCH_3^+$ geometries reported in Table 1.

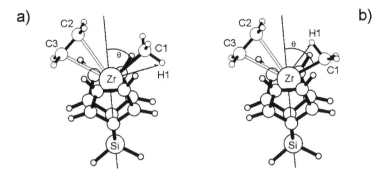

Figure5. Minimum energy geometries of the $H_2Si(Cp)_2ZrCH_3(C_2H_4)^+$ system starting from the olefin-free $H_2Si(Cp)_2ZrCH_3^+$ species without, part a, and with, part b, a α-agostic interaction.

As for the olefin-free species, two different geometries can be localized. The most stable is derived from coordination of the olefin to the α-agostic olefin-free $H_2Si(Cp)_2ZrCH_3^+$ species, with the geometry obtained from ethene coordination to the $H_2Si(Cp)_2ZrCH_3^+$ without agostic interactions roughly 1-2 kcal/mol higher in energy, as indicated by the lower ethene uptake energies. The Zr–H distances and the θ angle in the structures with a α-agostic interaction are rather similar, whatever methodological approach is used. In the structures without agostic interactions, instead, the θ angle is comprised between 50° and 65°, the larger values corresponding to the pure BP86 and BPW91 functionals. In all cases the olefin is unsymmetrically coordinated to the metal atom, with the ethene C atom closest to the CH_3 group roughly 0.1–0.3 Å farther away from the metal atom. The pure BP86 and BPW91 functionals, and the MP2 geometry show slightly shorter Zr–C(ethene) distances.

The HF and all the DFT uptake energies are in the relatively small range of 18.3–22.9 kcal/mol. The MP2 uptake energy, instead, is remarkably higher, 30.7 and 32.0 kcal/mol, depending on the particular geometry considered. In the case of the B3LYP and MP2 geometries we also evaluated the influence of the BSSE on the binding energy with the counterpoise approach of Boys and Bernardi [51]. The BSSE is definitely large, roughly 8 and 13 kcal/mol at the B3LYP and MP2 level, respectively, and strongly reduces the B3LYP and MP2 uptake energies from roughly 22 and 31 kcal/mol to roughly 14 and 18 kcal/mol, respectively. Interestingly, after BSSE correction the difference between the B3LYP and MP2 uptake energies is reduced from 11 to 4 kcal/mol, only. These results indicate that BSSE corrections must be considered for realistic olefin binding energies.

Table 3. Geometries and ethene uptake energies for the $H_2Si(Cp)_2ZrCH_3(C_2H_4)^+$ species. The angle θ and the atoms numbering scheme are defined in Figure 6. The B3LYP and MP2 BSSE corrected uptake energies are reported in parenthesis. In both geometries, the ethene uptake energy is calculated with respect to the most stable $H_2Si(Cp)_2ZrCH_3^+$ geometry.

QM Level	θ (deg)	Zr–H2 (Å)	Zr–C2 (Å)	Zr–C3 (Å)	E(uptake) (kcal/mol)
		No agostic interaction			
BP86	63.9	3.00	2.61	2.91	21.8
BPW91	65.4	3.00	2.63	2.92	20.0
BLYP	59.8	2.98	2.68	2.96	19.9
B3LYP	55.4	2.93	2.68	2.95	21.9 (13.8)
B1LYP	54.8	2.91	2.69	2.96	21.6
HF	43.1	2.82	2.90	2.96	18.3
MP2	50.4	2.93	2.64	2.88	30.7 (17.4)
		α-agostic interaction			
BP86	66.7	2.28	2.68	2.86	22.9
BPW91	66.9	2.29	2.69	2.88	20.9
BLYP	65.4	2.34	2.74	2.93	20.4
B3LYP	64.9	2.35	2.73	2.92	22.1 (14.6)
B1LYP	64.5	2.36	2.75	2.93	21.8
MP2	68.3	2.26	2.68	2.83	32.0 (19.1)

Before concluding this section, it has to be reminded that all the above calculations represent gas-phase processes without solvent and/or counterion effects, and without unfavorable entropic contributions. Recent studies have shown that they modify substantially the olefin-coordination energetics.

With regard to solvent effects, for cationic group 4 catalytic systems the solvent/metallocene interaction is mainly electrostatic (as the olefin/metallocene interaction). Therefore, it is reasonable to expect similar solvent and olefin coordination energies. In fact, Rytter and co-workers calculated that toluene coordinates with one C–C bond to the $Cp_2ZrC_3H_7^+$ species, and the binding energy of 4 kcal/mol is only 2.5 kcal/mol lower than the ethene binding energy to the same Zr-catalyst [49], while Cavallo et al. calculated binding energies around 8–10 kcal/mol for the coordination of benzene to some bis-Cp and substituted bis-Cp systems [52, 53].

With regard to the presence of the counterion, it has to be considered that it is the species which coordinates more strongly to the metallocene, due to its negative charge. Gas-phase heterolytic breakage of the metallocene$^+$-counterion$^-$ ion-pair was calculated to be roughly 70–100 kcal/mol, depending on the particular ion-pair considered [40, 54]. Solvent effects, usually accounted for with continuum models, reduce the energy required to dissociate the ion-pair [54], but still it remains a highly endothermic process. Thus, several groups started to replace the olefin coordination step, usually depicted as metallocene$^+$ + C_2H_4 → metallocene$(C_2H_4)^+$, with the reaction metallocene$^+$/An$^-$ + C_2H_4 → metallocene$(C_2H_4)^+$/An$^-$ (An$^-$ = counterion).

Fusco and co-workers performed the first simulations of this type, and showed that ethene coordination to associated ion-pairs of the type $Cp_2MtCH_3^+$/An$^-$ (An$^-$ = $Cl_2Al(CH_3)_2^-$ or $Cl_2Al[O(AlMe_3)AlMeH]_2^-$ as model for MAO) to give rise to olefin-coordinated species with the olefin sandwiched between the metallocene and the counterion is an endothermic process of roughly 10 kcal/mol [55, 56]. Very similar conclusions were obtained by Ziegler and co-workers, which modeled ethene and toluene (as solvent) coordination to the $Cp_2TiCH_3^+$/An$^-$, $CpMt(CH_3)_2^+$/An$^-$ and $H_2Si(Cp)(NH)TiCH_3^+$/An$^-$ systems (An$^-$ = μ-CH_3 coordinated $CH_3B(C_6F_5)_3^-$) [54].

Lanza, Fragalà and Marks modeled coordination of ethene to the CGC $H_2Si(Cp)(^tBuN)TiCH_3^+$/An$^-$ system (An$^-$ = $CH_3B(C_6F_5)_3^-$ μ-CH_3^- coordinated to the Ti atom) from the side opposite to the counterion through MP2 single point energetics on the HF optimized geometries [57]. Their calculations clearly indicated that the olefin-bound intermediate is only a shallow minimum energy situation. Along this line, Ziegler and co-workers performed static and dynamic BP86 simulations of the ethene insertion into the $Cp_2Zr_2CH_5^+$/$CH_3B(C_6F_5)_3^-$ ion-pair, and found that olefin coordination occurs with a large barrier [58].

Finally, the uptake energy values only represent a contribution to the total free energy of coordination. In fact, an always unfavorable uptake entropy has to be accounted for. Although few experimental data are available, it is reasonable to assume that the $-T\Delta S$ contribution to the free energy of olefin coordination to group 4 metallocenes at room temperature is close to the 10 kcal/mol value observed at 300 K for Ni and Pd compounds [59]. The few computational data also suggest a $-T\Delta S$ contribution close to 10 kcal/mol [38, 60]. As a consequence, olefin uptake energies higher than 10 kcal/mol are required to form stable olefin complexes in the gas-phase.

Whether the metal-olefin complex is a real chemical species, or the olefin undergoes direct insertion into the metal-carbon bond has been a matter of debate for many years. However, some experimental studies of the last years established that such species do exist [61-64], although as transient species. Moderately stable olefin adducts have been obtained when the olefin is tethered to the metal [61-66] or to the Cp ligands [67]. The experimental $\Delta G^{\ddagger}$ values for metal olefin dissociation are close to 10 kcal/mol [64, 65, 67]. Probably, the presence of the tether reduces strongly the entropy gain that favors the olefin dissociation, inducing the so-called "chelation effect". Upper bounds to the olefin uptake energy can be obtained by measurements of the π–σ–π processes in fluxional allyl derivatives of group 3 [68] and group 4 [69] metallocenes. Again, $\Delta G^{\ddagger}$ values close to 10 kcal/mol were observed. Very recent NMR experiments of Casey and co-workers on propene coordination/dissociation from the neutral group 3 $(C_5Me_5)Y_2CH_2CH_2CH(CH_3)_2$ system resulted in the rather lower propene binding ΔH° of 5 kcal/mol [70].

3.3   Insertion.

The insertion reaction of a coordinated olefin into the Mt–C σ-bond, where Mt is a group 4 metallocene, or a model of it, has been the subject of several theoretical studies [26, 32-36, 39-41, 43, 46, 47, 49, 71]. All authors agree that the insertion reaction occurs through a slipping of the olefin towards the first C atom of the growing chain, and that the four centers transition state assumes an almost planar geometry. For a model based on the analogous group 3 Cp_2ScCH_3 system, the transition state is only slightly more advanced relative to the one for the cationic zirconocene [43,50].

The electronics behind the insertion reaction is generally explained in terms of a simple three-orbitals four-electrons scheme. Hoffmann and Lauher early recognized that this is an easy reaction for d^0 complexes, and the relevant role played by the olefin π^* orbital in determining the insertion barrier [26]. According to them, the empty π^* orbital of the olefin can stabilize high energy occupied d orbitals of the metal in the olefin complex, but this stabilization is lost as the insertion reaction approaches the transition state. The net effect is an energy increase of the metal d orbitals involved in the d–π^* back-donation to the olefin π^* orbital. Since for d^0 systems this back-donation does not occur, d^0 systems were predicted to be barrierless, whereas a substantial barrier was predicted for d^n (n > 0) systems [26].

A similar picture was suggested by the DFT calculations of Ziegler and co-workers [72]. In agreement with Hoffmann and Lauher, for d^0 systems the lowest unoccupied molecular orbital (LUMO) of the olefin complex (see Figure 7) chiefly corresponds to a bonding $d-\pi^*$ metal-olefin interaction. In the transition state, the occupied sp^3 orbital of the first C atom of the growing chain (the one bonded to the metal) and the occupied π orbital of the olefin form an energetically unfavorable bonding/antibonding combination, with the latter corresponding to the highest occupied molecular orbital (HOMO). If the empty π^* orbital of the olefin mixes in, the antibonding character of the HOMO is transformed in a substantially more stable nonbonding HOMO, while the energy of the LUMO rises due to the $\pi-\pi^*$ mixing. Again, for d^0 systems the insertion reaction is substantially barrierless –since the LUMO energy does not contribute to the total energy–, whereas for d^n (n > 0) systems it is not, since for these systems the HOMO corresponds to the high energy LUMO of d^0 systems.

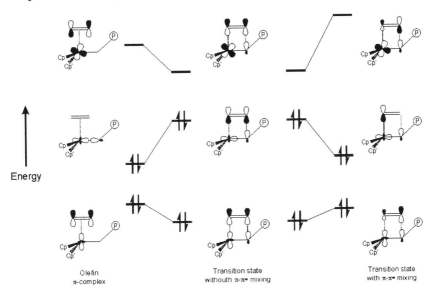

Figure 7. Molecular orbitals diagram of the mixing process involved in the insertion of ethene into a Mt-C(alkyl) bond for a generic d^0 neutral group 3 or cationic group 4 catalyst.

To quantify this point, Ziegler and co-workers compared the insertion barrier for ethene insertion into the cationic d^0 $Cp_2TiC_2H_5^+$ and neutral d^1 $Cp_2TiC_2H_5$ systems. The insertion barrier for the neutral d^1 system is roughly 20 times higher than the insertion barrier for the cationic d^0 system [72]. Finally, it is worth noting that Hoffmann and Ziegler predicted that d^1 and d^2 complexes can be suitable polymerization catalysts if other ligands can accept the d electrons in orbitals orthogonal to the π^* olefin orbital –which corresponds to a reduction of the relevance of the d-π^* interaction– [26, 72], while Ziegler also noted that if the occupied metal d orbitals are lower in energy, e.g. for late transition metals, the destabilization due to the disruption of the metal to olefin π^* orbital back-donation is smaller, and hence low insertion barriers are again possible [72].

The presence of a favorable α-agostic interaction which stabilize the transition state is another point of convergence between various authors [32-36, 39, 46, 49, 50, 73, 74]. Before continuing, it is worth noting that a short Zr–H(α) distance (indicative of a α-agostic interaction) is almost inevitable as the sp^3 orbital of the C atom of the growing chain bonded to Zr tilts away from the Zr–C axis to be oriented towards the closest C atom of the olefin, giving rise to the bonding interactions with the olefin itself. According to Janiak, the α-agostic stabilization becomes important through an increase in electron deficiency of the metal, that switches from a formally 16e$^-$ Zr in the olefin-coordinated reactant, to the formally 14e$^-$ Zr in the insertion product [73]. Similar ideas were developed by Grubbs and Coates, who also made a nice relationship between the hyperconjugative stabilization by β-hydrogen atoms of substrate undergoing nucleophilic substitution reactions in organic chemistry, and the agostic stabilization by α-hydrogen atoms in Ziegler-Natta catalysis [75].

Regarding the height of the insertion barrier, the situation is much more controversial, since pure density functionals and some MP2 calculations suggest that this a barrierless reaction, or it occurs with a negligible barrier. HF, hybrid density functionals and several post-HF calculations, instead, suggest a barrier in the range of 5–10 kcal/mol, roughly.

For instance, the static BP86 calculations of Ziegler and co-workers on the reaction of ethene with the $Cp_2ZrCH_3^+$ system predicts a barrier of 0.8 kcal/mol [50], whereas the reaction with the $Cp_2ZrCH_2CH_3^+$ system occurs with the negligible barrier of 0.2 kcal/mol [46]. Ahlrichs and co-workers investigated the $Cp_2TiCH_3(C_2H_4)^+$ system and found a considerable energy barrier and a transition state only without inclusion of electron correlation [35]. At the MP2 level, they found that the insertion reaction occurs on a very flat, downhill potential energy surface. However, the basis set they used (193 basis functions) was not particularly extended, and this can influence the MP2 energetics. When group 3 metals are considered, the situation is rather similar, since for ethene insertion on Cp_2ScCH_3, the static DFT calculations of Ziegler and co-workers predicted almost negligible insertion barriers, < 3 kcal/mol [43, 50]. For the insertion reactions C_2H_4 + $H_2Si(Cp)_2MtCH_3^+$ (Mt = Ti, Zr, Hf) Morokuma, and co-workers performed calculations at different levels of *ab initio* theory [39]. Their best estimate for the insertion barrier was 6.7 kcal/mol at the RQCISD level of theory. However, they also reported that the predicted barrier sensibly depends on the particular level of theory, and that larger basis sets were needed for accurate RQCISD calculations. Lanza and Fragalà investigated the $Cp_2MtCH_3(C_2H_4)^+$ (Mt = Ti and Zr) at the HF level with single point MP2 and CCSD calculations on the HF geometries. They found HF insertion barriers close to 20 kcal/mol for both metals, while the single point MP2 and CCSD insertion barriers are equal to 2.4 and 9.9 kcal/mol for Ti, and equal to 8.0 and 11.7 kcal/mol for Zr, respectively. However, for the Ti system they also calculated an insertion barrier of 1.2 kcal/mol if the geometry optimizations were performed at the MP2 level, and polarization functions were included on the C and H atoms [40].

First principles molecular dynamics simulations also resulted in processes with low free energy barriers. Meier and co-workers investigated ethene insertion on the $H_2Si(Cp)_2TiCH_3^+$ system with the Car-Parrinello method [41], and the insertion proceeded in a barrierless fashion, while Ziegler investigated the same reaction with the $Cp_2ZrC_2H_5^+$ system and a 5 kcal/mol free energy barrier was predicted. However, they also warned that quite longer simulation times were needed for quantitative predictions [48].

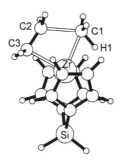

Figure 8. Transition state for the insertion of ethene with the $H_2Si(Cp)_2ZrCH_3^+$ system.

Since the situation about the height of the insertion barrier is not so clear, we performed a systematic comparison of the performances of different computational approaches in determining insertion barriers and geometries, with the aim to offer a further contribution to the discussion. The insertion transition state was located with different pure and hybrid DFT functionals, and at the HF and MP2 level of theory. The main geometrical parameters of the transition state for the insertion reaction of ethene into the Zr–C bond of the $H_2Si(Cp)_2ZrCH_3^+$ species are reported in Table 4.

Table 4. Geometries and activation barriers for the insertion reaction of ethene into the Zr–CH$_3$ bond of the H$_2$Si(Cp)$_2$ZrCH$_3^+$ species. The angle θ and the atoms numbering scheme are defined in Figure 8.

QM Level	Zr–C1	Zr–C3	C1–C2	Zr–H1	$E^‡$
	(Å)	(Å)	(Å)	(Å)	(kcal/mol
BP86	2.25	2.44	2.31	2.14	0.2
BPW91	2.25	2.43	2.30	2.14	0.2
BLYP	2.32	2.39	2.17	2.14	2.3
B3LYP	2.29	2.38	2.18	2.14	2.9
B1LYP	2.30	2.38	2.17	2.14	5.0
HF	2.34	2.35	2.14	2.20	16.3
MP2	2.27	2.39	2.19	2.10	2.8

The BP86 and BPW91 transition states occur relatively early, since to these functionals correspond the longest distances, roughly 2.30 Å, for the emerging C–C bond. The three BnLYP functionals and the MP2 approach predict that the transition state is reached later, when the emerging C–C bond is close to 2.17 Å. The latest transition state is predicted by the simple HF theory, at a distance of 2.14 Å. As reasonable, to longer C1–C2 distances in the transition state correspond shorter and longer Zr–C1 and Zr–C3 distances which correspond to the Zr–C σ-bonds which are going to be broken and formed, respectively. All the transition states are characterized by the presence of a strong α-agostic interaction, with Zr–H1 distances around 2.14 Å. The only exception is represented by the HF transition state which shows a longer Zr–H1 distance, around 2.20 Å.

With regard to the insertion barriers, the BP86 and BPW91 functionals predict extremely low barrier, around 0.2 kcal/mol only. The BnLYP functionals predict relatively higher insertion barriers, and the higher is the amount of HF exchange the higher is the insertion barrier. The HF predicts a remarkably high and unrealistic barrier (16.3 kcal/mol), in agreement with previous results [35, 39, 40]. At the MP2 level the insertion barrier is 2.9 kcal/mol, very close to the B3LYP value. The BP86 geometry we calculated is very similar to that calculated by Ziegler and co-workers for the same system, although in their geometry optimizations the BP86 gradient corrections were not included. The barrier to insertion they calculated, roughly 1 kcal/mol, is slightly larger than the one we predict with the BP86 functional.

In order to investigate the basis set effects on the insertion barrier, single point MP2 calculations on the MP2/MIDI-SVP geometries are reported in Table 5.

Table 5. Post-HF activation barriers for the insertion reaction of ethene into the Zr–CH3 bond of the $H_2Si(Cp)_2ZrCH_3^+$ species. All the reported insertion barriers were obtained through single point calculations on the MP2 geometries of Tables 3 and 4 (corresponding to run 3 in this Table). In the valence calculations the 1s orbitals on the C atoms, the orbitals up to 2p on the Si atom and up to the 3d on the Zr atom where not included in the active orbitals space. In the full MP2 calculations all occupied orbitals were correlated.

Run	Level of theory	Metal	$H_2Si(Cp)_2$	C_2H_4 and CH_3	$E^{\ddagger}$ (kcal/m
1	MP2-Valence	MIDI	SV	SV	6.0
2	MP2-Valence	MIDI	SV	SVP	2.8
3	MP2-Valence	MIDI	SVP	SVP	2.8
4	MP2-Full	MIDI	SVP	SVP	0.7
5	MP2-Valence	LANL2DZ	SVP	SVP	4.9
6	MP2-Valence	LANL2DZ	6-31G	6-31G	8.5
7	MP2-Valence	LANL2DZ	6-31G	6-31G(d,p)	5.6
8	MP2-Valence	LANL2DZ	6-31G(d,p)	6-31G(d,p)	5.9
9	MP2-Valence	LANL2DZ	6-31G	6-311G(d,p)	2.4
10	MP3-Valence	LANL2DZ	6-31G	6-31G(d,p)	11.2
11	MP4-Valence	LANL2DZ	6-31G	6-31G(d,p)	10.3
12	CCSD-Valence	LANL2DZ	6-31G	6-31G(d,p)	10.8
13	CCSD(T)-Valence	LANL2DZ	6-31G	6-31G(d,p)	8.7

Firstly, inclusion of polarization functions on the C and H atoms of the reactive groups (CH_3^- and C_2H_4) reduces considerably the insertion barrier (compare runs 1 and 2 as well as runs 6 and 7) and seems to be mandatory. Instead, inclusion of polarization functions on the ancillary $H_2Si(Cp)_2$ ligand has a negligible effect on the calculated insertion barrier (compare runs 2 and 3 as well as runs 7 and 8). Extension of the basis set on the reactive groups lowers further the insertion barrier (compare runs 7 and 9). Both the MIDI basis set on Zr, and the SVP basis set on the remaining atoms decrease the insertion barrier (compare runs 3, 5 and 8). Finally, the extension of the active orbitals space to include all the occupied orbitals reduces sensibly the insertion barrier (compare runs 3 and 4).

In Table 5 the insertion barrier at levels of theory higher than MP2 are also reported (runs 10-13). The MP3 and MP4 insertion barriers are both remarkably higher than the MP2 barrier. The CCSD insertion barrier also is quite larger than the MP2 barrier (5.2 kcal/mol above), but the perturbative inclusion of triple excitations in the couple cluster calculations reduces considerably the CCSD barrier, which is 8.7 kcal/mol (3.1 kcal/mol above the MP2 insertion barrier). The insertion barriers reported in Table 5 can be used to obtain a further approximation of the insertion barrier. In fact, the CCSD(T) barrier of 8.7 kcal/mol should be lowered by roughly 3 kcal/mol if the 6-311G(d,p) basis set would be used on the reacting C and H atoms (compare runs 7 and 9), and should be lowered by roughly 2 kcal/mol if all the occupied orbitals would be included in the active orbitals space (compare runs 3 and 4). When these corrections are added to the CCSD(T) barrier, the best estimate we obtain amounts to 3–4 kcal/mol roughly, if the LANL2DZ and 6-31G(d,p) basis sets are considered, and should be lower if the MIDI and SVP basis sets would be used.

With regards to the mono-Cp CGC catalysts, the BP86 calculations of Ziegler and co-workers on the $H_2Si(Cp)(NH)MtCH_3^+$ (Mt = Ti, Zr and Hf) model systems predicted a barrier of 3.8, 5.1 and 5.8 kcal/mol for Ti, Zr and Hf, respectively [44]. The HF and single points MP2 and CCSD calculations of Lanza, Fragalà and Marks on the $H_2Si(Cp)(^tBuN)MtCH_3^+$ (Mt = Ti and Zr) systems resulted in insertion barriers of 18.1, 7.9 and 11.6 kcal/mol for Ti, and of 22.2, 14.6 and 17.0 kcal/mol for Zr, respectively [40]. The authors remarked that these barriers are certainly overestimated, since inclusion of polarization functions on the C and H atoms, and geometry optimizations at the MP2 level reduced the insertion barrier for the Ti system to 3.5 kcal/mol only, in rather good agreement with the BP86 value of Ziegler.

As for the final state after the insertion step, all authors do agree that the α-agostic interaction occurring in the transition state evolves into a γ-agostic one. Moreover, the γ-agostic is usually predicted to correspond to the kinetic product that, through a conformational rearrangement of low energy, could evolve into the thermodynamic β-agostic product [33, 34, 36, 37, 39-41, 46, 71].

When the insertion reaction takes place on models of growing chain longer than simple methyl groups, reaction path due to different orientations/rearrangements of the growing chain have to be considered. Ziegler and co-workers performed static and dynamics DFT calculations for ethene insertion on $Cp_2ZrC_2H_5^+$ systems, while Rytter, Ystenes and co-workers as well as Ruiz-López and co-workers performed DFT calculations for ethene insertion on the $Cp_2ZrC_3H_7^+$ and $(Me_5Cp)_2ZrC_3H_7^+$ systems considering both frontside and backside ethene approach to the catalyst (corresponding to ethene approach from the C–H agostic bond side, or from the from Zr–C σ-bond side, respectively) [46, 49, 71]. The frontside insertion requires a rearrangement of the growing chain from a β-agostic to an α-agostic orientation. This barrier was predicted to be rather low. Once the growing chain adopts a α-agostic geometry, the insertion is almost barrierless according to the BP86 calculations of Ziegler [46], lower than 2.5 kcal/mol and equal to 1.6 kcal/mol according to the BLYP calculations of Rytter and co-workers [49], and of Ruiz-López and co-workers [71].

As for the backside approach, the insertion can occur without any particular rearrangement. The activation barrier is about 7 kcal/mol high, and is slightly assisted by a β-agostic interaction [46,71]. Støvneng and Rytter, and more recently Rytter, Ystenes and co-workers studied the propagation step on a γ-agostic growing chain for ethene polymerization on the $Cp_2ZrC_3H_7^+$ system [49, 74]. The last authors investigated the same reaction path with the analogous $(Me_5Cp)_2ZrC_3H_7^+$ system also. They found that insertion can occur easily also in presence of the γ-agostic interaction, with activation barriers lower than 3 kcal/mol. Finally, Ystenes proposed the so-called "trigger mechanism" [76]. This mechanism involves a two-monomer transition state, where the entering of a new monomer unit triggers the insertion of the already complexed monomer. Specific calculations to support this model are not available.

A clear experimental estimate of the intrinsic reaction barrier to olefin insertion is still missing. The NMR analysis of Erker and co-workers estimated the intrinsic activation barrier for 1-olefin insertion into the Zr–C bond of the $(MeCp)_2Zr(\mu-C_4H_6-borate\ betaine)$ to be about 10–11 kcal/mol [69, 77]. Very recent NMR experiments of Casey and co-workers on the propene insertion into the Y–C σ-bond of the neutral group 3 $(C_5Me_5)YCH_2CH_2CH(CH_3)_2$ system resulted in a $\Delta G^{\ddagger}$ of 11.5 kcal/mol [70].

With regards to the mono-Cp CGC catalysts, the BP86 static and dynamic calculations of Ziegler and co-workers on the ethene reaction with the $H_2Si(Cp)(NH)TiC_3H_7^+$ system indicated that interconversion between geometries with different agostic interactions (α, β and γ) is rather easy. Ethene coordination to the preferred γ-agostic olefin-free species is likely to lead to the most stable β-agostic olefin-bound species. The insertion reaction occurs after a rearrangment of the growing chain from a β-agostic to a α-agostic geometry. The static energy barrier they calculated for the insertion reaction is 4.7 kcal/mol, while the dynamic free energy insertion barrier was estimated to be 5.8 kcal/mol.

All the calculations reported above indicate that the insertion reaction is a rather easy process, and usually underestimate the experimental apparent propagation barrier. This led several groups to include solvent and/or counterion in the computer modeling of the insertion reaction. The aspects regarding olefin coordination were discussed in Section 3.2 and thus will be not repeated here. With regard to the insertion reaction, Lanza, Fragalà and Marks modeled the insertion of ethene with the CGC $H_2Si(Cp)(^tBuN)TiCH_3^+$ system in the presence of a $CH_3B(C_6F_5)_3^-$ counterion μ-coordinated to the Ti atom through the CH_3^- group through MP2 single point energetics on the HF optimized geometries [57]. Their calculations clearly indicated that the ethene-bound intermediate is only a shallow minimum energy situation along the reaction coordinate, when the counterion is considered, and the insertion barrier raises from 7.9 kcal/mol (gas-phase simulations at the same level of theory) [40] to roughly 14 kcal/mol in the presence of the counterion

[57]. Along this line, Ziegler and co-workers performed static and dynamic BP86 simulations of the ethene insertion into the $Cp_2Zr_2CH_5^+/CH_3B(C_6F_5)_3^-$ ion-pair [58]. In the olefin-free ion-pair the ethyl group which simulates the growing chain shows no agostic interactions with the Zr atom. Ethene coordination occurs from the side opposite to the anion, in agreement with the simulations of the CGC-catalysts of Lanza et al. After olefin coordination, which occurs with a large barrier, the insertion step has a low energy barrier. The overall barrier, from separated reactants to the transition state, was calculated to be 11.6 kcal/mol, and the largest part of the barrier was suggested to be related to the olefin complexation [58].

The last results described are a strong indication that any computer modeling of the activity of early transition metal catalysts for the polymerization of olefins probably requires the inclusion of the counterion in the simulations.

4.   REGIOSELECTIVITY

The polymerization of 1-olefins introduces the problems of regioselectivity and stereoselectivity. In this section we will focus on the origin of the regioselectivity of the insertion reaction (primary vs. secondary 1-propene insertion, see Figure 9), while the origin of the stereoselectivity will be discussed in the next section.

Before continuing, it has to be noted that the energy difference between the secondary and primary propene insertion, ΔE_{regio}, can be considered composed by two main contributions, electronic and steric. The steric contribution to ΔE_{regio}, due to steric interaction between the monomer, the growing chain and the ligand skeleton, was modeled successfully through simple molecular mechanics calculations [78-80], and was reviewed recently [11,24]. For this reason in the following we will focus only on the electronic contribution to ΔE_{regio}.

Figure 9. Representation of primary (regioregular) and secondary (regioirregular) propene insertions into the Mt–C bond of group 4 polymerization catalysts, parts a and b, respectively.

Regarding the origin for the electronic preference for primary versus secondary propene insertion, the *ab initio* calculations of Morokuma and co-workers on approximated transition state geometries for primary and secondary propene insertion into the Ti–methyl σ-bond of the system $Cl_2TiCH_3(propene)^+$ indicated that secondary propene insertion was disfavored by 4.6 kcal/mol relative to primary propene insertion [33]. The latter is essentially stabilized by favorable electrostatic attraction and less serious exchange repulsion, in agreement with the experimental results and the Markovnikov rule of organic chemistry. Their Mulliken analysis on the transition state for ethene insertion into the Ti–methyl σ-bond indicated that the ethene C atom that is going to be bonded to the metal atom is more negatively charged relatively to the ethene C atom that is going to be bonded to the methyl group. Thus, they argued that the additional methyl group of the propene would give a more favorable electrostatic interaction with the C atom of the olefin closer to the methyl group than to the metal atom. That is, primary insertion is favored over secondary insertion. Moreover, in the transition state for the secondary insertion, the propene methyl group is closer to the additional metal ligands than for primary insertion, causing a larger exchange repulsion.

To confirm these conclusions with more realistic models, we performed BP86 calculations on the primary and secondary propene insertion into the $Zr–CH_3^+$ σ-bond of the $H_2Si(Cp)_2ZrCH_3(propene)^+$ system. The transition states for primary and secondary propene insertion are reported in Figure 10.

Firstly, the presence of the methyl group of the propene molecule does not modify substantially the geometry of the transition state, which remains substantially planar, and both of them are stabilized by a strong α-agostic interaction. However, the transition state occurs slightly earlier in the primary than in the secondary propene insertion, as evidenced by the slightly less formed C–C bond in the transition state for the primary insertion. The length of the breaking Zr–C bond is the same in both transition states, whereas the forming Zr–C bond is considerably longer in the transition state leading to secondary propene insertion.

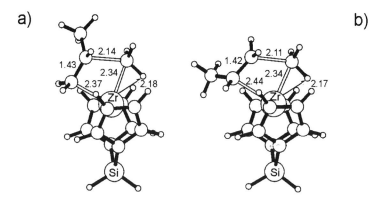

Figure 10. Transition states for the primary, part a, and secondary, part b, insertion of propene into the Zr–CH$_3$ bond of the H$_2$Si(Cp)$_2$ZrCH$_3^+$ system.

Of relevance, however, is the fact that the transition state leading to secondary propene insertion is 3.6 kcal/mol higher in energy than the transition state leading to primary insertion. This results confirms the pioneering conclusions of Morokuma and co-workers, and are in substantial agreement with the high preference experimentally observed for primary propene insertion with group 4 metallocenes. In fact, in most polypropylene samples obtained by these catalytic systems the regioirregularities are usually in the range of 0-2% [11].

5.   STEREOCHEMISTRY

In the previous section we have established that propene insertion into a metal-carbon bond with group 4 polymerization catalysts is mostly primary, while in this section we focus on the choice of the propene enantioface (upon coordination a prochiral olefin as propene gives rise to not superposable *re* and *si* coordinations, see ref. 81) which defines the stereochemistry of each insertion (the catalyst stereoselectivity). Since every propene insertion, whatever its orientation, creates a new stereogenic center, the catalyst stereospecificity (which defines the stereoregularity or tacticity of the polymer produced) is determined by the stereochemical relationship between the stereogenic carbon atoms in the polymer chain. Multiple insertions of the same enantioface produce a polymer chain with chiral centers of the same configuration, i.e. an isotactic polymer (Figure 11a). Multiple insertions of alternating enantiofaces produce a polymer chain with chiral centers of alternating configuration, i.e. a syndiotactic polymer (Figure 11b). Random enantioface insertions produce a polymer chain with no configurational regularity, i.e. an atactic polymer (Figure 11c). While both isotactic and syndiotactic polypropylenes are partially crystalline materials with relatively high melting points (up to 160-170 °C for isotactic and ~150 °C for syndiotactic polypropylene), atactic polypropylene is a fully amorphous polymer, since it lacks long-range stereochemical regularity.

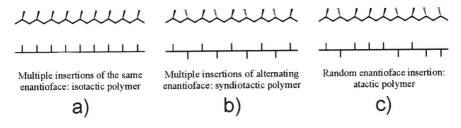

Multiple insertions of the same
enantioface: isotactic polymer
a)

Multiple insertions of alternating
enantioface: syndiotactic polymer
b)

Random enantioface insertion:
atactic polymer
c)

Figure 11. Schematic representation of iso-, syndio- and atactic polymers, parts a, b and c, respectively. Chain segments are shown in their trans-planar and modified Fisher projections.

A necessary (but not sufficient) prerequisite for models of catalysts for the stereospecific polymerization of 1-olefins polymerization, is the stereoselectivity of each monomer insertion step. The possible origin of stereoselectivity in this class of systems was investigated through simple molecular mechanics calculations [11, 14, 24, 32, 52, 78-80, 82-86].

Since molecular mechanics cannot be used to calculate the energy of transition states, suitable models were adopted. These models are extremely similar to the π-olefin complex with an orientation of the growing chain rather similar to that adopted when a α-agostic interaction is present. They were often called pre-insertion intermediates because the insertion transition state could be reached from these intermediates with a minimal displacement of the reacting atoms.

All the molecular modeling studies performed indicate that for group 4 metallocenes (independently of their structure and symmetry), when a substantial stereoselectivity is calculated for primary monomer insertion, this is not due to direct interactions of the π-ligands of the chiral metallocene with the monomer. Instead, the origin of stereoselectivity is connected to a chiral orientation of the growing chain determined by its interactions with the π-ligands of the chiral metallocene. It is the chirally oriented growing chain which discriminates between the two prochiral faces of the propene monomer. In short, the stereoselectivity is mainly due to non-bonded energy interactions of the methyl group of the chirally coordinated monomer with the chirally oriented growing chain.

According to the scheme of Figure 12, in the framework of a regular chain-migratory mechanism the C_2 and some C_S symmetric metallocenes lead to iso- and syndiotactic polymers, respectively. In fact, for the C_2 symmetric systems the same propene enantioface is enchained at each insertion step, whereas for the C_S symmetric systems there is a regular alternation of the propene enantioface which is enchained.

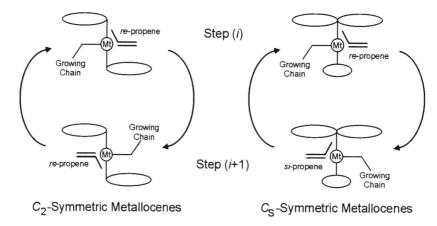

Step (*i*)

Step (*i+1*)

C$_2$-Symmetric Metallocenes   C$_S$-Symmetric Metallocenes

Figure 12. Scheme of stereospecific 1-olefins polymerization with generic C_2 and C_S symmetric metallocenes. In the framework of a regular chain migratory mechanism, the C_2 and C_S symmetric catalysts lead to iso- and syndiotactic polymers, respectively. In fact, multiple insertions of the same enantioface occur with C_2 symmetric metallocenes, while multiple insertions of alternating enantiofaces occur with C_S metallocenes.

The molecular mechanics calculations used to develop this mechanism of stereoselectivity assume the applicability of the Hammond postulate to this case (i.e., the energy difference between the transition states for the insertion reaction can be estimated as the energy difference between the corresponding π-olefin complexes). The validity of these assumption is confirmed by the large amount of experimental results which have been rationalized with this kind of calculations [11, 12, 14, 24]. Nevertheless, in this final section we present a quantum mechanics validation of this mechanism through full DFT localization of the transition states for the insertion reaction of propene into the Zr-isobutyl bond of two typical stereospecific homogeneous catalysts. The first system is based on the *racemic*-dimethylsilyl-bis-1-indenyl zirconocene, with a C_2 symmetry of coordination of the aromatic ligand, and the actual catalyst leads to a highly isotactic polymer. The second system is based on the dimethylsilyl-cyclopentadienyl-9-fluorenyl zirconocene, with a C_S symmetry of coordination of the aromatic ligand, and the corresponding catalyst leads to a highly syndiotactic polymer.

The models for the isospecific catalyst coordinate to a Zr atom a propene molecule, an isobutyl group (simulating a primary growing chain) and the stereorigid (bridged) bis-indenyl (R,R) coordinated π-ligand (to define the configuration of coordinated aromatic ligands it is customary to use the notation (*R*) or (*S*) according to the Cahn-Ingold-Prelog rules [87] as extended by Schlögl [88]).

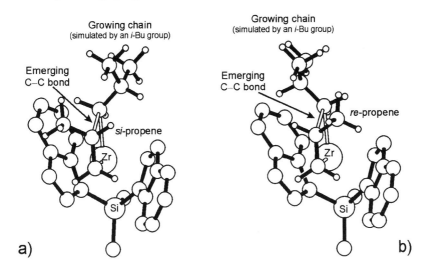

Figure 13. Transition states for propene insertion into the Zr–isobutyl bond of the *racemic-* dimethylsilyl-bis-1-indenyl zirconocene with a (*R,R*) coordination of the aromatic ligand. C_2 is the overall symmetry of the metallocene, while *re* and *si* is the chirality of coordination of the propene molecule in the transition states of parts a and b, respectively.

In the transition state of Figure 13a the growing chain is pushed away from the bulky bis-indenyl ligand, and is oriented in a uncrowded part of space, whereas in the transition state of Figure 13b the growing chain interacts repulsively with the indenyl group on the left. Both transition states present an *anti* orientation of the methyl group of the reacting propene molecule and of the C_β atom as well as the following of the growing chain, i.e. they are on opposite sides of the plane defined by the four centers transition state. Transition states with a *cis* orientation of the propene methyl group and of the growing chain (i.e. on the same side of the plane defined by the four centers transition state) were always calculated to be of higher energy.

The repulsive interactions between the growing chain and the π-ligand in the structure of Figure 13b are at the origin of the energy difference between the two diastereoisomeric transition states. In fact, the structure reported in Figure 13a is calculated to be 3.5 kcal/mol more stable than the structure of Figure 13b. According to the mechanism of migratory insertion of Figure 12, in the successive insertion step the relative coordination position of the growing chain and of the propene molecule are switched. However, due to the symmetry of the catalysts they are identical and thus at each insertion step the same enantioface of the propene is inserted. That is, an isotactic polymer is produced. Incidentally, for the (*R,R*) coordination of the bis-indenyl ligand (as in Figures 13a and 13b) the insertion of the *re* enantioface is favored. Of course, for the (*S,S*) coordination of the bis-indenyl ligand the insertion of the *si* enantioface is favored.

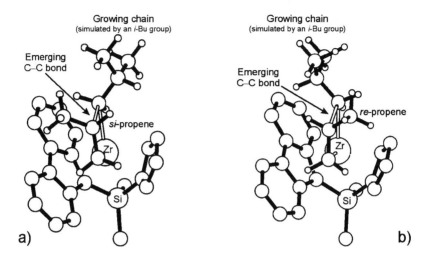

Figure 14. Transition states for propene insertion into the Zr–isobutyl bond of the dimethylsilyl-cyclopentadienyl-9-fluorenyl zirconocene with a *R* configuration at the metal atom. C_S is the overall symmetry of the metallocene, while *re* and *si* is the chirality of coordination of the propene molecule in the transition states of parts a and b, respectively.

The models for the syndiospecific catalyst coordinate to a Zr atom a propene molecule, an isobutyl group (simulating a primary growing chain) and the bridged dimethylsilyl-cyclopentadienyl-9-fluorenyl π-ligand. In this case the coordination of the π-ligand is achiral. The metal atom, instead, is chiral after coordination of the monomer and of the growing chain, since four different ligands are coordinated to it, and the standard Cahn-Ingold-Prelog CIP nomenclature can be used [87].

In the transition state of Figure 14a the growing chain is pushed away from the bulky fluorenyl group, and is oriented in a uncrowded part of space, whereas in the transition state of Figure 14b the growing chain interacts repulsively with the fluorenyl group. As for the transition states leading to an isotactic polymer, both transition states of Figure 14 present an *anti* orientation of the methyl group of the reacting propene molecule and of the growing chain.

Again, the repulsive interactions between the growing chain and the π-ligand in the structure of Figure 14b is at the origin of the energy difference between the two diastereoisomeric transition states. In fact, the structure reported in Figure 14a is calculated to be 2.4 kcal/mol more stable than the structure of Figure 14b. According to the mechanism of migratory insertion of Figure 12, in the successive insertion step the relative coordination position of the growing chain and of the propene molecule are switched. Due to the symmetry of the catalyst this corresponds to have enantiomeric transition states and thus there is a regular alternance in the insertion of the two enantiofaces of the propene. That is, a syndiotactic polymer is produced. Incidentally, for the *R* configuration at the metal atom (as in Figures 14a and 14b) the insertion of the *re* enantioface is favored. Of course, for the *S* configuration at the metal the insertion of the *si* enantioface is favored.

In the framework of the regular chain migratory mechanism of Figure 12, the enantioface selectivities we have calculated for the C_2- and C_S-symmetric catalysts of Figures 13 and 14, explain the iso- and syndiospecificity experimentally found for the corresponding real catalysts [89-91].

6.   CONCLUSIONS

The advent of single-site homogeneous catalysts for the polymerization of olefins initiated a revolution whose effects in ordinary life are at the beginning. New polymeric materials with new properties have been discovered, and the fields of applications of these catalysts are expanding continuously. In the last years an impressive understanding of the relationship between the catalysts-structure and the properties of the produced polymer has been achieved. Computational chemists certainly gave a strong contribution, and in the previous sections we reported about the status of the art in the field, and we showed the major contributions gave by a molecular modeling approach. It is now possible to model successfully the whole propagation process, and very important details of it, as the regio- and stereoselectivity of the insertion step.

Nevertheless, there is still much work to do in this field. The inclusion of solvent and/or counterions is just at the beginning, and solvent effects have been included with continuum models only. In the next years we will probably arrive to dynamically simulate the whole polymerization process in the presence of the counterion and of explicit solvent molecules. As for the experimental issues which have been not rationalized yet computationally, we remark that still it is not easy to model the relative activity of different catalysts, and even to predict if a certain catalyst will show any activity at all. Moreover, copolymerizations still represent an untackled problem. However, considering the pace at which the understanding of once obscure facts progressed it is not difficult to predict that also these challenges will be positively solved.

APPENDIX: COMPUTATIONAL DETAILS

Stationary points on the potential energy surface were calculated with the Gaussian 98 package [92]. Minima were localized by full optimization of the starting structures, while the transition states for the insertion reaction were approached through a linear transit procedure which started from the olefin-coordinated intermediate. Full transition state searches were started from the structures corresponding to the maximum of the energy along the linear transit paths. The real nature of these structures as first order saddle points was confirmed by frequency calculations which resulted in only one imaginary frequence. Geometry optimizations were performed at different level of theory. As for methods based on the wavefunction, the classical *ab initio* Hartree-Fock (HF) and Møller-Plesset perturbative theory up to the second order (MP2) were utilized. For the DFT calculations the following functionals were used; 1) BP86, which corresponds to the Becke's [93] exchange and Perdew's 1986 [94, 95] correlation functionals; 2) BPW91, which corresponds to the Becke's [93] exchange and to the Perdew-Wang's 1991 correlation functionals [96, 97]; 3) BLYP, which corresponds to the Becke's [93] exchange and to the Lee-Yang-Parr's correlation functional [98]; 4) B3LYP, which corresponds to the Becke's three parameter hybrid exchange [99] and to the Lee-Yang-Parr's correlation functionals [98]; 5) B1LYP, which corresponds to the Becke's one parameter hybrid exchange [100] and to the Lee-Yang-Parr's correlation functional [98] as implemented by Adamo and Barone [101].

Single point coupled cluster calculations with inclusion of single, double and perturbatively connected triple excitations, CCSD(T) [102, 103], were performed on the B3LYP geometries.

With regard to the basis set, the all electrons MIDI basis set of Huzinaga was used on the Zr atom [104], while the valence double-ζ basis set agumented with a polarization function (p on the H atoms and pure d functions on the C and Si atoms) of Ahlrichs, denoted as SVP [105], was used on all the atoms. The polarization functions, were also included on all the atoms of the $H_2Si(Cp)_2$ ligand. For the single point CCSD(T) calculations we used the ECP LANL2DZ basis set on Zr [106-108], and the 6-31G basis set agumented with a polarization function (p on the H atoms and cartesians d functions on the C), denoted as 6-31G(d,p) [109-113], on the H and C atoms of the ethene and of the CH_3 groups used to simulate the growing chain. For all the H, C and Si atoms of the $H_2Si(Cp)_2$ ligand the 6-31G basis set was used. If not explicitly stated, the MP2 [114] and CCSD(T) calculations were performed within the frozen core approximation, which corresponds to exclude the 1s orbital of the C atoms, the orbitals up to 2p on the Si atom, and the orbitals up to 3d on the Zr atom from the active orbitals space. To investigate the effect of the basis set on the MP2 insertion barrier we performed a series of single-point MP2 where we utilized also the valence double-ζ basis of Ahlrichs with no polarization functions, denoted as SV, and the 6-311G basis set agumented with a polarization function (p on the H atoms and cartesians d functions on the C), denoted as 6-311G(d,p) [115,116].

The calculations relative to the regioselectivity and stereoselectivity of the propene insertion reactions (Sections 4 and 5) were performed with the Amsterdam Density Functional (ADF) package [117-119]. The electronic configuration of the molecular systems were described by a triple-ζ STO basis set on zirconium for 4s, 4p, 4d, 5s, 5p (ADF basis set IV). Double-ζ STO basis sets were used for silicon (3s,3p), carbon (2s,2p) and hydrogen (1s), augmented with a single 3d, 3d, and 2p function, respectively (ADF basis set III). The inner shells on zirconium (including 3d), silicon (including 2p), and carbon (1s), were treated within the frozen core approximation. Energies and geometries were evaluated by using the local exchange-correlation potential by Vosko *et al* [120], augmented in a self-consistent manner with Becke's [93] exchange gradient correction and Perdew's [94,95] correlation gradient correction.

ACKNOWLEDGEMENTS

I thank P. Corradini (Università di Napoli) and G. Guerra (Università di Salerno) for their teaching, and T. Ziegler (University of Calgary) who introduced me to the world of DFT. This work was supported by the MURST of Italy (PRIN-2000) and by Basell Polyolefins. We thank the CIMCF of the Università Federico II di Napoli for technical support.

REFERENCES

1. Ziegler, K. *Nobel Lectures in Chemistry, 1963-1970*; Elsevier, 1972, pp 6.
2. Natta, G. *Nobel Lectures in Chemistry, 1963-1970*; Elsevier, 1972, pp 27.
3. Natta, G.; Mazzanti, G. *Tetrahedron* **1960**, *8*, 86.
4. Long, W. P.; Breslow, D. S. *J. Am. Chem. Soc.* **1960**, *82*, 1953.
5. Long, W. P.; Breslow, D. S. *Liebigs Ann. Chem.* **1975**, 463.
6. Andresen, A.; Cordes, H. G.; Herwig, J.; Kaminsky, W.; Merck, A.; Mottweiler, R.; Pein, J.; Sinn, H.; Vollmer, H. J. *Angew. Chem. Int. Ed. Engl.* **1976**, *15*, 630.
7. Ewen, J. A. *J. Am. Chem. Soc.* **1984**, *106*, 6355.
8. Ewen, J. A.; Jones, R. L.; Razavi, A.; Ferrara, J. D. *J. Am. Chem. Soc.* **1988**, *110*, 6255.
9. Rieger, B.; Jany, G.; Fawzi, R.; Steimann, M. *Organometallics* **1994**, *13*, 647.
10. Brintzinger, H. H.; Fischer, D.; Mülhaupt, R.; Rieger, B.; Waymouth, R. M. *Angew. Chem. Int. Ed. Engl.* **1995**, *34*, 1143.
11. Resconi, L.; Cavallo, L.; Fait, A.; Piemontesi, F. *Chem. Rev.* **2000**, *100*, 1253.
12. Coates, G. W. *Chem. Rev.* **2000**, *100*, 1223.
13. Alt, H. G.; Köppl, A. *Chem. Rev.* **2000**, *100*, 1205.
14. Angermund, K.; Fink, G.; Jensen, V. R.; Kleinschmidt, R. *Chem. Rev.* **2000**, *100*, 1457.
15. Armstrong, D. R.; Perkins, P. G.; Stweart, J. J. P. *J. Chem. Soc. Dalton Trans.* **1972**, 1972.
16. Giunchi, G.; Clementi, E.; Ruiz-Vizcaya, M. E.; Novaro, O. *Chem. Phys. Lett.* **1977**, *49*, 8.
17. Novaro, O.; Blaisten--Barojas, E.; Clementi, E.; Giunchi, E.; Ruiz-Vizcaya, M. E. *J. Chem. Phys.* **1978**, *68*, 2337.
18. Fujimoto, H.; Yamasaki, T.; Mizutani, H.; Koga, N. *J. Am. Chem. Soc.* **1985**, *107*, 6157.
19. Corradini, P.; Barone, V.; Fusco, R.; Guerra, G. *Eur. Polym. J.* **1979**, *15*, 1133.
20. Corradini, P.; Barone, V.; Fusco, R.; Guerra, G. *Eur. Polym. J.* **1979**, *15*, 133.
21. Corradini, P.; Barone, V.; Guerra, G. *Macromolecules* **1982**, *15*, 1242.
22. Corradini, P.; Barone, V.; Fusco, R.; Guerra, G. *J. Catal.* **1982**, *77*, 32.
23. Corradini, P.; Guerra, G.; Barone, V. *Eur. Polym. J.* **1984**, *20*, 1177.
24. Corradini, L.; Cavallo, L.; Guerra, G. In: *Metallocene Based Polyolefins, Preparation, Properties and Technology*; Scheirs, J. and Kaminsky, W., Eds.; John Wiley & Sons: New York, 2000; Vol. 2, pp 3.
25. Guerra, G.; Cavallo, L.; Corradini, P. *Top. Stereochem.* **2002**, in press.

26. Lauher, J. W.; Hoffmann, R. *J. Am. Chem. Soc.* **1976**, *98*, 1729.
27. Breslow, D. S.; Newburg, N. R. *J. Am. Chem. Soc.* **1959**, *81*, 81.
28. Cossee, P. *Tetrahedron Lett.* **1960**, *17*, 17.
29. Cossee, P. *J. Catal.* **1964**, *3*, 80.
30. Brookhart, M.; Green, M. L. H. *J. Organomet. Chem.* **1983**, *250*, 395.
31. Laverty, D. T.; Rooney, J. J. *J. Chem. Soc., Faraday Trans. 1* **1983**, *79*, 869.
32. Castonguay, L. A.; Rappé, A. K. *J. Am. Chem. Soc.* **1992**, *114*, 5832.
33. Kawamura-Kuribayashi, H.; Koga, N.; Morokuma, K. *J. Am. Chem. Soc.* **1992**, *114*, 8687.
34. Axe, F. U.; Coffin, J. M. *J. Phys. Chem.* **1994**, *98*, 2567.
35. Weiss, H.; Ehrig, M.; Ahlrichs, R. *J. Am. Chem. Soc.* **1994**, *116*, 4919.
36. Jensen, V. R.; Børve, K. J. *J. Comput. Chem.* **1998**, *19*, 947.
37. Bierwagen, E. P.; Bercaw, J. E.; Goddard, W. A., III *J. Am. Chem. Soc.* **1994**, *116*, 1481.
38. Margl, P. M.; Deng, L.; Ziegler, T. *Organometallics* **1998**, *17*, 933.
39. Yoshida, T.; Koga, N.; Morokuma, K. *Organometallics* **1995**, *14*, 746.
40. Lanza, G.; Fragalà, I. L.; Marks, T. J. *Organometallics* **2001**, *20*, 4006.
41. Meier, R. J.; van Doremaele, G. H. J.; Iarlori, S.; Buda, F. *J. Am. Chem. Soc.* **1994**, 7274.
42. Yang, X.; Stern, C. L.; Marks, T. J. *J. Am. Chem. Soc.* **1991**, *113*, 3623.
43. Woo, T. K.; Fan, L.; Ziegler, T. *Organometallics* **1994**, *13*, 2252.
44. Fan, L.; Harrison, D.; Woo, T. K.; Ziegler, T. *Organoemtallics* **1995**, *14*, 2018.
45. Ziegler, T.; Tschinke, V.; Versluis, L.; Baerends, E. J. *Polyhedron* **1988**, *7*, 1625.
46. Lohrenz, J. C. W.; Woo, T. K.; Ziegler, T. *J. Am. Chem. Soc.* **1995**, *117*, 12793.
47. Jolly, C. A.; Marynick, D. S. *J. Am. Chem. Soc.* **1989**, *111*, 7968.
48. Margl, P.; Lohrenz, J. C. W.; Ziegler, T.; Blöchl, P. E. *J. Am. Chem. Soc.* **1996**, *118*, 4434.
49. Thorshaug, K.; Støvneng, J. A.; Rytter, E.; Ystenes, M. *Macromolecules* **1998**, *31*, 7149.
50. Woo, T. K.; Fan; Ziegler, T. *Organometallics* **1994**, *13*, 432.
51. Boys, S. F.; Bernardi, F. *Mol. Phys.* **1970**, *19*, 553.
52. Guerra, G.; Longo, P.; Corradini, P.; Cavallo, L. *J. Am. Chem. Soc.* **1999**, *121*, 8651.
53. Longo, P.; Grisi, F.; Guerra, G.; Cavallo, L. *Macromolecules* **2000**, *33*, 4647.
54. Chan, M. S. W.; Vanka, K.; Pye, C. C.; Ziegler, T. *Organometallics* **1999**, *18*, 4624.
55. Fusco, R.; Longo, L.; Masi, F.; Garbassi, F. *Macromolecules* **1997**, *30*, 7673.
56. Fusco, R.; Longo, L.; Proto, A.; Masi, F.; Garbassi, F. *Macromol. Rapid Commun.* **1998**, *19*, 257.
57. Lanza, G.; Fragalà, I. L.; Marks, T. J. *J. Am. Chem. Soc.* **1998**, *120*, 8257.
58. Chan, M. S. W.; Ziegler, T. *Organometallics* **2000**, *19*, 5182.
59. Rix, F. C.; Brookhart, M.; White, P. S. *J. Am. Chem. Soc.* **1996**, *118*, 4746-4764.
60. Musaev, D. G.; Froese, R. D. J.; Svensson, M.; Morokuma, K. *J. Am. Chem. Soc.* **1997**, *119*, 367.
61. Casey, C. P.; Hallenbeck, S. L.; Pollock, D. W.; Landis, C. R. *J. Am. Chem. Soc.* **1995**, *117*, 9770.
62. Casey, C. P.; Carpenetti, D. W., II; Sakurai, H. *J. Am. Chem. Soc.* **1999**, *120*, 9483.
63. Witte, P. T.; Meetsma, A.; Hessen, B. *J. Am. Chem. Soc.* **1997**, *119*, 10561.
64. Wu, Z.; Jordan, R. F. *J. Am. Chem. Soc.* **1995**, *117*, 5867.
65. Casey, C. P.; Hallenbeck, S. L.; Wright, J. M.; Landis, C. R. *J. Am. Chem. Soc.* **1997**, *119*, 9680.
66. Casey, C. P.; Carpenetti, D. W., II,; Sakurai, H. *Organometallics* **2001**, *20*.
67. Galakhov, M. V.; Heinz, G.; Royo, P. *Chem. Commun.* **1998**, 17.

68. Abrams, M. B.; Yoder, J. C.; Loeber, C.; Day, M. W.; Bercaw, J. E. *Organometallics* **1999**, *18*, 1389.

69. Karl, J.; Dahlmann, M.; Erker, G.; Bergander, K. *J. Am. Chem. Soc.* **1998**, *120*, 5643.

70. Casey, C. P.; Lee, T.-L.; Tunge, J. A.; Carpenetti, D. W., II *J. Am. Chem. Soc.* **2001**, *123*, 10762.

71. Petitjean, L.; Pattou, D.; Ruiz-López, M. F. *J. Phys. Chem. B* **1999**, *103*, 27.

72. Margl, P. M.; Deng, L.; Ziegler, T. *J. Am. Chem. Soc.* **1998**, *120*, 5517.

73. Janiak, C. *J. Organomet. Chem.* **1993**, *452*, 63.

74. Støvneng, J. A.; Rytter, E. *J. Organomet. Chem.* **1996**, *519*, 277.

75. Grubbs, R. H.; Coates, G. W. *Acc. Chem. Res.* **1996**, *29*, 85.

76. Ystenes, M. *J. Catal.* **1991**, *129*, 383.

77. Erker, G. *Acc. Chem. Res.* **2001**, *34*, 309.

78. Guerra, G.; Cavallo, L.; Moscardi, G.; Vacatello, M.; Corradini, P. *J. Am. Chem. Soc.* **1994**, *116*, 2988.

79. Guerra, G.; Cavallo, L.; Corradini, P.; Longo, P.; Resconi, L. *J. Am. Chem. Soc.* **1997**, *119*, 4394.

80. Toto, M.; Cavallo, L.; Corradini, P.; Guerra, G.; Resconi, L. *Macromolecules* **1998**, *31*, 3431.

81. Hanson, K. R. *J. Am. Chem. Soc* **1966**, *88*, 2731.

82. Corradini, P.; Busico, V.; Guerra, G. *Comprehensive polymer science*; Allen, G. and Bevington, J. C., Ed.; Pergamon Press: Oxford, 1989; Vol. 4, pp 29-50.

83. Corradini, P.; Guerra, G.; Vacatello, M.; Villani, V. *Gazz. Chim. Ital.* **1988**, *118*, 173.

84. Cavallo, L.; Corradini, P.; Guerra, G.; Vacatello, M. *Macromolecules* **1991**, *24*, 1784.

85. Cavallo, L.; Corradini, P.; Guerra, G.; Vacatello, M. *Polymer* **1991**, *32*, 1329.

86. van der Leek, K.; Angermund, K.; Reffke, M.; Kleinschmidt, R.; Goretzki, R.; Fink, G. *Chem. Eur. J.* **1997**, *3*, 585.

87. Cahn, R. S.; Ingold, C.; Prelog, V. *Angew. Chem., Int. Ed. Engl.* **1966**, *5*, 385.

88. Schlögl, K. *Top. Stereochem.* **1966**, *1*, 39.

89. Spaleck, W.; Antberg, M.; Aulbach, M.; Bachmann, B.; Dolle, V.; Haftka, S.; Küber, F.; Rohrmann, J.; Winter, A. In: *Ziegler Catalysts*; Fink, G., Mülhaupt, R. and Brintzinger, H.-H., Eds.; Springer-Verlag: Berlin, 1995, pp 83.

90. Resconi, L.; Piemontesi, F.; Camurati, I.; Sudmeijer, O.; Nifant'ev, I. E.; Ivchenko, P. V.; Kuz'mina, L. G. *J. Am. Chem. Soc.* **1998**, *120*, 2308.

91. Schaverien, C. J.; Ernst, R.; Terlouw, W.; Schut, P.; Sudmeijer, O.; Budzelaar, P. H. M. *J. Mol. Cat. A: Chem.* **1998**, *128*, 245.

92. Frisch, M. J.; Trucks, G. W.; Schlegel, H. B.; Scuseria, G. E.; Robb, M. A.; Cheeseman, J. R.; Zakrzewski, V. G.; Montgomery, J. A., Jr.; Stratmann, R. E.; Burant, J. C.; Dapprich, S.; Millam, J. M.; Daniels, A. D.; Kudin, K. N.; Strain, M. C.; Farkas, O.; Tomasi, J.; Barone, V.; Cossi, M.; Cammi, R.; Mennucci, B.; Pomelli, C.; Adamo, C.; Clifford, S.; Ochterski, J.; Petersson, G. A.; Ayala, P. Y.; Cui, Q.; Morokuma, K.; Malick, D. K.; Rabuck, A. D.; Raghavachari, K.; Foresman, J. B.; Cioslowski, J.; Ortiz, J. V.; Baboul, A. G.; Stefanov, B. B.; Liu, G.; Liashenko, A.; Piskorz, P.; Komaromi, I.; Gomperts, R.; Martin, R. L.; Fox, D. J.; Keith, T.; Al-Laham, M. A.; Peng, C. Y.; Nanayakkara, A.; Gonzalez, C.; Challacombe, M.; Gill, P. M. W.; Johnson, B. G.; Chen, W.; Wong, M. W.; Andres, J. L.; Head-Gordon, M.; Replogle, E. S.; Pople, J. A. *Gaussian 98 A-9*; Gaussian, Inc.: Pittsburgh PA, 1998.

93. Becke, A. *Phys. Rev. A* **1988**, *38*, 3098.

94. Perdew, J. P. *Phys. Rev. B* **1986**, *33*, 8822.

95. Perdew, J. P. *Phys. Rev. B* **1986**, *34*, 7406.

96. Perdew, J. P.; Burke, K.; Wang, Y. *Phys. Rev. B* **1996**, *54*, 16533.
97. Perdew, J. P. In: *Electronic Structure of Solids '91*; Ziesche, P. and Eschrig, H., Eds.; Akademie Verlag: Berlin, 1991, pp 11.
98. Lee, C.; Yang, W.; Parr, R. G. *Phys. Rev. B* **1988**, *37*, 785.
99. Becke, A. D. *J. Chem. Phys.* **1993**, *98*, 5648.
100. Becke, A. D. *J. Chem. Phys.* **1996**, *104*, 1040.
101. Adamo, C.; Barone, V. *Chem. Phys. Lett.* **1997**, *274*, 242.
102. Raghavachari, K.; Trucks, G. W.; Pople, J. A.; Head-Gordon, M. *Chem. Phys. Lett.* **1989**, *157*, 479.
103. Purvis, G. D.; Bartlett, R. J. *J. Chem. Phys.* **1982**, *76*, 1910.
104. Huzinaga, S.; Andzelm, J.; Klobukowsi, M.; Radzio-Andzelm, E.; Sakai, Y.; Tatewaki, H. *Gaussian Basis Sets for Molecular Calculations*; Elsevier: Amsterdam, 1984.
105. Schaefer, A.; Horn, H.; Ahlrichs, R. *J. Chem. Phys.* **1992**, *97*, 2571.
106. Hay, P. J.; Wadt, W. R. *J. Chem. Phys.* **1985**, *82*, 270.
107. Hay, P. J.; Wadt, W. R. *J. Chem. Phys.* **1985**, *82*, 284.
108. Hay, P. J.; Wadt, W. R. *J. Chem. Phys.* **1985**, *82*, 299.
109. Ditchfield, R.; Hehre, W. J.; Pople, J. A. *J. Chem. Phys.* **1971**, *54*, 724.
110. Hehre, W.; Ditchfield, J. R.; Pople, J. A. *J. Chem. Phys.* **1972**, *56*, 2257.
111. Hariharan, P. C.; Pople, J. A. *Mol. Phys.* **1974**, *27*, 209.
112. Gordon, M. S. *Chem. Phys. Lett.* **1980**, *76*, 163.
113. Hariharan, P. C.; Pople, J. A. *Theo. Chim. Acta.* **1973**, *28*, 213.
114. Møller, C.; Plesset, M. S. *Phys. Rev.* **1934**, *46*, 618.
115. Krishnan, R.; Binkley, J. S.; Seeger, R.; Pople, J. A. *J. Chem. Phys.* **1980**, *72*, 650.
116. McLean, A. D.; Chandler, G. S. *J. Chem. Phys.* **1980**, *72*, 5639.
117. ADF 2000 Users Manual, Vrije Universiteit Amsterdam, Amsterdam, The Netherlands.
118. Baerends, E. J.; Ellis, D. E.; Ros, P. *Chem. Phys.* **1973**, *2*, 41.
119. Fonseca Guerra, C.; Snijders, J. G.; te Velde, G.; Baerends, E. J. *Theor. Chem. Acc.* **1998**, *99*, 391.
120. Vosko, S. H.; Wilk, L.; Nusair, M. *Can. J. Phys.* **1980**, *58*, 1200.

Chapter 3

The Key Steps in Olefin Polymerization Catalyzed by Late Transition Metals

Artur Michalak[1,2] and Tom Ziegler[1,*]

[1]*Department of Chemistry, University of Calgary, University Drive 2500, Calgary, Alberta, Canada T2N 1N4;* [2]*Department of Theoretical Chemistry, Faculty of Chemistry, Jagiellonian University, R. Ingardena 3, 30-060 Cracow, Poland*

Abstract: Brookhart and coworkers have recently developed Ni(II) and Pd(II) bis-imine based catalysts of the type $(ArN=C(R)-C(R)=NAr)M-CH_3^+$ for olefin polymerization. We discuss in the first part how bulky Ar and R groups can be used to enhance chain growth and suppress chain termination for this type of catalyst. The second part concentrates on the ability of bis-imine catalysts to produce branched polymers and the way in which the branching can be controlled by changing the substituents Ar and R. Consideration is finally given to other types of late transition metal catalysts.

Keywords: Density functional theory, olefin polymerization, β-hydride elimination, stochastic approach, branching.

1. INTRODUCTION

Brookhart and coworkers [1] have recently developed Ni(II) and Pd(II) bis-imine based catalysts of the type $(ArN=C(R)-C(R)=NAr)M-CH_3^+$ (**1a** of Figure 1) that are promising alternatives to both Ziegler-Natta systems and metallocene catalysts for olefin polymerization. Traditionally, such late metal catalysts are found to produce dimers or extremely low molecular weight oligomers due to the favorability of the β-elimination chain termination process [2].

57

F. Maseras and A. Lledós (eds.), Computational Modeling of Homogeneous Catalysis, 57–78.
© 2002 *Kluwer Academic Publishers. Printed in the Netherlands.*

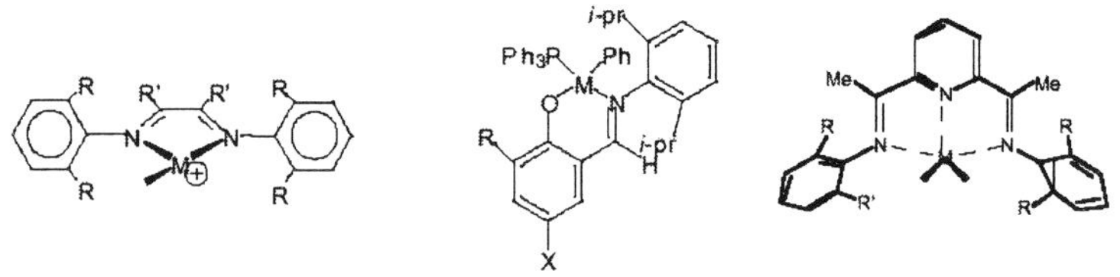

la M = Ni(II),Pd(II)   1b M = Ni(II),Pd(II)   1c M=Fe(II),Co(II)

Bisimine catalyst   Salicylaldiminato   Imino Pyridine
(Brookhart)   catalyst (Grubbs)   Catalyst
                                                        (Brookhart/Vernon)

Figure 1. Late transition metal catalysts for olefin polymerization of current interest

With the Brookhart systems very high molecular weight polymers can be produced. They also exhibit high activities which are competitive with those of commercial metallocene [3] catalysts. Not only can these catalysts convert ethylene into high molecular weight polyethylene, but the polymers also exhibit a controlled level of short chain branching. NMR studies which indicate the presence of multiple methine, methylene and methyl signals suggests branches of variable length with methyl branches predominating.[1] The extent of the branching is a function of temperature, monomer concentration and catalyst structure. Thus, by simply varying these parameters, polymers which are highly branched or virtually linear can be tailored.

Brookhart's group has studied the mechanistic details of the polymerization including the role of the bulky substituents on the bis-imine ligands.[1,2] Three main processes are thought to dominate the polymerization chemistry of these catalyst systems, namely propagation, chain branching and chain termination (Figure 2). Following cocatalyst activation of the precatalyst, a bis-imine methyl cation is formed. The first insertion of ethylene yields a bis-imine alkyl cation which upon uptake of another ethylene molecule produces a metal alkyl olefin π-complex. This π-complex has been established by NMR studies[1] to be the catalytic resting state of the system. The chain propagation cycle is depicted in Figure 2a. The first step involves the insertion of the co-ordinated olefin moiety to form a metal alkyl cationic species. Rapid uptake of monomer returns the system to the initial resting state π-complex. The unique short chain branching observed with these catalysts is proposed to occur via an alkyl chain isomerization process as sketched in Figure 2c.

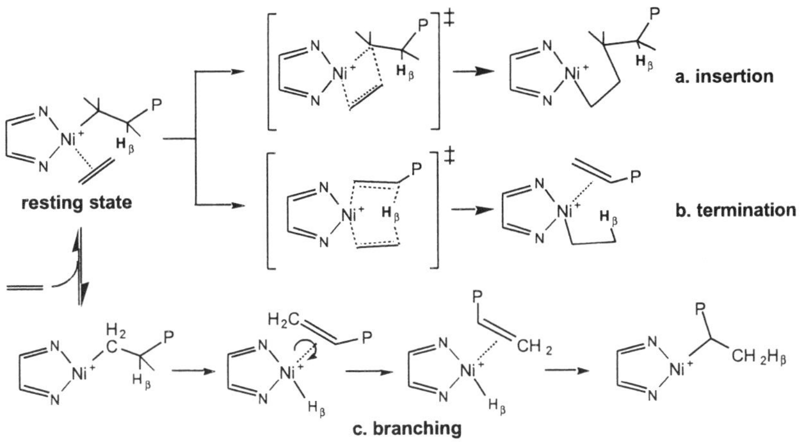

Figure 2. Proposed reaction mechanism for (a) insertion, (b) chain termination and (c) chain branching in the case of the Brookhart Ni-bis-imine polymerization catalyst. Large bulky substituents have been removed for clarity

In this proposed process, β-hydride elimination first yields a putative hydride olefin π-complex. Rotation of the π-coordinated olefin moiety about its co-ordination axis, followed by reinsertion produces a secondary carbon unit and therefore a branching point. Consecutive repetitions of this process allows the metal center to migrate down the polymer chain, thus producing longer chain branches. Chain termination occurs via monomer assisted β-hydrogen elimination, either in a fully concerted fashion as illustrated in Figure 2b or in a multistep associative mechanism as implicated by Johnson[1] *et al.*

Similar Ni and Pd catalysts developed by Keim [4] and others [5, 6] which do not posses the bulky ligand systems have been used to produce dimers or extremely low molecular weight oligomers. Brookhart has suggested[1] that the bulky aryl ligands act to preferentially block the axial sites of the metal center as illustrated by Figure 3. This feature in the catalyst system must in some way act to retard the chain termination process relative to the propagation process, thereby allowing these catalysts to produce high molecular weight polymers.

More recently the groups of Brookhart [7] and Gibson [8] have investigated the catalytic potential of iron(II) and cobalt(II) complexes with tridentate pyridine bis-imine ligands (**1c** of Figure 1). They find that especially the iron(II) system can produce high-density polyethylene in good yields when bulky ortho-substituted aryl groups are attached to the imine-nitrogens. The new catalysts have polymerization activities comparable to, or even higher than, those of metallocenes under similar conditions. They exhibit further great potential for controlling polymer properties by external parameters such as pressure and temperature.

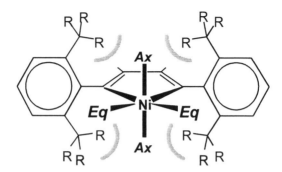

Figure 3. Axial(*Ax*) and equatorial(*Eq*) coordination sites of the metal center and their potential steric interactions with the bulky substituents

Most recently, Grubbs' group demonstrated that some neutral salicylaldiminato nickel(II) complexes, whose skeleton structure appears as **1b** in Figure 1, show catalytic activities rivaling those of the bisimine complexes [9]. This potentially opens the door to a new class of catalysts as the active sites derived from these nickel complexes are neutral, thus reducing the ion-pairing problems encountered in the current catalysts.

Theoretical studies have been carried out on all the late transition metal catalysts **1a** [10-13] , **1b** [14] and **1c** [15] in Figure 1. It is not the objective here to review all the computational results. We shall instead describe the general mechanistic insight that has been gained from the theoretical studies with the main emphasis on Brookhart's bis-imine catalysts. The experimental work on late transition metal olefin polymerization catalysts has been reviewed recently by Ittel [16] et al.

2.   STERIC CONTROL OF MOLECULAR WEIGHT

We discuss in this section how steric bulk can be used to increase molecular weight by enhancing the rate of insertion (Figure 1) and decrease the rate of termination. We shall demonstrate this point by first discussing the barriers of insertion/termination for the generic bis-imine system $(HN=C(H)-C(H)=NH)M-R^+$ in which we have replaced the aryl rings of **1a** in Figure 1 with hydrogens. These barriers will be compared to those obtained from calculations on the full system in order to gauge the influence of steric bulk.

The uptake of an ethylene molecule by (HN=C(H)-C(H)=NH)NiC$_3$H$_7^+$, **2a**, is exothermic [13a] by 19.9 kcal/mol and results in the formation of a π–complex in which the ethylene molecule is situated in an axial position above the N-Ni-N coordination plane, **3a** of Figure 4. The π–complexation energy is higher than for early d^0 transition metal complexes because of the additional metal-to-olefin back donation. In fact, the π–complex is so stable that according to experimental studies [16] becomes the resting state for the catalytic system. For (HN=C(H)-C(H)=NH)PdC$_3$H$_7^+$, **2b**, one finds [13c] a similar ethylene complexation energy of 18.8 kcal/mol.

Insertion of ethylene into the Ni-C bond in **3a** leads to the alkyl complex **4a** via the transition state TS[**3a-4a**] with a barrier [13a] of 17.5 kcal/mol relative to **3a**. It is worth to note that in TS[**3a-4a**] both ethylene and the α–carbon of the growing (propyl) chain are situated in the N-Ni-N plane. For the corresponding palladium complex the insertion barrier [13c] is somewhat higher at 19.9 kcal/mol.

The termination process of Figure 1 takes place from **3a** by transfer of a β–hydrogen on the growing chain to a carbon on the incoming ethylene monomer. The result (**5a** of Figure 4) is a new (ethyl) growing chain and a complexed olefinic (propylene) unit made up of the old (propyl) growing chain. The olefinic unit might subsequently dissociate, thus giving rise to chain termination. The transition state for this process TS[**4a–5a**] has a barrier [13a] of 9.7 kcal/mol. It is important to note that in TS[**4a –5a**] (Figure 4) we have two groups in the axial position over and below the N-Ni-N coordination plane, see also Figure 3. One group is the incoming ethylene and the other the α–carbon of the growing (propyl) chain.

It is also important to note [13a] that for the generic catalyst, termination has a much lower barrier than insertion. Thus (HN=C(H)-C(H)=NH)NiC$_3$H$_7^+$ is not going to be an efficient olefin polymerization catalyst. Rather, **2a,** will at best be able to produce small oligomers of ethylene. This is in line with the experimental observation [16] that only bis-imines with bulky substituents are able to function as polymerization catalysts whereas less encumbered systems works as oligomerization catalysts.

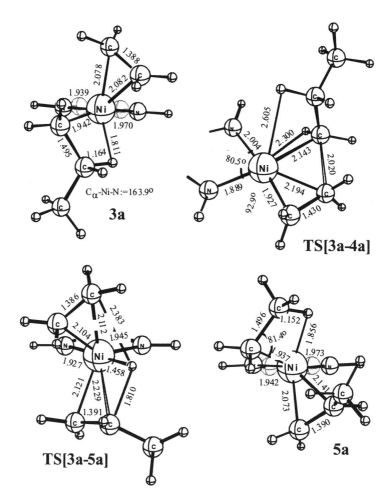

Figure 4. Structures resulting from ethylene insertion and chain termination due to the generic catalyst (HN=C(H)-C(H)=NH)PdC$_3$H$_7^+$. Ethylene complex (**3a**); insertion transition state (TS[**3a-4a**]); termination transition state (TS[**3a-5a**]); new olefin product from termination process (**5a**).

We shall now discuss how the introduction of steric bulk influences both insertion and termination. Our discussion will be based on calculations [13b] involving **1a** of Figure 1 in which M=Ni, R= *i*-Pr, and R' = H. We shall label the different species involved by Roman numerals to distinguish them from the corresponding generic systems labeled by Arabic numerals. In the actual calculations on **Ia** (or **IIa**) use was made of a combined quantum mechanical (QM) and molecular mechanical (MM) scheme [13b] in which the core ((HN=C(H)-C(H)=NH)NiC$_3$H$_7^+$) was represented by QM and the remaining part by MM. Quite similar calculations have been carried out by Morokuma [11e] et al.

The uptake of olefin by **IIa** (Figure 5) leads to the ethylene complex **IIIa** of Figure 5 with a complexation energy [13b] of 14.7 kcal/mol. The ethylene complexation energy is reduced compared to the generic system 3a by 5.2 kcal/mol. This is understandable since the ethylene molecule in **IIIa** is sitting in one of the axial positions and thus encumbered sterically by the bulky phenyl groups, Figures 3 and 5. Further studies have shown that the uptake energy can be influenced by a few kcal/mol by changing the substituents R and R' on 1a of Figure 1. It has also been demonstrated that the steric bulk on the bis-imine nitrogens gives rise to a free energy of activation for the uptake due to entropic effects as the ethylene has to approach a narrow channel in order to bind to the metal.

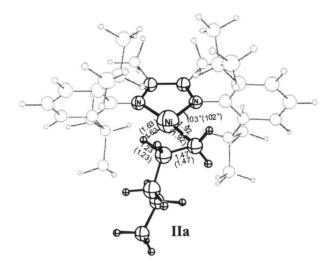

Figure 5. QM/MM representation of structure **IIa**.

The subsequent insertion of ethylene into the Ni-C bond transforms **IIIa** into the pentyl complex **IVa** via the insertion transition state TS[**IIIa – Va**] of Figure 6. However it is interesting to note that the barrier of insertion has been reduced from 17.5 kcal/mol for the generic system (TS[**3a-4a**]) to 13.2 kcal/mol for the real system [13b] (TS[**IIIa – IVa**]). The reduction is not so much due to a change in the relative energies of the transition states since both the monomer and the α–carbon of the growing chain are situated in the N-Ni-N coordination plane (Figure 6) where they are relatively unencumbered by the bulky phenyl groups. Instead it is the steric destabilization of the **IIIa** resting-state relatively to **3a** that is responsible for the reduction in the insertion barrier for the real system. Thus, by introducing steric bulk one is able to enhance the polymerization activity of the Brookhart based Ni(II)-bis-imine complexes. Similar conclusion have been reached for the homologous Pd(II) systems [1]. However, the effect is not as pronounced as the steric congestion around the palladium center is less pronounced due to the longer Pd-N bonds.

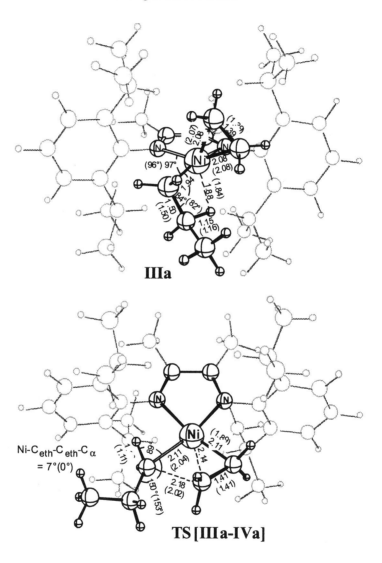

Figure 6. Ethylene complex (**IIIa**) and insertion transition state TS[**IIIa-IVa**] for the polymerization process involving **1a** (or **IIa**)

It is worth to note that the calculated insertion barrier [13b] of 13.2 kcal/mol recently has been confirmed by Brookhart [16] et al with an experimental estimate of 12.0 kcal/mol. A similar good agreement betweeen theory and experiment has also been obtained for the palladium [13f] system.

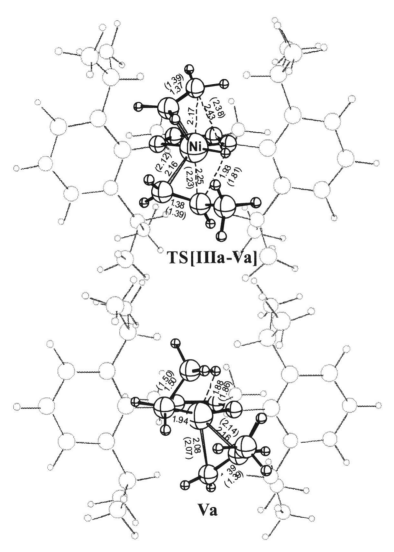

Figure 7. Transition state TS(**IVa-Va**) and product **Va** process involving hydride transfer process from original ethylene complex IVa.

The termination process leading from **IIIa** over TS[**IIIa-Va**] to **Va** by the transfer of a β–hydrogen on the growing chain to a carbon on the incoming ethylene monomer is calculated to have a barrier of 18.6 kcal/mol. This is a substantial increase compared to the generic system with a barrier (TS[**4a** – **5a**]) of 9.7 kcal/mol. The increase can be understood by observing that the transition state TS[**IIIa-Va**] has the ethylene molecule in one axial position and the α–carbon of the growing chain in the other. It is thus understandable that introducing bulky aryl groups considerably destabilizes TS[**IIIa-Va**] compared to TS[**4a** –**5a**] with the result that termination becomes less feasible that propagation for the real system.

The discussion here illustrates that it should be possible by gradually increasing the size of the substituents on the bis-imine nitrogen atoms to proceed from catalysts for oligomerization to polymerization catalysts. For the recent Grubbs catalyst [9] (1b of Figure 1) steric bulk is also required [14] to obtain high molecular weight polymers. For the Fe(II) and Co(II) catalysts [7, 8] with tridentate pyridine bis-imine ligands (**1c** of Figure 1), steric bulk is required to destabilize the resting state in the form of a olefin complex and lower [15a-b] the barrier of insertion sufficiently. The Fe(II) system seems further to carry out insertion and termination on energy surfaces with different [15a,d] spin-multiplicity.

3.   CONTROL OF THE POLYMER STRUCTURE

The third process in Figure 2 involves the possible isomerization (branching) of a growing chain. The isomerization is mediated first by the migration of a β–hydrogen on the alkyl chain of **IIa** to the metal center thus producing a hydrido olefin complex. The β–hydrogen elimination is followed by a 180° rotation of the olefin unit and a subsequent insertion of the olefin into the M-H bond resulting in an isomerization (branching) of the growing chain. The principle is illustrated by the isomerization of the n-propyl chain in **IIa** via the transition state TS[**IIa-VIa**] of Figure 8 to produce the iso-propyl complex **VI**. The internal barrier (relative to **IIa**) for the process is 15.3 kcal/mol [13b]. For the generic $C(H)=NH)Ni(n-C_3H_7)^+$ system the isomerization barrier is [13a] somewhat lower at 12.8 kcal/mol. This is understandable since the olefin unit has more space for its rotation.

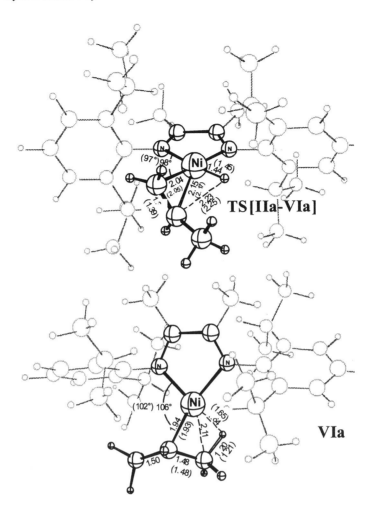

Figure 8. Transition state **TS[IIa-VIa]** and product **VIa** from the isomerization of **IIa.**

In the case of the Pd-bis-imine catalyst the isomerization process has a barrier of 5.8 kcal/mol [13c] for the generic system and 6.1 kcal/mol [16] for the real catalyst. The lower barrier is a result of the longer Pd-N distance (compared to the Ni-N bond length) which provides space for rotation of the olefin unit, even when the bis-imine nitrogens are attached to bulky substituents.

The fact that the internal barrier of isomerization is much lower for the palladium than for the nickel system makes the former a more likely candidates for producing branched polymers. We shall in the following illustrate how palladium catalysts can be used to produce different branched structures by changing the steric bulk around the metal center. The principle will be illustrated in connection with the polymerization of propylene.

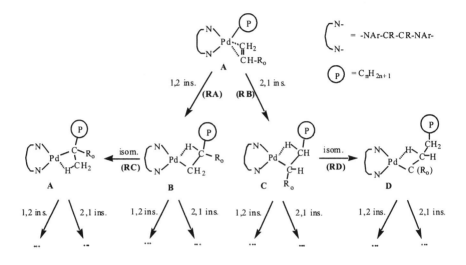

Figure 9. Polymerization of propylene ($R_0 = CH_3$) by the Pd-bis-imine Brookhart catalysts.

3.1   Stochastic Model for Polymer Growth

The propylene polymerization is shown in Figure 9. In the case of Brookhart catalyst, the resting state is an olefin π-complex, **A**, starting from which the polymer chain may grow via two alternative insertion paths. After the 1,2-insertion [reaction **(A)**] the β-agostic alkyl complex **B** with the unsubstituted olefin carbon attached to the metal is formed. On the other hand, the 2,1-insertion [reaction **(B)**] leads to the formation of the complex **C**, with substituted olefin carbon linked to the metal. Both insertion reactions introduce the branch R_o. Prior to a formation of subsequent olefin complexes and the following insertion reactions, however, the polymer chain may isomerize via reactions **(C)** and **(D)**, resulting in the complexes **A** and **D**, with tertiary and primary carbon forming bond to the palladium atom, respectively. The isomerization reaction **(C)** introduces an additional methyl branch, while the chain straightening isomerization reaction **(D)** shortens, and eventually removes a branch, leading to the linear polymer chain. Moreover, not only the number of branches, but also their microstructure [17] can be controlled by changes in reaction conditions and the catalyst structure. The outcome of the complex polymer growth and branching in Figure 9 is best modelled by the kind of stochastic simulations [18] described in the following. The outcome of such simulations are predictions about the polymer structures. A few initial steps in a typical simulation are presented in Figure 10.

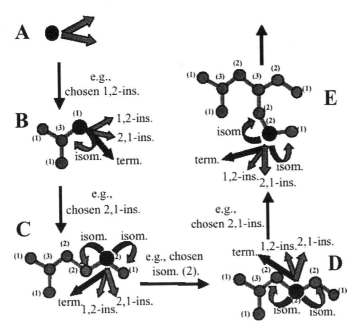

Figure 10. Stochastic simulations. In the starting structure (**A**) one C atom is attached to the catalyst; only two events are possible: propylene capture followed by the 1,2- or 2,1-insertion. For the structure **B** four events are taken into account: isomerization to the tertiary carbon, 1,2- and 2,1-insertions, and a termination. For the structures **C**, **D** and **E** five events are considered: two isomerizations, two insertions and a termination. The probabilities of these events are equal for the structures **C** and **E** (in both cases two different isomerizations lead to a primary or secondary carbon at the metal), and different for the structure **D** (for which both isomerizations lead to the structure with a secondary carbon attached to the metal). For clarity, the numbers [(1), (2), and (3)] labeling different atom types (primary, secondary, and tertiary, respectively) are shown.

Initially, one carbon atom (a methyl group) is attached to the metal of the catalyst (**A** in Figure 1). In the first step, it will capture and insert a propylene molecule via either 1,2- or 2,1-insertion route. Thus, one of these insertion events is stochastically chosen; this choice, however, is not totally random but weighted by the probabilities of the two reactions. Here the relative probabilities are proportional to the relative rates. Now, if one assumes that the 1,2-insertion has happened in the first step, i.e. the iso-butyl group is attached to the catalyst (**B**) after insertion. At this stage four different elementary events are possible: two alternative insertion routes (1,2- and 2,1-) proceeded by the capture of olefin, the termination reaction, and the isomerization reaction that would lead to a tert-butyl group attached to the metal center. If, for instance, the 2,1-insertion happened, a heptyl group would be attached to the catalyst by its secondary carbon atom (**C**); thus, five reactive events would be possible (two insertions, a termination, and two isomerizations), one of them would be stochastically chosen in a next step, etc.

In the stochastic model one assigns different probabilities for similar events starting from or/and leading to structures with a carbon atom of different character being attached to a metal. For example, the 1,2-insertion starting from a primary carbon is not equivalent to the 1,2-insertion starting from the secondary carbon. Similarly, the isomerizations starting from a primary, secondary, and tertiary carbon are not equivalent in general, and also the two isomerizations starting from a secondary carbon may be inequivalent, e.g. if one of them leads to a primary and another – to a secondary or tertiary carbon at the metal (as at stage **C** and **E**). In other words, one takes into account in this model: three different 1,2- insertions, three different 2,1-insertion, three different terminations (each starting from $1°$-, $2°$-, and $3°$-carbon), and nine different isomerizations (starting from and leading to $1°$-, $2°$-, and $3°$-carbon).

It should be emphasised here as well, that at different stages the *absolute* probabilities of equivalent events may be different, since they depend on the probabilities of all the other events (because of the probability normalisation). For example, at the stages **C** and **D** in Figure 1, the secondary carbon is attached to the metal, and five reactive events are possible. However, at the stage **C** the isomerization reactions are inequivalent (one leads to the primary carbon and another - to the secondary carbon), while at the stage **D** they are equivalent. As a result, the *absolute* probabilities of *all the events* at the stage **C** will differ from those at the stage **D**.

Figure 11 summarises the way in which the stochastic probabilities are generated from the rates of the different reactions. The basic assumption here is that the *relative* probabilities of elementary reactions at the microscopic level, π_i/π_j, are equal to their *relative* reaction rates (macroscopic), r_i/r_j. Thus, the *relative* reaction rates (eq.1 of Figure 11) for all pairs of the considered reactive events together with the probability normalization condition (eq. 2 of Figure 11) constitute the system of equations that can be solved for the *absolute* probabilities of all the events at a given stage. With this assumption, one can use the experimentally determined reaction rates or the theoretically calculated relative rate constants, obtained from the energetics of the elementary reactions with the standard Eyring exponential equation. The Eyring equation introduces as well a temperature dependence of all the relative probabilities (as in eq. 3 of Figure 11).

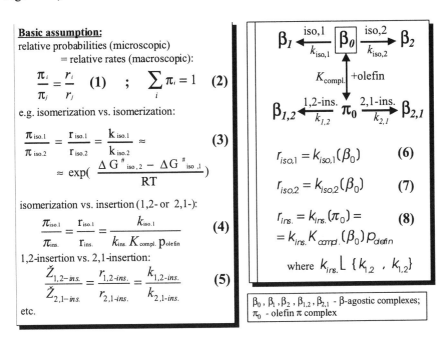

Figure 11. Basic equations used in the stochastic model.

Let us now have a closer look at three basic types of the relative probabilities appearing in the model: for an isomerization vs. another isomerization, the 1,2-insertion vs. 2,1-insertion, and an isomerization vs. an insertion. The right-hand part of Figure 11 summarizes the equations for the macroscopic reaction rates for the alternative reactive events starting from an alkyl complex β_0; let us assume that the secondary carbon atom is attached to the metal, so that two isomerization reactions have to be considered.

The isomerization reactions are first-order in concentration of the initial alkyl complex, $[\beta_0]$ (eqs 6, 7 of Figure 11). Thus, the relative probability for the two isomerizations (eq.3) is given by the ratio of their rate constants, $k_{iso,1}/k_{iso,2}$, (as equal to the relative rate, $r_{iso,1}/r_{iso,2}$), and at given temperature it can be calculated from the isomerization barriers $\Delta G^{\#}_{iso,1}$ and $\Delta G^{\#}_{iso,1}$, as in eq. 3.

From alkyl complex β_0, the olefin can be captured to form the π-complex π_0, and inserted via 1,2- or 2,1-insertion route. In the model applied here we consider the olefin capture and its insertion as one reactive event i.e. we assume a pre-equilibrium between the alkyl and olefin complexes, described by an equilibrium constant $K_{compl.} = [\pi_0] / [\beta_0] = \exp(\Delta G_{compl.} / RT)$. This corresponds to neglecting the barrier for the monomer capture. Such an approach is valid for the late-transition metal complexes, e.g. the diimine catalysts studied in the present work, where the resting state of the catalyst is a very stable olefin π-complex [16] and the olefin capture barrier and the related π-complex dissociation barrier is much lower than the insertion barriers. This assumption allows one to speed up the simulation: otherwise many olefin capture/dissociation steps, not important for the final result of the simulation, would be happening before insertion takes place. It follows from the above considerations that the insertion rate is given by eq. 8, and the equation for the isomerization vs. insertion relative probability (eq. 4) includes the isomerization and insertion rate constants, the equilibrium constant, $K_{compl.}$, and the olefin pressure, $p_{olefin,}$. Finally, the relative probability for the two alternative insertions is given by eq. 4.; it depends on the two rate constants ratio only.

It is important to emphasize here that the model in such a form allows one to simulate the influence of the reaction conditions. The temperature dependence of all the relative probabilities appears in the exponential expressions for the rate constants and the equilibrium constants. The olefin pressure influences the isomerization-insertion relative probabilities. As a result, both, temperature and olefin pressure influence the values of the *absolute* probabilities for *all* the reactive events considered. In the following use has been made of calculated reaction rates [13f] to evaluate all stochastic probabilities, unless otherwise stated.

3.2   Control by Steric Bulk

In Figure 12 examples of the polymer structures together with the values characterizing the structure (number of branches / 1000 C; % of atoms in the main chain; % of atoms in primary branches; average ratio of isomerization and insertion events) are listed for the catalysts **1-7** (for numbering see Scheme 1), while Table 1 collects the values of probabilities of the reactive events at the selected stages of polymer growth. Let us first discuss the results of the simulations for the propylene polymerization catalyzed by the Pd-based bis-imine catalyst with $R=CH_3$, $Ar=C_6H_3(i\text{-}Pr)_2$ (**6**), as for this system there exist experimental data [16]. At T=298K and for p=1 (arbitrary units) the simulations lead to the structure characterized by an average number of 238 branches / 1000 C atom in the chain. The experimentally determined value for this catalyst is 213 br./1000C [1c]. Thus, the agreement between the two values seems to be quite good, especially when one takes into account the simplicity of the model, and notices that the 'ideal' polymerization without isomerizations would produce 333.3 br./1000C.

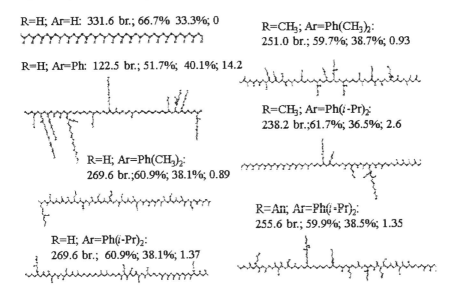

Figure 12. Examples of the polymer structures obtained with different catalysts (T=298K, p=1). The values above the plots denote: the average number of branches / 1000 C, % of atoms in the main chain, % of atoms in the primary branches, and the ratio between the number of isomerization and insertion steps. Different atom shadings are used to mark different types of branches (primary, secondary, etc.).

In the structures obtained from the simulations ca. 61.7 % of the carbon atoms belong to the main chain and c.a. 36.5 % to the primary branches, i.e. less than 2% appears in the branches of higher order. The lengths of the longest observed primary, secondary, tertiary branches are 28, 11, and 7, respectively. However, the average lengths of those branches are muchshorter: 1.6, 2.3, and 0.3. Only one quaternary methyl branch was obtained in all the simulations.

The polymerization process is characterized by an average probability ratio of isomerization vs. insertion steps of 2.6. A closer look at the simulation results shows that for this catalyst the insertions practically occur only at the primary carbon, the insertion from the secondary carbon happen very rarely. To illustrate this point, the values of the probabilities of alternative events may be helpful. If the primary carbon is attached to the metal, the probabilities of the 1,2-insertion, 2,1-insertion and the isomerization (to secondary or tertiary carbon) are 0.700, 0.286, and 0.014, respectively. If the secondary carbon, neighboring with the two secondary carbon atoms is attached to the metal, the corresponding values are 0.002 (1, 2- ins.), 0.001 (2,1-ins.), and 0.499 (two equivalent isomerizations). And if the secondary carbon, neighboring with one primary C and one secondary C is at the Pd center, the probablilities are: 0.002 (1,2-ins.), 0.001 (2,1-ins.), 0.265 (isomerization to primary C), 0.731 (isomerization to secondary C).

Table 1. The probabilities of the reactive events at selected stages of the polymer chain growth

	Probabilities									
	Pd-CH$_2$-...[1]			CH$_3$-CH(Pd)-CH$_2$-...[2]				...-CH$_2$-CH(Pd)-CH$_2$-...[3]		
	1,2- ins.	2,1- ins.	isom er.	1,2- ins.	2,1- ins.	isom. to 1°C	isom. to 2°C	1,2- ins.	2,1- ins.	both isom.
1	0.003	0.969	0.001	0.001	0.994	0.001	0.003	0.992	0.001	0.006
2	0.301	0.582	0.117	0.001	0.002	0.266	0.731	0.001	0.002	0.998
3	0.102	0.898	0.001	0.050	0.445	0.134	0.369	0.041	0.360	0.600
4	0.046	0.954	0.000	0.019	0.403	0.154	0.424	0.015	0.317	0.666
5	0.057	0.943	0.000	0.017	0.502	0.128	0.352	0.014	0.410	0.574
6	0.700	0.286	0.014	0.002	0.001	0.265	0.731	0.002	0.001	0.998
7	0.272	0.727	0.001	0.090	0.238	0.179	0.493	0.068	0.181	0.750

[1] primary carbon attached to the catalyst. [2] secondary carbon neighboring with a primary and secondary carbon attached to the catalyst. [3] secondary carbon neighboring with two secondary carbons atached to the catalyst.

Thus, the isomerization is not likely to happen after 1,2-insertion (when the primary carbon is attached to the catalyst), and usually occurs after 2,1-insertion (when the secondary carbon is at the catalyst). In any case, when the alternative isomerizations are considered, the one going along the chain is always preferred, compared to the isomerization leading to the primary carbon. However, the latter isomerization occurs often, since there is no insertions at the secondary carbon and its probability is quite high (0.265). This explains the relatively large ratio for isomerization/insertion steps of 2.6.

The above results lead to the conclusion that for this catalyst the overall number of branches is exclusively controlled by the relative ratio of 1,2- and 2,1- insertions; the first one always produces the branch, and the second one is always followed by chain running, eventually leading to a removal of one of the existing branches. Thus, from the probablilities of 1,2- and 2,1-insertions one can expect a removal of 28.6% of branches (95.3 branches / 1000C), i.e. the expected overall number of branches is 238, as observed from the simulations. The results presented here are in very good qualitative agreement with the experimental data: it has been concluded from the NMR analysis of the polpropylenes that for this catalyst insertion happens only at the primary carbons, all the 2,1- insertions lead to the removal of a branch, and that 1,2-insertion is the main route for a polymer growth [16].

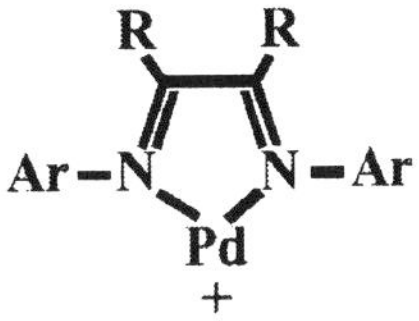

1: R = H;   Ar = H
2: R = H;   Ar = C$_6$H$_5$
3: R = H;   Ar = C$_6$H$_3$(CH$_3$)$_2$
4: R = H;   Ar = C$_6$H$_3$(*i*-Pr)$_2$
5: R = CH$_3$; Ar = C$_6$H$_3$(CH$_3$)$_2$
6: R = CH$_3$; Ar = C$_6$H$_3$(*i*-Pr)$_2$

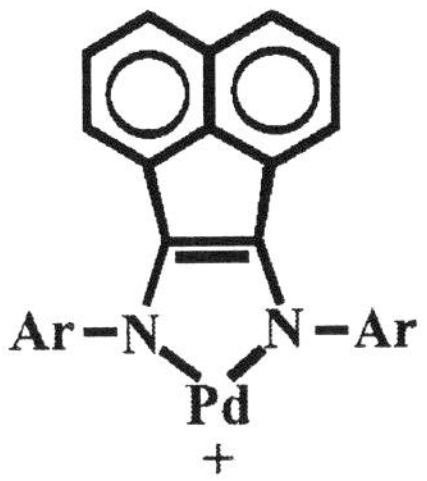

7: Ar = C$_6$H$_3$(*i*-Pr)$_2$

Figure 13. Different catalysts used in simulation

Let us now discuss the influence of the catalyst substituents on the polymer branching and its microstructure. In order to understand the influence of the steric bulk, the simulations have been performed for all the remaining catalyst of Figure 13. It should be mentioned here, however, that the less bulky catalysts, **1** and **2** [16], are not capable of polymerizing olefins. The termination barriers for these systems are very low, due to the lack of the steric bulk. Thus, the results obtained for the catalysts **1** and **2** are of purely theoretical character. They are useful, however, for understanding how the steric bulk can control the polyolefin branching. They also demonstrate how the microstructures can change with small modifications in the energetics of the polymerization cycle.

The results presented in Figure 12 demonstrate that for the generic catalyst model **1** (R=H, Ar=H) the chain grows by a regular sequence of 2,1-insertions. There is practically no isomerization, the average number of branches is 331.6 br./1000 C, and 66.7% of C belongs to the main chain, and the rest to the methyl branches. This comes from the fact that for the generic catalyst the 2,1- insertion is strongly preferred and the insertion barriers (calculated with respect to the separated reactants) are low compared to the isomerization: the probablilities of the 2,1-insertion at the primary and secondary C are 0.969 and 0.992-0.994, respectively.

The catalyst **2** is located on the opposite side of the range of branching patterns. Here, the lowest average number of branches of 122.5 br./1000 C is obtained. Only 51.7% of atoms belong to the main chain, and 40.1% to the primary branches, i.e. 8% of the carbon atoms is located in the branches of higher order. As in the example structure shown in Figure 12, the branches produced here are relatively long and often separated by a few methylene units. The longest observed primary, secondary, tertiary, and quaternary branches are built of 75, 28, 13, and 8 carbon atoms, while the average lengths of the primary, secondary, and tertiary branches are 4, 4, and 2.5, respectively. Thus, the structures obtained for the catalyst **2** are quite different from those for the systems **6** and **1**. Also, the average ratio of isomerization vs. insertion steps of 14.2 is the largest observed, and distinctly larger than those obtained for the remaining systems.

A comparison of the probabilities of the events (Table 1) for the catalysts **2** and **6** reveals that the basic difference between them is the reversed insertion regioselectivity: the 2,1- insertion is preferred for **2**, and the 1,2- insertion – for **6**. In both cases the insertions happen only at the primary carbons, and in both cases the isomerization starting from the primary carbon has much lower probability than any insertion. Therefore, it comes as no surprise that the observed number of branches is that low for the system **2**, since it is controlled only by the 1,2-/2,1- insertion ratio, as in system **6**. However, it is quite surprising and hard to predict without simulations that the average isomerization/insertion ratio and especially the resulting microstructures of the polymers are so different for the two systems. Just because of the reversed insertion regioselectivity!

A comparison of the results for the catalyst **3-7** (see Figure 12) shows that the increased steric bulk leads to a slight decrease in the global number of branches, and the increase in the insertion/isomerization ratio. For the systems **3** and **4** (with R=H) the average number of branches and the structural characteristics are almost identical. This arises from the fact that the energetics of the polymerization cycle is quite similar. This can be understood by observing that hydrogens as backbone substituent give the aryl rings flexibility to adjust their orientation and minimize a steric repulsion, when their substituents are increased in size [13e]. For the two systems with methyl backbone and different aryl substituents (catalysts **5** and **6**) the difference in the number of branches is more pronounced (251.0 br./ 1000C for **5**, and 238.3 br./1000 C for **6**).

The observed trend – a decrease in the number of branches with increased steric bulk – is quite surprising. One could expect the opposite trend, since the steric effects increase the ratio between 1,2- and 2,1- insertions, and intuitively, this should lead to an increase in the number of branches (less 2,1-insertions - less removed branches). However, for the systems **3**, **4**, **5**, and **7** the insertions at the secondary carbons happen with relatively large frequencies: for the systems **3**, **4**, and **5** the probabilities of the insertion starting from the secondary carbon are c.a. 0.4-0.5, and for the system **7** – c.a. 0.25-0.33. Since every insertion into the secondary carbon by definition adds a branch, the global number of branches for the systems **3-5** and **7** is larger than for the more bulky catalyst **6**, for which there are practically no insertions from the secondary carbons. An increase in the steric bulk leads to a decrease in the secondary-insertion probability, and eventually to a decrease in a number of branches, despite the larger fraction of the 1,2-insertions. Thus, the results show that for the systems **3-5** and **7** the branching is controlled by both, the 1,2-/2,1-insertion ratio, and the ration between the isomerization and insertions starting from the secondary carbon. This leads to the surprising results mentioned above. We would like to point out here again, that due to the lack of the experimental/theoretical results, the secondary insertion barriers for the real catalysts were *assumed* to be 1 kcal/mol higher than the primary insertion barriers (as *calculated* for the generic system). For the catalyst **6** this is *a posteriori* validated by the good agreement with experimental results. Unfortunately, there are no clear experimental data for propylene polymerization with the Pd-based catalyst for different substituents. For the related Ni-based systems, the reversed trend has been observed experimentally [16]: an increase in the number of branches with an increased in the steric bulk, as could be expected from the increased 1,2-insertion fraction. However, the Ni- and Pd-based systems are known to be very different, concerning the branching and the polymer microstructures [16]. Therefore, no conclusions can be drawn from those experimental data about the validity of our assumption

4. CONCLUDING REMARKS

Late transition metal complexes have over the past six years emerged as a promising new class of olefin polymerization catalysts thanks to the work of Brookhart and others [1-9, 16]. Theoretical calculations [10-15] have greatly helped in understanding the mechanistic aspects of these new catalysts. Especially, theoretical calculations have helped explain how steric bulk at the same time can enhance the rate of polymer growth and decrease the rate of chain termination. Theory has further helped to understand the way in which the new class of polymerization catalysts affords branched polymers. Based on this insight it has become possible to set up stochastic models that can predict what type of polymer will be produced by a certain ligand design for the late transition metal polymerization catalysts. The same type of catalysts can also be used to co-polymerize polar monomers with α-olefins [19]. This subject will be the subject of a forthcoming review.

ACKNOWLEDGMENT

This work has been supported by the National Sciences and Engineering Research Council of Canada (NSERC), Nova Chemical Research and Technology Corporation as well as donors of the Petroleum Research Fund, administered by the American Chemical Society (ACS-PRF No. 36543-AC3). We acknowledge the contributions from Liqun Deng, Peter Margl and Tom K. Woo.

REFERENCES

1   Johnson, L. K.; Killian, C. M.; Brookhart, M. *J. Am. Chem. Soc.* **1995**, *117*, 6414. (b) Johnson, L. K.; Mecking, S.; Brookhart, M. *J. Am. Chem. Soc.* **1996**, *118*, 267. (c) Killian, C. M.; Tempel, D. J.; Johnson, L. K.; Brookhart, M. *J. Am. Chem. Soc.* **1996**, *118*, 11664.

2   Wilke, G. *Angew. Chem., Int. Ed. Engl.* **1988**, 27, 185.

3   Haggin, J. *Chem. Eng. News* **1996**, *Feb 5*, 6.

4   Keim, W. *Angew. Chem., Int. Ed. Engl.* **1990**, 29, 235.

5   Abecywickrema, R.; Bennett, M. A.; Cavell, K. J.; Kony, M.; Masters, A. F.; Webb, A. G. *J. Chem. Soc., Dalton Trans.* **1993**, 59.

6   Brown, S. J.; Masters, A. F. *J. Organomet. Chem.* **1989**, *367*, 371

7   Small, B. L., Brookhart, M.; Bennett, A. M. A. *J. Am. Chem. Soc.* **1998**, 120. 4049.

8   Britovsek, G. J. P.; Bruce, M.; Gibson, V. C.;Kimberley, B. S.; Maddox, P. J.; Mastroianni, S.; McTavish, S. J.; Redshaw, C.; Solan, G. A.; Stromberg, S.; White, A. J. P.; Williams, D. J. *J. Am. Chem. Soc.* **1999**, *121*, 8728.

9   Wang, C.; Friedrish, S.; Younkin, T. R.; Li, R. T.; Grubbs, R. H.; Bansleben, D. A.; Day, M. W. *Organometallics* **1998**, *17*, 3149.

10 Dedieu, A. *Chem. Rev.* **2000**, *100*, 543.

11 Musaev, D. G.; Froese, R. D. J.; Morokuma, K. *J. Am. Chem. Soc.* **1997**, *119*, 367. (b) Musaev, D. G.; Svensson, M.; Morokuma, K.; Strömberg, S.; Zetterberg, K.; Siegbahn, P. *Organometallics* **1997**, *16*, 1933. (c) Musaev, D. G.; Froese, R. D. J.; Morokuma, K. *New. J. Chem.* **1997**, *22*, 1265. (d) Froese, R. D. J.; Musaev, D. G.; Morokuma, K. *J. Am. Chem. Soc.* **1998**, *120*, 1581. (e) Musaev, D. G.; Froese, R. D. J.; Morokuma, K. *Organometallics* **1998**, *17*, 1850. (f) Musaev, D. G.; Morokuma, K. *Top. Catal.* **1999**, *7*, 107.

12 Von Schenck, H.; Strömberg, S.; Zetterberg, K.; Ludwig, M.; Åkermark, B.; Svensson, M. *Organometallics* **2001**, *20*, 2813.

13 Deng, L.; Margl, P.; Ziegler, T. *J. Am. Chem. Soc.* **1997**, *119*, 1094. (b) Deng, L.; Woo, T. K.; Cavallo, L.; Margl, P.; Ziegler, T. *J. Am. Chem. Soc.* **1997**, *119*, 6177. (c) Michalak, A.; Ziegler, T. *Organometallics* **1999**, *18*, 3998. (d) Woo, T. K.; Ziegler, T. *J. Organomet. Chem.* **1999**, *591*, 204. (e) Woo, T. K.; Blöchl, P. E.; Ziegler, T. *J. Phys. Chem. A* **2000**, *104*, 121. (f) Michalak, A.; Ziegler, T. *Organometallics* **2000**, *19*, 1850.

14 Chan, M. S. W.; Deng, L.; Ziegler, T. *Organometallics* **2000**, *19*, 2741.

15 Deng, L.; Margl, P.; Ziegler, T. *J. Am.Chem.Soc.* **1999**, *121*, 6479. (b) Margl, P.; Deng, L.; Ziegler T. *Organometallics* **1999**; *18*, 5701. (c) Griffiths, E. A. H.; Britovsek, G. J. P.; Gibson, V.; Gould, I. R. *Chem. Commun.* **1999**, 1333. (d) Khoroshun,D.V.; Musaev, D.G.; Vreven, T.; Morokuma, K. *Organometallics* **2001**, *20*, 2007.

16 Ittel, S. D.; Johnson, L. K.; Brookhart, M. *Chem. Rev.* **2000**, *100*, 1169.

17 Hawker, C. J.; Frechet, J. M. J.; Grubbs, R. B.; Dao, J. *J.Am.Chem.Soc.* **1995**, *117*, 10763.

18 Michalak, A; Ziegler, T., work in progress.

19 Boffa,.L.S.; Novak,B.M. *Chem. Rev.* **2000**, 100, 1479.

Chapter 4

Hydrogenation of Carbon Dioxide

Shigeyoshi Sakaki[1,*] and Yasuo Musashi[2]
[1]Department of Molecular Engineering, Graduate School of Engineering, Kyoto University, Sakyo-ku, Kyoto 606-8501, Japan; [2]Information Processing Center, Kumamoto University, Kurokami 2-39-1, Kumamoto 860-8555 Japan

Abstract: Theoretical works on the hydrogenation of carbon dioxide catalysed by transition-metal complexes are reviewed. All the elementary processes in the catalytic cycle, such as insertion of carbon dioxide into the metal-hydride bond, reductive elimination of formic acid, and metathesis of the transition-metal formate intermediate with a dihydrogen molecule are discussed, based on recent theoretical works. Theoretically evaluated energy changes are compared among several possible catalytic cycles, and the most plausible reaction mechanism is proposed. Characteristic features of the catalytic cycle and elementary processes are also discussed.

Key words: hydrogenation of carbon dioxide, insertion of carbon dioxide into the metal-hydride bond, reductive elimination of formic acid, σ-bond metathesis

1.   INTRODUCTION

In this first section, we will describe the experimental knowledge on the hydrogenation of CO_2 to formic acid and the early theoretical works on this topic, which are concerned with the structure and formation of transition-metal CO_2 complexes.

1.1   Experimental Background

Transition-metal catalysed CO_2 fixation is among the most challenging subjects of research in organometallic and catalytic chemistry [1]. CO_2 is a naturally abundant resource of carbon, but its reactivity is limited by a very high thermodynamical stability. In this regard, many attempts have been made to convert CO_2 into useful chemical species.

F. Maseras and A. Lledós (eds.), Computational Modeling of Homogeneous Catalysis, 79–105.

$$(1a)$$

$$(1b)$$

$$(2)$$

$$H_2 \quad + \quad CO_2 \quad \longrightarrow \quad HCOOH \qquad (3)$$

Equations 1 to 3 show some of fixation reactions of carbon dioxide. Equations 1a and 1b present coupling reactions of CO_2 with diene, triene, and alkyne affording lactone and similar molecules [2], in a process catalyzed by low valent transition metal compounds such as nickel(0) and palladium(0) complexes. Another interesting CO_2 fixation reaction is co-polymerization of CO_2 and epoxide yielding polycarbonate (equation 2). This reaction is catalyzed by aluminum porphyrin and zinc diphenoxide [3].

Nevertheless, the hydrogenation of CO_2 by a dihydrogen molecule to formic acid (equation 3) has attracted much attention in recent years. Formic acid is indeed the starting material for a large variety of organic reaction. This hydrogenation process was originally suggested in the reaction of $RhCl(PPh_3)_3$ with CO_2 and H_2 [5]. In this first report, however, formic acid could not be detected, while $RhCl(CO)(PPh_3)_2$ was produced. The first report of CO_2 hydrogenation was presented by Inoue and his collaborators on 1976; they succeeded this reaction with $Pd(diphos)_2$, $Ni(diphos)_2$, $Pd(PPh_3)_4$, $RhCl(PPh_3)_2$, $RuH_2(PPh_3)_4$, and $IrH_3(PPh_3)_3$ in the presence of amines [6]. About 8 years later, the second report was made by Darensbourg and his collaborators, who performed hydrogenation of CO_2 to HCOOMe with $[MH(CO)_5]^-$ (M=Cr, Mo, or W) in the presence of H_2 (250psi) and CO_2 (250 psi) [7]. In 1989, Taqui Kahn and his collaborators [8] used $[Ru(EDTAH)Cl]^-$ as a catalyst and succeeded in hydrogenating CO_2 with somewhat large turnover numbers of about 180.

Between 1992 and 1994, three important works were reported in this field. Tsai and Nicholas carried out the hydrogenation of CO_2 with [Rh(NBD)(PMe$_2$Ph)$_3$]BF$_4$ (NBD = norbornadiene) as a catalyst [9] and detected spectroscopically [RhH$_2$(PMe$_2$Ph)$_3$(S)]BF$_4$ (S= H$_2$O or THF), [RhH$_2$(PMe$_2$Ph)$_4$]BF$_4$, [RhH(S)(PMe$_2$Ph)$_2$(η^2-O$_2$CH)]BF$_4$, and [RhH(S)$_{1,2}$(PMe$_2$Ph)$_{3,2}$(η^1-OCOH)]BF$_4$. From these observations, they proposed a catalytic cycle consisting of CO_2 insertion into the Rh(I)-H bond and the reductive elimination of HCOOH. On the same year, Leitner and collaborators reported the hydrogenation of CO_2 to formic acid with RhH(diphos)$_2$ (diphos=1,2-bis(diphenylphosphino)ethane or 1,3-bis(diphenylphosphino)propane) [10]. On 1994, Jessop, Ikariya, and Noyori.[11, 12] succeeded in performing a significantly efficient hydrogenation of CO_2 in super critical CO_2 (scCO$_2$) with ruthenium(II) complexes in the presence of triethylamine. The turnover numbers reached 7200 h^{-1} for formic acid formation [1c] and total turnover numbers reached 420000 for formamide formation [12b].

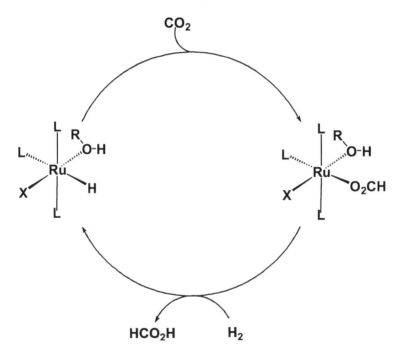

Figure 1. Reaction mechanism for the hydrogenation process of CO_2 proposed by Noyori and co-workers [12].

The reaction mechanism proposed by Noyori and co-workers [12] (Figure 1) is a good example of the generally accepted mechanism for this kind of processes. The reaction starts with the CO$_2$ insertion into the Ru(II)-H bond to afford a Ru(II)-formate intermediate. The final step seems to be σ-bond metathesis of the Ru(II)-formate intermediate with dihydrogen molecule, though it was not clearly described [12b] in the original formulation and could also be a combination of oxidative addition/reductive elimination. Very large turnover numbers (360000h^{-1}) were also reported with RuCl$_2$(dppe)$_2$ (dppe=1,2-bis(diphenylphosphino)ethane) by Kröcher et al. [13].

A slightly different catalytic system was reported by Lau and collaborators [14]. These authors used Ru(η^5: η^1-C$_5$H$_4$-(CH$_2$)$_3$NMe$_2$)(dppm) (dppm = bis(diphenylphosphino)methane) as a catalyst and proposed a modified reaction mechanism where H$_2$ undergoes heterolytic activation between the Ru center and the NMe$_2$ group in the ligand.

Hydrogenation of CO$_2$ can also be carried out in water [10c, 15]. In this case, the substrate is not CO$_2$ but HCO$_3^-$, and insertion of HCO$_3^-$ into the Ru-H bond is involved in the catalytic cycle. In our opinion, this step can be seen as a nucleophilic attack of the hydride ligand to the positively charged C atom of HCO$_3^-$.

1.2   Theoretical studies of CO$_2$ fixation reactions

In spite of the importance of CO$_2$ fixation reactions, relatively few theoretical works have been carried out on transition-metal CO$_2$ complexes

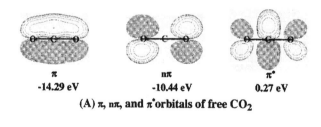

π	nπ	π*
-14.29 eV	-10.44 eV	0.27 eV

(A) π, nπ, and π*orbitals of free CO$_2$

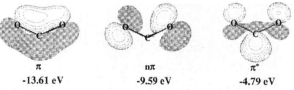

π	nπ	π*
-13.61 eV	-9.59 eV	-4.79 eV

(B) π, nπ, and π*orbitals of CO$_2$ with distorted structure

Figure 2. Frontier orbitals of CO$_2$.

and CO$_2$ fixation reactions. These results are nevertheless helpful for a better understanding of the hydrogenation of CO_2 to produce formic acid, which will be discussed in detail in the rest of the chapter.

First, we will discuss the frontier orbitals of CO_2. As shown in Figure 2, these are the π, non-bonding π ($n\pi$), and π^* orbitals. Since CO_2 is a Lewis acid, the π^* orbital is considered particularly important. It is worth noticing that the C p orbital contributes to the π^* orbital to a much greater extent than the O p orbital, and that the π^* orbital lowers its energy by the O-C-O bending distortion (see Figure 2B). The knowledge of these orbitals is necessary to understand and explain the geometry, bonding nature, and reaction behaviour of transition metal CO_2 complexes.

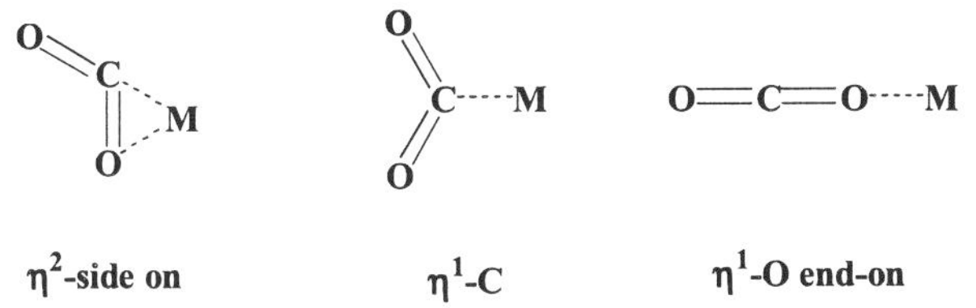

Figure 3. The three possible bonding modes of CO_2 in a transition metal complex.

Three coordination modes are considered possible in transition metal CO_2 complexes, as shown in Figure 3. They were theoretically investigated with semi-empirical [16, 17] and ab initio methods [18, 19]. A short summary of these results is collected in Table 1 [20], where the importance of the frontier orbitals of CO_2 for this interaction is apparent.

Table 1. Coordination mode of transition metal CO_2 complexes and electronic structure of the metal moiety [20].

	η^2- side-on	η^1-C	η^1-O end-on
Back-donation	strong when d_π is HOMO	strong when d_σ is HOMO	weak in general
Electrostatic	medium	unfavorable	strong for $M^{\delta+}$
Exchange repulsion	strong when d_σ is occupied	strong when d_π is occupied	strong when d_σ is occupied

Among the fixation reactions of CO_2 not directly related to hydrogenation, we will mention theoretical studies on two of them. The first is the transition metal-catalysed electrochemical reduction of CO_2 to CO, which was theoretically investigated with the ab initio SD-CI method. This reaction which had been previously observed by several authors [21, 22] was experimentally well characterized by Suavage and collaborators [23], who succeeded in carrying out the electrochemical reduction of CO_2 with Ni(cyclam)Cl$_2$. They proposed that the first step is the one-electron reduction of the Ni(II) complex to afford a Ni(I)-CO_2 complex with an η^1-C coordination form, and that the second step is another one-electron reduction of this $Ni(I)$-CO_2 species followed by proton addition to yield CO. SD-CI calculations of all the intermediates in this cycle clearly proved that the proposed reaction mechanism was reasonable and the intermediate, the $Ni(I)$-CO_2 complex, certainly took the η^1-C coordination form [24]. The preference for η^1-C over η^2-side-on coordination in this intermediate could be easily explained by the fact that the $Ni(I)$-cyclam species has a doubly occupied d_{z2} orbital at a high energy (Table 1).

The second fixation reaction that has been the subject of theoretical study is the nickel(0)-catalyzed coupling reaction of CO_2 with acetylene (equation 1), which was theoretically investigated with the ab initio SD-CI method [25]. The theoretical calculations clearly showed that if the nickel(0) moiety was eliminated from the reaction system, the activation barrier increased very much, and that the C-C bond formation between CO_2 and acetylene was accelerated by the charge-transfer from the nickel(0) d orbital to the orbital resulting from the combination between π^* orbitals of acetylene and CO_2. Actually, the HOMO contour map of $Ni(PH_3)(CO_2)(C_2H_2)$ clearly displays this orbital mixing in the transition state.

Though the theoretical studies presented in this section are relatively old and modern computational methods were not used, we consider that the essence of their conclusions is still valid now.

2.   THE ELEMENTARY STEPS IN THE CATALYTIC CYCLE OF CO$_2$ HYDROGENATION

As briefly discussed in section 1.1, and shown in Figure 1, the accepted mechanism for the catalytic cycle of hydrogenation of CO_2 to formic acid starts with the insertion of CO_2 into a metal-hydride bond. Then, there are two possible continuations. The first possibility is the reductive elimination of formic acid followed by the oxidative addition of dihydrogen molecule to the metal center. The second possible path goes through the σ-bond metathesis of a metal formate complex with a dihydrogen molecule. In this section, we will review theoretical investigations on each of these elementary processes, with the exception of oxidative addition of H_2 to the metal center, which has already been discussed in many reviews.

2.1 Insertion of CO_2 into the Metal-Hydride and Similar Bonds

Insertion reactions of CO_2 into the metal-hydride and metal-alkyl bonds are of considerable importance, since these reactions are involved not only in the catalytic cycle of the hydrogenation of CO_2 into formic acid but also in the catalytic cycle of co-polymerization of CO_2 and epoxide. In this regard, insertions of CO_2 into various metal-hydride, metal-alkyl, and similar bonds have been the subject of intense experimental investigation. For instance, CO_2 insertions into Cu(I)-CH_3, Cu(I)-OR, Cu(I)-alkyl [26-28], Ru(II)-H [29], Cr(0)-H, Mo(0)-H, W(0)-H [30], Ni(II)-H and Ni(II)-CH_3 bonds [31, 32] have been so far reported.

2.1.1 CO_2 and C_2H_4 Insertion into Cu(I)-H Bonds

CO_2 insertions into the Cu(I)-H and Cu(I)-CH_3 bonds (equations 4 and 5) were theoretically investigated and compared with the similar insertion reactions of ethylene into the Cu(I)-H and Cu(I)-CH_3 bonds (eqs 6 and 7) [33,34].

$$CuH(PH_3)_2 + CO_2 \rightarrow Cu(\eta^1\text{-OCOH})(PH_3)_2$$

$$\rightarrow Cu(\eta^2\text{-O}_2CH)(PH_3)_2 \qquad (4)$$

$$Cu(CH_3)(PH_3)_2 + CO_2 \rightarrow Cu(\eta^1\text{-OCOCH}_3)(PH_3)_2$$

$$\rightarrow Cu(\eta^2\text{-O}_2CCH_3)(PH_3)_2 \qquad (5)$$

$$CuH(PH_3)_2 + C_2H_4 \rightarrow Cu(C_2H_5)(PH_3)_2 \qquad (6)$$

$$Cu(CH_3)(PH_3)_2 + C_2H_4 \rightarrow Cu(C_2H_4CH_3)(PH_3)_2 \qquad (7)$$

Though only an alkyl complex is possible in the ethylene insertion, there are several possible products of the CO_2 insertion, such as η^1-OCOH, η^2-O_2CH, and η^1-COOH species, as shown in Figure 4. Thus, the following issues were investigated for the CO_2 insertion reactions: (1) Which species is more easily formed, the metal-formate (M-OCOH) or the metal-carboxylic acid (M-COOH)? (2) What are the most important interactions in the CO_2 insertion? (3) How different is the CO_2 insertion from the C_2H_4 insertion?

(4) Which bond, Cu(I)-H or Cu(I)-CH$_3$, is more reactive for the CO$_2$ insertion and why?

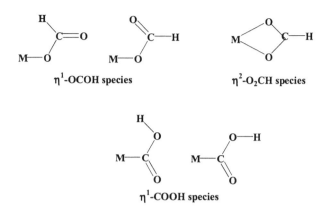

Figure 4. Possible products for the insertion of CO$_2$ into the M-H bond.

Structures involved in the process indicated in equation 4 were optimized with the Hartree-Fock (HF) method [34], as shown in Figure 5. First, CO$_2$ approaches the Cu(I) center, to afford the precursor complex **2A**. Since the Cu(I) center is not appropriate for the CO$_2$ coordination, the carbon dioxide fragment is far from the metal center in **2A**. In the transition state **3A**, the C-H distance is 1.678 Å, which is much longer than the usual C-H bond distance. The Cu-O^1 distance is still 2.5 Å, which is also much longer than that of the product (1.98 Å). The C-O^1 bond is slightly longer than the C-O^2 bond, which indicates that the C-O^1 and C-O^2 bonds have not changed yet to the C-O single and C-O double bonds, respectively. All these features indicate that the transition state is reactant-like. In the product, there are several possible structures, **4A**, **5A**, and **6A**. **4A** isomerises to **5A** in the optimisation calculation and **6A** is slightly more stable than **5A**.

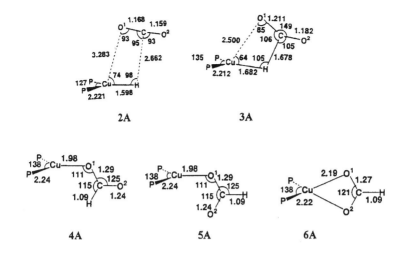

Figure 5. Optimized structures participating in the CO_2 insertion into the Cu-H bond of CuH(PH$_3$)$_2$. **2A** is the reactant, **3A** is the transition state, and **4A-6A** are products [34b].

The activation barrier and the reaction energy were evaluated with various computational methods, including MP4 (SDQ), SD-CI and CCD [34]. The reaction was found to be exothermic by a value of about 40 kcal/mol. The activation barriers were found to be 9.9 kcal/mol at the MP4(SDQ) level, 4.2 at the SD-CI level and 4.3 at the CCD level. Apparently, the MP4(SDQ) method yields a much larger activation barrier than SD-CI and CCD methods. Since the CCD and SD-CI methods present a similar activation barrier, these values were taken as the correct ones. From these calculations, it was concluded that CO_2 is easily inserted into the Cu(I)-H bond with a moderate activation barrier.

The detailed study of the molecular orbitals in the different species allowed a better understanding of the interactions under way. It was proved that the charge-transfer from the HOMO of the metal moiety to the π^* orbital of CO_2 is the most important interaction in the transition state and that the anti-bonding mixing of the π orbital of CO_2 also plays a significant role. The leading role of this HOMO-LUMO interaction also explains why the M-OCOH species is more easily formed than the M-COOH species.

The CO_2 insertion into the Cu(I)-H bond leading to Cu(η^1-COOH)(PH$_3$)$_2$ was compared with the insertion leading to Cu(η^1-OCOH)(PH$_3$)$_2$. Because the transition state (TS) of the CO_2 insertion leading to the Cu-(η^1-COOH) species could not be optimized with the Hartree-Fock method used in the geometry optimizations, the assumed TS-like structure was calculated with the O-H distance and the Cu-H-O angle arbitrarily taken to be 2.0 Å and 100 or 120 , respectively. This assumed TS-like structure was much more unstable than the transition state leading to the $Cu-(\eta^1-OCOH)$ species by about 28 kcal/mol. Thus, it could be clearly concluded that the CO_2 insertion leading to the $Cu-(\eta^1-COOH)$ is unfavourable. As mentioned above, these results are easily understood in terms of the electron distribution and the shape of the π^* orbital of CO_2, which is the most important for the interaction. Electrostatic interactions go also in the same direction. The product, $Cu(\eta^1-OCOH)(PH_3)_2$, is also much more stable than $Cu(\eta^1-COOH)(PH_3)_2$, for essentially the same reasons.

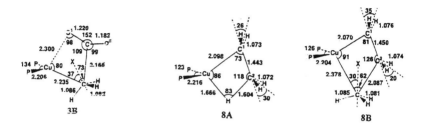

Figure 6. Transition states for the insertion of CO_2 into the $Cu-CH_3$ bond of $Cu(CH_3)(PH_3)_2$ (**3B**), the insertion of C_2H_4 into the $Cu-H$ bond of $CuH(PH_3)_2$ (**8A**), and the insertion of C_2H_4 into the $Cu-CH_3$ bond of $Cu(CH_3)(PH_3)_2$ (**8B**) [34].

Several interesting differences are observed between the insertion reactions of CO_2 and C_2H_4. The transition state of the C_2H_4 insertion reaction (**8A**, Figure 6) is much more product-like than that of the CO_2 insertion reaction (**3A**, Figure 5). The other important difference is that the CO_2 insertion needs a smaller activation barrier than the C_2H_4 insertion. While the carbon dioxide insertion had a barrier of 4.2 kcal/mol at the SD-CI level, the barrier for the ethylene insertion is 5.4 kcal/mol at this same level [34b]. This difference is easily interpreted in terms of the electrostatic and charge-transfer interactions. On one hand, the initial approach of the bond-forming atom to the Cu(I) center is more favourable in the case of CO_2, where it is a negatively charged O atom. On the other hand, the charge-transfer interaction between $CuH(PH_3)_2$ and CO_2 is much stronger than that between $CuH(PH_3)_2$ and C_2H_4 because the π^* orbital of CO_2 becomes much lower in energy than that of C_2H_4 after the bending distortion of CO_2.

A comparison was made also between the energy barriers for insertion of the CO_2 and C_2H_4 molecules in two different bonds: $Cu(I)-H$ and $Cu(I)-CH_3$. The barriers were lower for the case of $Cu(I)-H$ by values of 5.1 and 13.1 kcal/mol at the SD-CI level [34]. This difference arises from the fact that the valence orbital of the H ligand is a spherical 1s orbital and that of the CH_3 ligand is a directional sp^3 orbital. The lone pair from the CH_3 ligand must change its direction to form a new C-C bond and this has an energy cost. Similar differences were observed between the oxidative additions of H-H and H-CH$_3$ bonds [35] and between the insertions of C$_2$H$_4$ and C$_2$H$_2$ into the M-H and M-SiR$_3$ bonds [36]. Finally, the CO$_2$ insertion into the Cu(I)-OH bond was also analyzed, and found to proceed with no barrier [34b]. This is interpreted in terms that the OH group has two lone pair orbitals; one is used for bonding interaction with the Cu(I) center and the other can be used to form a new bonding interaction with CO$_2$.

2.1.2 CO$_2$ Insertion in Rh(III)-H, Rh(I)-H, Cu(I)-H and Ru(II)-H Bonds

We will discuss here the differences in the CO$_2$ insertion into a variety of M-H bonds. The energetics of these reactions are collected in Table 2.

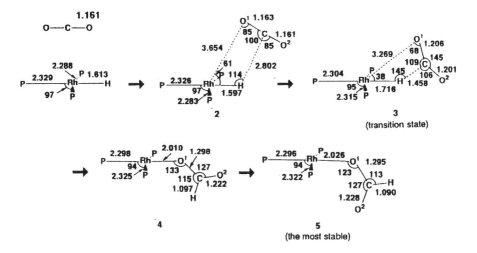

Figure 7. Optimized structures for the reaction of insertion of CO$_2$ into the Rh(I)-H bond of Rh(PH$_3$)$_3$ [37].

Table 2. Computed activation barrier (E_a, kcal/mol) and reaction energy (ΔE, kcal/mol) of the CO_2 insertion into Rh(III)-H, Rh(I)-H and Ru(II)-H bonds.

	E_a	ΔE	Ref.
RhH(PH$_3$)$_3$ + CO$_2$ → Rh(η^1-OCOH)PH$_3$)$_3$			
MP2//HF	24.6	-7.7	39a
MP3//HF	15.1	-27.9	39a
MP4(DQ)//HF	21.0	-17.8	39a
MP4(SDQ)//HF	20.5	-15.6	39a
SD-CI(DS)$^{a)}$ //HF	15.7	-26.2	39a
SD-CI(P)$^{b)}$ //HF	15.6	-24.2	39a
CCD//HF	15.9	-18.7	39a
RhH(PH$_3$)$_2$ + CO$_2$ → Rh(η^1-OCOH)(PH$_3$)$_3$			
MP2//MP2	4.3	-20.2	38
QCISD//MP2	-4.3	-24.5	38
[RhH$_2$(PH$_3$)$_3$]$^+$ + CO$_2$ → [RhH(η^2-O$_2$CH)(PH$_3$)$_3$]$^+$			
MP2//HF	52.9	-3.3	39a
MP3//HF	44.2	-13.9	39a
MP4(DQ)//HF	48.8	-8.8	39a
MP4(SDQ)//HF	46.8	-10.5	39a
SD-CI(DS)$^{a)}$ //HF	44.5	-13.7	39a
SD-CI(P)$^{b)}$ //HF	44.7	-13.7	39a
CCD//HF	45.8	-11.6	39a
MP4(SDQ)//MP2	53.8	-3.3	39b
[RhH$_2$(PH$_3$)$_2$(H$_2$O)]$^+$ + CO$_2$ → [RhH(η^2-O$_2$CH)(PH$_3$)$_2$(H$_2$O)]$^+$			
MP4(SDQ)//HF	24.0	-22.0	39a
MP4(SDQ)//MP2	24.5	-26.2	39b
RuH$_2$(PH$_3$)$_4$ + CO$_2$ → RuH(η^1-OCOH)(PH$_3$)$_4$			
MP2//B3LYP	41.8	8.7	46
MP3//B3LYP	25.8	-6.0	46
MP4(DQ)//B3LYP	37.2	4.3	46
MP4(SDQ)//B3LYP	37.0	4.1	46
CCSD//B3LYP	29.3	-2.3	46
CCSD(T)//B3LYP	29.7	-1.4	46
B3LYP//B3LYP	29.2	-0.3	41
RuH$_2$(PH$_3$)$_3$ + CO$_2$ → RuH(η^1-OCOH)(PH$_3$)$_3$			
MP2//B3LYP	19.6	7.6	46
MP3//B3LYP	6.0	8.4	46
MP4(DQ)//B3LYP	14.0	0.3	46
MP4(SDQ)//B3LYP	14.2	1.3	46
CCSD//B3LYP	10.3	4.9	46
CCSD(T)//B3LYP	11.0	-2.8	46
B3LYP//B3LYP	10.3	-4.0	41

The CO_2 insertion into the Rh(I)-H bond of the square planar RhH(PH$_3$)$_3$ complex was investigated with the Hartree-Fock method for geometry optimisation and MP2-MP4(SDQ) and CCD methods for energy evaluation [37]. Results are shown in Figure 7. CO_2 approaches the H ligand without interaction with the Rh(I) center. The activation barrier was evaluated with various computational methods, as listed in Table 2. Its value considerably fluctuates around MP2 and MP3 methods, but SD-CI and CCD methods present almost the same value of about 16 kcal/mol. Interestingly, this value is much larger than that into the Cu(I)-H bond [37]. This is interpreted in terms of the stronger exchange repulsion between $RhH(PH_3)_3$ and CO_2. The doubly occupied d_π orbital of the Rh(I) center gives rise to exchange repulsion with the π and non-bonding π orbitals of CO_2. This exchange repulsion is larger than that of the Cu(I) system because the 4d orbital of Rh(I) expands to a much greater extent than does the 3d orbital of Cu(I). This effect is reflected in the transition state geometry, where the Rh-H-O angle (145°) is larger than in the corresponding copper system (Cu-H-O, 106°). The Rh-O distance is also very long in the transition state.

More recently, the CO_2 insertion into the Rh(I)-H bond of the T-shape $RhH(PH_3)_2$ system has been theoretically investigated with MP2//MP2 and QCISD(T)//MP2 methods [38]. As can be seen in Table 2, this insertion reaction requires a much smaller activation barrier (4.3 kcal/mol with MP2//MP2 and –4.3 kcal/mol with QCISD(T)//MP2) than the insertion reaction in $RhH(PH_3)_3$. The reason is in the coordination number of rhodium. In $RhH(PH_3)_2$, the d^8 configuration of Rh(I) means that the d_σ orbital is unoccupied. As CO_2 approaches this empty coordination site, it does not give rise to exchange repulsion with a doubly occupied d orbital. As a result, the Rh-H-O angle is much smaller than that of the transition state of the insertion reaction of $RhH(PH_3)_3$ and the Rh-O bond is almost formed in the transition state.

The CO_2 insertion into the Rh(III)-H bond of the square pyramidal complex $[RhH_2(PH_3)_3]^+$ was investigated with the MP2 method for geometry optimisation and the MP2-MP4(SDQ), SD-CI, and CCD//HF methods for energy evaluation [39b]. Because of the strong trans-influence of the H ligand, the position trans to it is empty and CO_2 approaches this empty site, as shown in Figure 8. After the transition state, the η^2-OCOH species is directly formed instead of the η^1-OCOH that was found in other cases. This is probably because the Rh(III) center has a d^6 electron configuration and it tends to take a six-coordinate structure. The activation barrier, fluctuating moderately between the MP2 and MP3 values, presents a very large value over 40 kcal/mol (Table 2). One reason for this high barrier is the strong trans-influence of the H (hydride) ligand, which is trans to the formate being created. Actually, the activation barrier of the CO_2 insertion into the Rh(III)-H bond becomes much lower when H_2O coordinates with the Rh(III) center at a position trans to CO_2 (see Table 2) [39b]. However, even in this case the activation barrier is still higher than that of the CO_2 insertion into the Rh(I)-H bond.

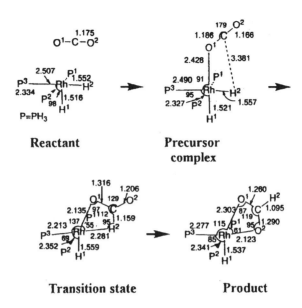

Figure 8. Optimised structures for the reaction of insertion of CO_2 into the Rh(III)-H bond of $RhH_2(PH_3)_3$ [37].

We recently investigated [40] the reason why CO_2 is inserted into the Rh(I)-H bond with a significantly lower barrier than into the Rh(III)-H bond, as shown in Table 2. As discussed above, charge-transfer from the metal-hydride moiety to the π^* orbital of CO_2 is very important in the CO_2 insertion reaction, and, at the same time, the metal-formate moiety is very much stabilized by the donation of electrons from the metal fragment. Since the Rh(I) center is more electron-rich than Rh(III), the charge-transfer from the Rh(I)-H moiety to the π^* orbital of CO_2 is favored, and the formate moiety is provided with sufficient electrons. Consequently, CO_2 is more easily inserted into the Rh(I)-H bond than into the Rh(III)-H bond.

CO_2 insertion into the Ru(II)-H bond of cis-$RuH_2(PH_3)_n$ (n=3 or 4) was theoretically investigated as a part of investigation of a full catalytic cycle [41]. Structures were optimised with the DFT-based B3LYP method, and the activation barriers and reaction energies were evaluated with MP2-MP4(SDQ), CCSD(T), and B3LYP methods, the results being shown in Table 2. The energies fluctuate considerably within the MP methods but CCSD(T) and B3LYP present similar values to each other. This suggests that the B3LYP method is reliable for the CO_2 insertion reaction. Two reaction paths were considered from cis-$RuH_2(PH_3)_4$, one of them involving

PH₃ dissociation. This phosphine dissociation costs 24.8 kcal/mol (B3LYP), but activates very much the complex towards the reaction. CO_2 insertion requires an activation barrier of 29.3 kcal/mol for cis-$RuH_2(PH_3)_4$, but only 10.3 kcal/mol for cis-$RuH_2(PH_3)_3$ (B3LYP). The reason of the significant difference between cis-$RuH_2(PH_3)_3$ and cis-$RuH_2(PH_3)_4$ is found by inspecting the Ru-O distance in the transition state. In cis-$RuH_2(PH_3)_4$, the Rh-O distance is very long, because this complex is coordinatively saturated, and the O atom cannot approach the Ru center.

It is noticeable that these activation barriers for CO_2 insertion into Ru(II)-H bonds are lower than those for the Rh(III)-H bond discussed above. This can be interpreted in terms of electron density at the metal center, which in the Ru(II) system must be larger than in Rh(III) but smaller than in the Rh(I) system [40].

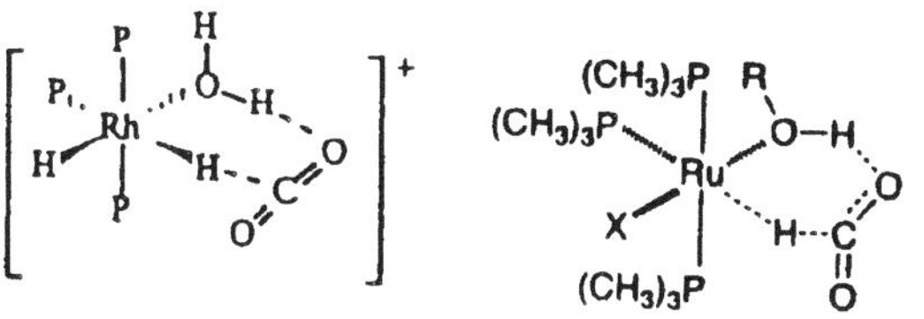

Figure 9. Proposed intermediates for the water or alcohol assisted insertion of carbon dioxide into metal-hydride bonds.

A last topic to discuss concerning the insertion reaction of CO_2 concerns the possible acceleration of this reaction by the presence of acidic protons from solvents or other ligands. Acceleration of the CO_2 insertion into the Rh(III)-H and Ru(II)-H bonds by water and alcohol was experimentally proposed, as shown in Figure 9 [9, 12b]. Although this possibility has not been computationally investigated yet, a very related topic has been the subject of a recent theoretical study. The role of a protonated amine-arm was experimentally proposed in the hydrogenation of CO_2 with Ru(dppm)(Cp-$(CH_2)_2$-NMe₂) [14], as mentioned in the Introduction, and has been the subject of recent DFT study [42]. Several paths were considered, but the more interesting aspect in connection with carbon dioxide insertion can be obtained from the comparison of the two paths represented by the transition states in Figure 10.

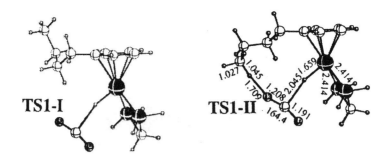

Figure 10. Transition states corresponding to two representative pathways of insertion of CO_2 into the Ru(II)-H bond of Ru(η^5-C_5H_4-(CH_2)$_3$NMe_2H$^+$)(dppm) (dppm = bis(dimethylphosphino)methane). [42] **TS1-I** in which the amine arm does not interact with CO_2 could not be optimized, and its geometry was taken to be the same as **TS1-II** without the orientation of amine arm. In **TS1-II**, the ammine arm interacts with the O atom of CO_2.

When the protonated amine-arm does not participate in the reaction (**TS1-I**), the reaction is very endothermic. However, the activation barrier decreases to 2.6 kcal/mol and the endothermicity of the reaction becomes very small (only 0.2 kcal/mol), if the protonated amine-arm participates in the reaction (**TS1-II**). In this transition state the O-H distance is 1.709Å, typical of a hydrogen bond. The origin of the acceleration was explained in terms of the enhancement of polarization of CO_2 induced by the protonated amine arm, which increases the positive charge in the carbon atom. This polarization favors the electrostatic interaction between hydride and CO_2, and since this polarization increases the contribution of the C p orbital in the π^* orbital of CO_2, it favors also the charge-transfer from the H ligand to the π^* orbital of CO_2. In other paths starting with direct interaction between CO_2 and Ru the effect of the protonated amine was much smaller.

2.2   Reductive Elimination of Formic Acid from Transition-Metal Formate Complexes

As shown in Figure 1, the next step in the catalytic cycle of carbon dioxide hydrogenation is either reductive elimination of formic acid from the transition-metal formate hydride complex or σ-bond metathesis between the transition-metal formate complex and dihydrogen molecule. In this section, we will discuss the reductive elimination process. Activation barriers and reaction energies for different reactions of this type are collected in Table 3.

Table 3. Computed activation barrier (E_a, kcal/mol) and reaction energy (ΔE, kcal/mol) of the reductive elimination of H-OCOH from formate hydride complexes.

	E_a	ΔE	Ref.
RhH$_2$(η^1-OCOH)(PH$_3$)$_2$ → RhH(HCOOH)(PH$_3$)$_2$			
MP2//MP2	23.6	13.0	38
QCISD(T)//MP2	24.7	11.5	38
RuH(η^1-OCOH)(PH$_3$)$_3$ → Ru(PH$_3$)$_3$ + HCOOH			
3-center reductive elimination			
MP2//B3LYP	36.6	-	46
MP3//B3LYP	50.3	-	46
MP4(DQ)//B3LYP	42.3	-	46
MP4(SDQ)//B3LYP	40.4	-	46
CCSD//B3LYP	47.6	-	46
CCSD(T)//B3LYP	45.1	-	46
B3LYP//B3LYP	39.8	41.4	46
5-center reductive elimination			
MP2//B3LYP	12.7	12.3	46
MP3//B3LYP	29.2	32.5	46
MP4(DQ)//B3LYP	19.3	18.1	46
MP4(SDQ)//B3LYP	16.1	13.9	46
CCSD//B3LYP	21.1	25.4	46
CCSD(T)//B3LYP	21.7	23.1	46
B3LYP//B3LYP	17.5	17.8	46

The reductive elimination of formic acid from RhH$_2$(η^1-OCOH)(PH$_3$)$_2$ was theoretically investigated first by Dedieu and his collaborators [38]. A three-center transition state was optimised, in which the Rh-O and Rh-H distances lengthened by about 0.2 Å and 0.1 Å, respectively, and the O-H distance was 1.25 Å. The activation barrier was evaluated to be 23.6 kcal/mol with the MP2 method and 24.7 kcal/mol with the QCISD(T) method (Table 3). From these results, it was concluded that the reductive elimination was difficult.

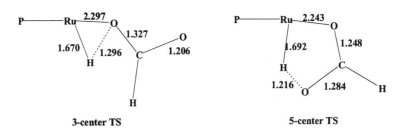

Figure 11. Schematic representation of the transition states for the two different pathways in the reductive elimination of H-OCOH from $RuH(\eta^1\text{-OCOH})(PH_3)_3$ [41]. Bond distances are in Å. Two PH_3 ligands which are perpendicular to this plane are omitted here for brevity.

The reductive elimination of formic acid from $RuH(\eta^1\text{-OCOH})(PH_3)_n$ (n=3, 4) was also recently investigated theoretically by Musashi and Sakaki [41]. They examined not only a three-center transition state but also a five-center one, as shown in Figure 11. In the three-center transition state, the Ru-O and Ru-H distances lengthen by 0.23Å and 0.11Å, respectively, and the O-H distance (1.37Å) is still much longer than the usual O-H bond. These geometrical features are essentially the same as those previouly reported [38] for the reductive elimination in $RhH_2(\eta^1\text{-OCOH})(PH_3)_2$. The activation barrier was evaluated with various computational methods, as listed in Table 3. The barrier is 39.8 kcal/mol at the B3LYP level. This very high activation barrier clearly indicates that the three-center reductive elimination is difficult.

Things are however different for the five-center elimination. In the transition state of this process (Figure 11), the Ru-O distance increases moderately by 0.16 Å, while the Ru-H distance lengthens considerably by 0.3 Å. These features suggest that this reductive elimination can be viewed as a hydrogen transfer from the Ru(II) center to the formate ligand. The activation barrier is 17.5 kcal/mol when n=3 and 25.5 kcal/mol when n=4 (B3LYP). Thus, the five-center pathway is much favored over the three-center one, at least for these particular $RuH(\eta^1\text{-OCOH})(PH_3)_n$ complexes. The reason for this difference can be understood from the orbitals presented in Figure 12. In the three-center transition state, the HOMO of $\eta^1\text{-OCOH}$, that is used for the Ru-O bonding interaction, must change its direction toward the H ligand to form a new O-H bond. In the five-center transition state, this directional change is not necessary.

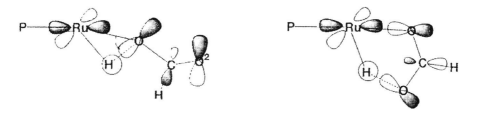

Figure 12. Orbital diagram explaining why the 3-center reductive elimination (left side) has a higher barrier than the 5-center process (right side).

2.3 Metathesis of a Transition-Metal Formate Complex with a Dihydrogen Molecule:

The other reaction path to obtain formic acid from the transition metal formate complex is metathesis with a dihydrogen molecule. This reaction course has been proposed experimentally, but no clear evidence has been reported so far. Energetics of this reaction from different complexes and with a variety of methods are collected in Table 4.

Table 4. Computed activation barrier (E_a, kcal/mol) and reaction energy (ΔE, kcal/mol) of the σ-bond metathesis reactions of metal formate complexes with the dihydrogen molecule.

	E_a	ΔE	Ref.
RhH$_2$(η^1-OCOH)(PH$_3$)$_2$ → RhH(HCOOH)(PH$_3$)$_2$			
MP2//MP2	11.4	-3.4	38
QCISD(T)//MP2	14.8	0.2	38
RuH(η^1-OCOH)(PH$_3$)$_3$ → Ru(HCOOH)(PH$_3$)$_3$			
4-center σ-bond metathesis			
MP2//B3LYP	27.9	-	46
MP3//B3LYP	27.5	-	46
MP4(DQ)//B3LYP	27.7	-	46
MP4(SDQ)//B3LYP	28.6	-	46
CCSD//B3LYP	28.1	-	46
CCSD(T)//B3LYP	28.3	-	46
B3LYP//B3LYP	24.8	15.9	41
6-center σ-bond metathesis			
MP2//B3LYP	8.5		46
MP3//B3LYP	9.8		46
MP4(DQ)//B3LYP	9.2		46
MP4(SDQ)//B3LYP	9.3		46
CCSD//B3LYP	9.8		46
CCSD(T)//B3LYP	9.3		46
B3LYP//B3LYP	8.2	-0.4	41

Dedieu and his collaborators theoretically investigated in detail the metathesis with complex $Rh(\eta^1\text{-OCOH})(PH_3)_2$ [38]. The first step of this reaction is coordination of the dihydrogen molecule to the metal center, a step that takes place easily because the 3-coordinate $Rh(\eta^1\text{-OCOH})(PH_3)_2$ complex has an empty coordination site. The binding energy is, however, very small; 3.2 kcal/mol at the MP2//MP2 level and 5.7 kcal/mol at the QCISD(T)//MP2 level. In the transition state, the H-H distance lengthens to 1.10Å and the O-H distance is still 1.34 Å which is considerably longer than the usual O-H bond. The Rh-O distance lengthens to about 2.2 Å, which is slightly longer than that of the reactant and almost the same as that of the product. The activation barrier was evaluated to be 11.4 kcal/mol with the MP2//MP2 method and 14.8 kcal/mol with the QCISD(T)//MP2 method (Table 4). This metathesis process could be explained with simple molecular orbital arguments.

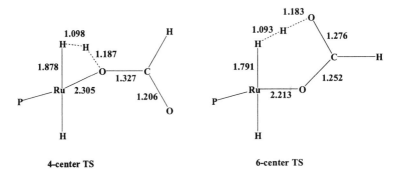

4-center TS   **6-center TS**

Figure 13. Schematic representation of the transition states for the two different pathways in the four-centered (left) and six-centered metathesis (right) of $RuH(\eta^1\text{-OCOH})(PH_3)_3$ with a dihydrogen molecule [41]. Bond distances are in Å.

The metathesis reaction between a ruthenium formate complex, $RuH(\eta^1\text{-OCOH})(PH_3)_3$, and a dihydrogen molecule was recently investigated with a B3LYP method for geometry optimisation and MP2-MP4(SDQ), CCSD(T), and B3LYP methods for energy evaluation [41]. As in the case of reductive elimination discussed in the previous section, two different mechanisms can be considered, the usual one involving a four-center transition state, and an additional one involving the two oxygen atoms of CO_2, leading in this case to a six-center transition state. Both transition states are shown in Figure 13. The four-center transition state has a geometry similar to that previously reported for $Rh(\eta^1\text{-OCOH})(PH_3)_2$ [38]. The activation barrier is calculated to be about 28 kcal/mol with MP2-MP4(SDQ) and CCSD(T) methods, while the B3LYP method presents a slightly smaller value of 25 kcal/mol, as listed in Table 4.

To cause the six-center metathesis, the formate moiety must take the structure in which the terminal O atom of η^1-OCOH is in the same side as H_2, as shown in Figure 13. Since $RuH(\eta^1\text{-OCOH})(PH_3)_3$ does not take this structure just after the CO_2 insertion, an isomerization consisting of a rotation around the Ru-O bond must take place first. This process has a moderate barrier of 8.5kcal/mol (B3LYP). The next step is the six-center metathesis itself. This metathesis occurs through the transition state shown in Figure 13. In this geometry the hydrogen atom being transferred is at almost an intermediate position between the oxygen and the other hydrogen atom. Another important feature of this structure is that the hydrogen atom being transferred becomes more positively charged while the other one gains negative charge. In other words, the metathesis involves the heterolytic H-H bond fission. This transition state is very similar to that of the C-H activations of methane and benzene by $Pd(\eta^2\text{-OCOH})_2$ [43]. As listed in Table 4, all the computational methods present almost the same activation barrier between 8 and 10 kcal/mol for this process. These activation barriers are much lower than those of the four-centered σ-bond metathesis presented above. This significant difference between four-center and six-center metatheses is easily understood in terms of the direction change of the HOMO of formate. As shown in Figure 14, the HOMO must change its direction toward the H atom in the four-center metathesis, but not in the six-center process.

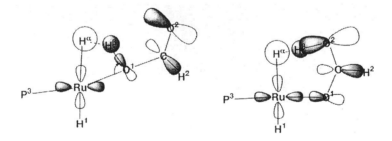

Figure 14. Orbital diagram explaining why the 4-center σ-bond metathesis (left side) has a higher barrier than the 6-center process (right side).

Another factor to be investigated in the metathesis process is the effect of bases in the reaction media. Bases such as triethylamine are added in the experimental conditions to stabilize the formic acid product because otherwise the product is thermodynamically less stable than the separate carbon dioxide and dihydrogen reactants. As discussed above, the σ-bond methathesis involves the heterolytic H-H bond fission, which would be accelerated by the presence of the base. This effect was theoretically investigated in the four-center σ-bond metathesis between $Rh(\eta^1$-

OCOH)(PH$_3$)$_2$ and a dihydrogen molecule [44]. The introduction of a NH$_3$ molecule in the system converts this single-step process in a three-step process, with three transition states. The energy change occurs smoothly and the highest barrier is only 2.1 kcal/mol (MP2//MP2) which is much smaller than the barrier in the absence of NH$_3$ (11.4 kcal/mol, Table 4). This means that the four-center metathesis occurs with nearly no barrier in the presence of base. However, the base effects are likely overestimated in this calculations. The experimental triethylamine system cannot do all the interactions ammonia does, and polar or protic solvents should weaken the interaction between the base and the complex.

After the metathesis, HCOOH must eliminate from the metal center, but this elimination gives rise to a significantly large destabilization energy. For instance, the destabilization energy is 11.3 kcal/mol in the RuH$_2$(PH$_3$)$_3$-catalyzed hydrogenation of CO$_2$. However, the fact that a recent theoretical study with the DFT/SCRF (polarizable continuum model) method indicates that CO$_2$ coordinates with the Rh center to form RhH(PH$_3$)$_2$(CO$_2$) [45] suggests that super critical CO$_2$ can assist the elimination of formic acid from the metal center.

3.   CATALYTIC CYCLE OF HYDROGENATION OF CO$_2$ TO FORMIC ACID

After discussing in detail each step of the hydrogenation process, we will comment here on the computational studies of the full catalytic cycle.

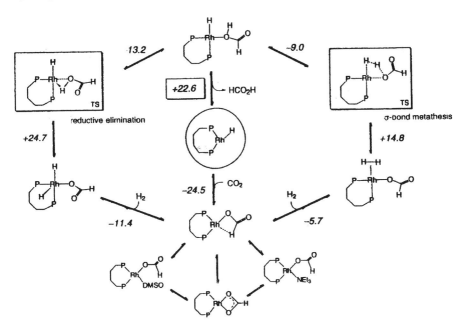

Figure 15. Computed cycle for the hydrogenation of CO_2 catalyzed by $RhH(PH_3)_2$ [38]. MP2 relative energies are in kcal/mol.

Dedieu and his collaborators reported the calculations on the full catalytic cycle of the reaction with $RhH(PH_3)_2$, which are summarized in Figure 15 [38]. Apparently, the oxidative addition of H_2 followed by the reductive elimination of HCOOH requires a much higher activation barrier (24.7 kcal/mol; MP2//MP2) than the σ-bond metathesis (14.8 kcal/mol). Thus, it was concluded that the hydrogenation proceeds through CO_2 insertion into the Rh(I)-H bond followed by the σ-bond metathesis. Though the elimination of formic acid from the Rh(I) center gives rise to a significantly large destabilization energy (22.6 kcal/mol), an amine molecule from the medium can stabilize the formic acid by formation of salt. This was the first report providing theoretical support to the reaction mechanism involving the σ-bond metathesis of the transition-metal formate complex with a dihydrogen molecule. According to these calculations, the rate-determining step is precisely the σ-bond metathesis.

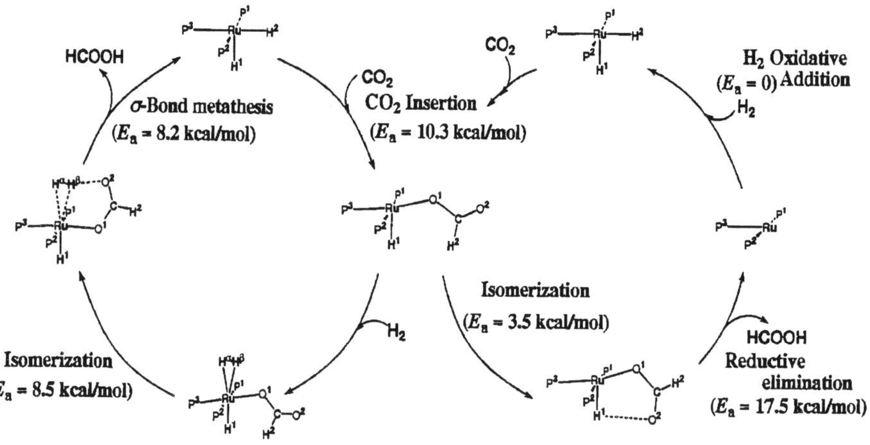

Figure 16. Computed cycle for the hydrogenation of CO_2 catalyzed by $RuH_2(PH_3)_3$ [41]. B3LYP relative energies are in kcal/mol.

The full catalytic cycle of the Ru-catalysed hydrogenation of CO_2 was theoretically investigated by Musashi and Sakaki [41], with the results summarized in Figure 16. Two kinds of active species were examined: the coordinatively saturated $RuH_2(PH_3)_4$ and the coordinatively unsaturated $RuH_2(PH_3)_3$. The barrier for CO_2 insertion into the Ru-H bond was very high for the case of $RuH_2(PH_3)_4$, thus the coordinatively unsaturated complex was found to be the active one, giving rise to the cycle shown in the Figure. As discussed in the previous section, either the five-centered reductive elimination or the six-centered σ-bond metathesis can occur after the CO_2 insertion. In both cases, $RuH(\eta^1\text{-OCOH})(PH_3)_3$ must isomerize to the structure in which these processes can take place. The isomerization needs a moderate activation barrier of 8.5 kcal/mol in the reaction path involving metathesis and 3.5 kcal/mol in the reaction path involving reductive elimination (B3LYP). However, the reductive elimination requires an activation barrier of 17.5 kcal/mol, considerably higher than the 8.2 kcal/mol of the σ-bond metathesis. Thus, it was concluded that the Ru-catalyzed hydrogenation proceeds through the path shown in the left part of Figure 16: CO_2 insertion into the Ru(II)-H bond, isomerization of the ruthenium formate hydride complex, and six-center σ-bond metathesis of the ruthenium formate hydride complex with a dihydrogen molecule. The active species is the coordinatively unsaturated complex $RuH_2(PH_3)_3$, and the rate-determining step is the CO_2 insertion. These conclusions are nuanced by the incorporation of solvent effects. A polar solvent decreases the activation barrier of the CO_2 insertion reaction and, as a result, in super-critical CO_2 the isomerization becomes the rate-determining step [46].

4. CONCLUSIONS AND PERSPECTIVES FOR FUTURE CALCULATIONS

The full catalytic cycle of transition metal catalysed hydrogenation of CO_2 to formic acid has been theoretically investigated, and valuable knowledge has been obtained on the reaction mechanism, the active species, and the rate-determining step. However, a number issues still remain to be investigated theoretically. The first of them is the importance of solvent effects. For instance, it is well known that the hydrogenation of CO_2 is efficiently catalysed by the ruthenium(II) complexes in super-critical CO_2, [11, 12]. Though solvent effects were theoretically investigated in a pioneering work by Tomasi et al. [45], their role on all the elementary steps has not been investigated yet. The second issue is the possibility to introduce the real phosphines in the calculation instead of modeling them as PH_3. The ONIOM method proposed by Morokuma's group [47] is an attractive way to take account of the effect of the real ligands. A third issue concerns the reliability of computational methods, which should be examined in detail. Though the DFT method with the B3LYP functional is not bad, there are still some differences between DFT and CCSD(T) methods. Of course, the final goal is to be able to predict what combination of transition metal element and ligands will provide efficient catalysis for this very important reaction.

REFERENCES

1. Braunstein, P.; Matt. D.; Nobel, D. *Chem. Rev.* **1988**, *88*, 747. b) Behr, A. *Angew., Chem., Int. Ed. Engl.,* **1988**, *27*, 661. c) Jessop, P. G.; Ikariya, T.; Noyori, R. *Chem. Rev.* **1995**, *95*, 259. d) Darensbourg, D. J.; Holtcamp, M. W. *Coord. Chem. Rev.*, **1996**, *153*, 155. e) Walther, D.; Rubens, M.; Rau, S. *Coord. Chem. Rev.* **1999**, *182*, 67.
2. For instance; a) Sasaki, Y.; Inoue, Y.; Hashimoto, H. *J. Chem. Soc., Chem. Commun.* **1976**, 605. b) Inoue, Y.; Itoh, Y.; Kazama, H.; Hashimoto, H. *Bull. Chem. Soc. Jpn.* **1980**, *53*, 3329. c) Behr, A.; He, R.; Gross, S.; Milchereit, A. *Angew. Chem., Int. Ed. Engl.* **1987**, *26*, 571. d) Behr, A.; Kanne, U. *J. Organomet. Chem.* **1986**, *317*, C41. e) Hoberg, H.; Gross, S.; Milchereit, A. *Angew. Chem., Int. Ed. Engl.,* **1987**, *26*, 571. f) Tsuda, T.; Morikawa, S.; Sumiya, R.; Saegusa, T. *J. Org. Chem.* **1988**, *53*, 3140. g) Tsuda, T.; Maruta, K.; Kitaike, Y. *J. Am. Chem. Soc.* **1992**, *114*, 1498. h) Derien, S.; Dunach, E.; Perichon, J. *J. Am. Chem. Soc.* **1991**, *113*, 8447.

3. For instance, a) Aida, T.; Inoue, S. *J. Am. Chem. Soc.* **1983**, *105*, 1304. b) Darensbourg, D. J.; Niezgoda, S. A.; Draper, J. D.; Reibenspies, J. H. *J. Am. Chem. Soc.* **1998**, *120*, 4690.

4. Kozima, F.; Aida, T.; Inoue, S. *J. Am. Chem. Soc.* **1986**, *108*, 391. b) Komatsu, M.; Aida, T.; Inoue, S. *J. Am. Chem. Soc.* **1991**, *113*, 8492. c) Sugimoto, H.; Kimura, T.; Inoue, S. *J. Am. Chem. Soc.* **1999**, *121*, 2325. d) Paddock, R. L.; Nguyen S. T. *J. Am. Chem. Soc.* **2001**, *123*, 11498.

5. Koinuma, H.; Yoshida, Y.; Hirai, H. *Chem. Lett.* **1975**, 1223.

6. Inoue, Y.; Izumida, H.; Sasaki, S.; Hashimoto, H. *Chem. Lett.* **1976**, 863.

7. Darensbourg, D. J.; Ovalles, C. O. *J. Am. Chem. Soc.* **1984**, *106*, 3750, and **1987**, *109*, 3330.

8. Taqui Khan, M. M.; Halligudi, S. B.; Shukla, S. *J. Mol. Catal.*, **1989**, *57*, 47.

9. Tsai, J. C.; Nicholas, K. H. *J. Am. Chem. Soc.* **1992**, *114*, 5117.

10. Graf, E.; Leitner, W. *J. Chem. Soc. Chem. Commun.* **1992**, 623. b) Burgemeister, T.; Kastner, F.; Leitner, W. *Angew. Chem., Int. Ed. Engl.* **1993**, *32*, 739. c) Gassner, F.; Leitner, W. *J. Chem. Soc., Chem. Commun.* **1993**, 1465.

11. Jessop, P. G.; Ikariya, T.; Noyori, R. *Nature*, **1994**, *368*, 231.

12. Jessop, P. G.; Hsiao, Y.; Ikariya, T.; Noyori, R. *J. Am. Chem. Soc.* **1994**, *116*, 8851. b) Jessop, P. G.; Ikariya, T.; Noyori, R. *J. Am. Chem. Soc.* **1996**, *118*, 344.

13. Kröcher, O.; Köppel, R. A.; Baiker, A. *J. Chem. Soc., Chem. Commun.* **1997**, 453.

14. Chu, H. S.; Lau, C. P.; Wong, K. Y. *Organometallics*, **1998**, *17*, 2768.

15. Laurenczy, G.; Joo, F.; Nádasdi, L. *Inorg. Chem.* **2000**, *39*, 5083.

16. Sakaki, S.; Kudou, N.; Ohyoshi, A. *Inorg. Chem.* **1977**, *16*, 202.

17. Mealli, C.; Hoffmann, R.; Stockis, A. *Inorg. Chem.* **1984**, *23*, 56.

18. Sakaki, S.; Kitaura, K.; Morokuma, K. *Inorg. Chem.* **1982**, *21*, 760. b) Sakaki, S.; Dedieu, A. *Inorg. Chem.* **1987**, *26*, 3278. c) Sakaki, S.; Aizawa, T.; Koga, N.; Morokuma, K.; Ohkubo, K. *Inorg. Chem.* **1989**, *28*, 103. d) Sakaki, S.; Koga, N.; Morokuma, K. *Inorg. Chem.* **1990**, *29*, 3110.

19. Jegat, C.; Fouassier, M.; Tranquille, M.; Mascetti, J.; Tomasi, I.; Aresta, M.; Ingold, F.; Dedieu, A. *Inorg. Chem.* **1993**, *32*, 1279. b) Dedieu, A.; Ingold, F. *Angew. Chem., Int. Ed. Engl.* **1989**, *28*, 1694.

20. Sakaki, S. in *Stereochemistry of Organometallic and Inorganic Compounds*, Vol. 4, Bernal, I, Ed.. Elsevier, Amsterdam, 1990, p95.

21. For instance, Meshitsuka, S.; Ichikawa, M.; Tamaru, K. *J. Chem. Soc., Chem. Commun.* **1974**, 158.

22. Hiratsuka, K.; Takahashi, K.; Sasaki, H.; Toshima, S. *Chem. Lett.* **1977**, 1137. Fishcher, B.; Eisenberg, R. *J. Am. Chem. Soc.* **1980**, *102*, 7363. Tezuka, M.; Yajima, T.; Tsuchiya, A.; Matsumoto, Y.; Uchida, Y.; Hidai, M. *J. Am. Chem. Soc.* **1982**, *104*, 6834.

23. Beley, M.; Collin, J. –P.; Ruppert, R.; Sauvage, J. –P. *J. Chem. Soc., Chem. Commun.* **1984**, 1315. b) Beley, M.; Collin, J. –P.; Ruppert, R.; Sauvage, J. –P. *J. Am. Chem. Soc.* **1986**, *108*, 7461.

24. Sakaki, S. *J. Am. Chem. Soc.*, **1990**, *112*, 7813, and *J. Am. Chem. Soc.* **1992**, *114*, 2055.

25. Sakaki, S.; Mine, K.; Taguchi, D.; Arai, T. *Bull. Chem. Soc. Jpn.* **1993**, *66*, 3289. b) Sakaki, S.; Mine, K.; Hamada, T.; Arai, T. *Bull. Chem. Soc. Jpn.* **1995**, *68*, 1873.

26. Miyashita, A.; Yamamoto, A. *J. Organomet. Chem.* **1973**, *49*, C57. b) Ikariya, T.; Yamamoto, A. *J. Organomet. Chem.* **1974**, *72*, 145. c) Miyashita, A.; Yamamoto, A. *J. Organomet. Chem.* **1976**, *113*, 187.

27. Tsuda, T.; Saegusa, T. *Inorg. Chem.* **1972**, *11*, 2561. b) Tsuda, T.; Sanada, S.; Ueda, K.; Saegusa, T. *Inorg. Chem.* **1976**, *15*, 2329. c) Tsuda, T.; Chujo, Y.; Saegusa, T. *J. Am. Chem. Soc.* **1978**, *100*, 630.

28. Bianchini, C.; Ghilardi, C. A.; Meli, A.; Midollini, S.; Orlandini, A. *Inorg. Chem.* **1985**, *24*, 924.

29. Komiya, S.; Yamamoto, A. *Bull. Chem. Soc. Jpn.* **1976**, *49*, 784. b) Komiya, S.; Yamamoto, A. *J. Organomet. Chem.* **1972**, *46*, C58.

30. Darensbourg, D. J.; Rokicki, A.; Darensbourg, M. Y. *J. Am. Chem. Soc.* **1981**, *103*, 3223. b) Darensbourg, D. J.; Rokicki, A. *J. Am. Chem. Soc.* **1982**, *104*, 349. c) Slater, S. G.; Lush, R.; Schumann, B. F.; Darensbourg, M. *Organometallics*, **1982**, *1*, 1662. d) Darensbourg, D. J.; Pala, M.; Waller, J. *Organometallics*, **1983**, *2*, 1285. e) Darensbourg, D. J.; Kudaroski, R. *J. Am. Chem. Soc.* **1984**, *106*, 3672. f) Darensbourg, D. J.; Hanckel, R. K.; Bauch, C. G.; Pala, M.; Simmons, D.; White, J. N. *J. Am. Chem. Soc.* **1985**, *107*, 7463. g) Darensbourg, D. J.; Scanchez, K. M.; Reibenspies, J. H.; Rheingold, A. L. *J. Am. Chem. Soc.* **1989**, *111*, 7094. h) Darensbourg, D. J.; Wiegreffe, H. P.; Wiegreffe, P. W. *J. Am. Chem. Soc.* **1990**, *112*, 9252.

31. Darensbourg, D. J.; Darensbourg, M. Y.; Goh, L. Y.; Ludvig, M.; Wiegreffe, P. *J. Am. Chem. Soc.* **1987**, *109*, 7539.

32. Carmona, E.; Gutiérrez-Puebla, E.; Marín, J. M.; Monge, A.; Paneque, M.; Poveda, M. L.; Ruiz, C. *J. Am. Chem. Soc.* **1989**, *111*, 2883.

33. Sakaki, S.; Ohkubo, K. *Inorg. Chem.* **1988**, *27*, 2020. b) Sakaki, S.; Ohkubo, K. *Inorg. Chem.* **1989**, *28*, 2583.

34. Sakaki, S.; Ohkubo, K. *Organometallics*, **1989**, *8*, 2970. b) Sakaki, S.; Musashi, Y. *Inorg. Chem.* **1995**, *34*, 1914.

35. Blomberg, M. R. A.; Brandemark, U.; Siegbahn, P. E. M. *J. Am. Chem. Soc.* **1984**, *105*, 5557. b) Saillard, J. Y.; Hoffmann, R. *J. Am. Chem. Soc.* **1984**, *106*, 2006. c) Low, J. J.; Goddard, W. A. *J. Am. Chem. Soc.* **1986**, *108*, 6115.

36. Sakaki, S.; Ogawa, M.; Musashi, Y.; Arai, T. *J. Am. Chem. Soc.* **1994**, *116*, 7258. b) Sakaki, S.; Mizoe, N.; Sugimoto, M. *Organometallics*, **1998**, *17*, 2510.

37. Sakaki, S.; Musashi, Y. *J. Chem. Soc., Dalton Trans.* **1994**, 3047.

38. Hutschka, F.; Dedieu, A.; Eichberger, M.; Fornika, R.; Leitner, W. *J. Am. Chem. Soc.*, **1997**, *119*, 4432.

39. Sakaki, S.; Musashi, Y. *Int. J. Quant. Chem.* **1996**, *57*, 481. b) Musashi, Y.; Sakaki, S. *J. Chem. Soc., Dalton Trans.* **1998**, 577.

40. Musashi, Y.; Sakaki, S. *J. Am. Chem. Soc.*, to be published.

41. Musashi, Y.; Sakaki, S. *J. Am. Chem. Soc.* **2000**, *122*, 3867.

42. Matsubara, T. *Organometallics*, **2001**, *20*, 19.

43. Biswas, B. Sugimoto, M.; Sakaki, S. *Organometallics*,

44. Hutschka, F.; Dedieu, A. *J. Chem. Soc., Dalton Trans.* **1997**, 1899.

45. Pomelli, C. S.; Tomasi, J.; Solà, M. *Organometallics*, **1998**, *17*, 3164.

46. Musashi, Y.; Sakaki, S. to be published.

47. Dapprich, S.; Komáromi, I.; Byun, K-S.; Morokuma, K. *J. Mol. Struct. (Theochem)*, **1999**, *461-462*, 1.

Chapter 5

Catalytic Enantioselective Hydrogenation of Alkenes

Steven Feldgus and Clark R. Landis*
Department of Chemistry, University of Wisconsin-Madison, 1101 University Avenue, Madison, WI 53706, USA

Abstract: Computations utilizing the ONIOM scheme have been used to model key features of the the Rh(chiral bisphosphine)-catalyzed enantioselective hydrogenation of prochiral enamides. Extensive computations and comparison with detailed mechanistic studies demonstrate that major features of the catalytic mechanism are reproduced by computation. These features include (1) relative stabilities of diastereomeric catalyst-enamide adducts (2) the anti "lock-and-key" motif, in which the majority of catalytic flux is carried by the less stable catalyst-enamide adduct (3) reversal of the sense of enantioselection when the enamide substrate bears a bulky, electron-donating group in the α-position. Based on the computational results, simple models demonstrate how specific steric effects of the chiral ligand environment co-mingle with orbital interactions between the catalyst and substrate to result in distinctive, and often surprising, mechanistic features of enantioselective hydrogenation reactions.

Key words: ONIOM, hydrogenation, enantioselectivity, asymmetric catalysis, DFT, reaction mechanism, chiral phosphine, ab initio, valence bond, oxidative addition, migratory insertion, reductive elimination.

1.   INTRODUCTION

As recently recognized by the Nobel Chemistry award committee, the conceptualization, development, and commercial application of enantioselective, homogeneous hydrogenation of alkenes represents a landmark achievement in modern chemistry. Further elaboration of asymmetric hydrogenation catalysts by Noyori, Burk, and others has created a robust and technologically important set of catalytic asymmetric synthetic techniques. As frequently occurs in science, these new technologies have spawned new areas of fundamental research. Soon after the development of

F. Maseras and A. Lledós (eds.), Computational Modeling of Homogeneous Catalysis, 107–135.
© 2002 Kluwer Academic Publishers. Printed in the Netherlands.*

practical asymmetric hydrogenation methods, research from the groups of Brown and Halpern painted a detailed, and surprising, picture of the overall mechanism of hydrogenation. Viewed at low resolution, this picture describes a two-state model of catalysis in which the more populated diastereomeric state corresponding to the best fit of the catalyst and substrate is relatively unreactive. As a result, all productive catalysis funnels through the small amount of catalyst in the less populated diastereomeric state. We have coined the phrase "anti-lock-and-key motif" to describe this mechanistic model. Viewed at higher resolution the empirically determined mechanism of asymmetric hydrogenation offers detailed insight into the origin of unusual pressure and temperature effects (e.g., the observation of increased enantioselectivity with increased temperature) observed for the hydrogenation of dehydroamino acid derivatives.

It is the detail of Landis and Halpern's kinetic model for enantioselective hydrogenation that affords the luxury of asking even more intimate questions about these precedent-setting catalytic transformations. What is the origin of the stereoselectivity? How does the "anti-lock-and-key" behavior arise? Why is the selectivity of the reaction so strongly dependent on the electronic properties of the olefin substituents? Because catalysis is purely a kinetic phenomenon, such questions invariably call upon discussion of transition state structures and energetics. Empirical data provide only crude boundaries for these discussions. However, the remarkable contemporaneous development of fast ab initio methods and fast computing hardware allow such questions to be addressed efficiently with computation. In this chapter, we discuss the computational approaches we have used to study the behavior of asymmetric hydrogenation catalysts and summarize the insights elucidated by these studies.

2.   HISTORICAL BACKGROUND

Wilkinson and coworkers reported the first well-characterized homogeneous transition metal catalyst for alkene hydrogenation nearly thirty five years ago [1]. Although Wilkinson's $Rh(PPh_3)_3Cl$ catalyst was not enantioselective, it opened the door for new attempts at generating asymmetric hydrogenation catalysts, a goal that was very difficult to realize with existing heterogeneous catalysts. The first successful homogeneous, asymmetric hydrogenations were performed by Büthe et al. and Knowles et al. with chiral monophosphine ligands [2, 3]. The approach of Knowles et al. combined two technologies that had just emerged: Wilkinson's hydrogenation catalysts and Mislow's synthesis of resolved, chiral phosphines. However these monophosphine ligands either gave fairly low enantiomeric excesses or required long reaction times [4]. Kagan and Dang achieved greater success with their chelating diphosphine ligand, DIOP (**1**) [5, 6]. Other effective chelating disphosphines followed, such as CHIRAPHOS [7] (**2**), BINAP [8] (**3**), and DiPAMP [9, 10] (**4**). Monsanto's L-Dopa synthesis, the first industrial application of a catalytic asymmetric hydrogenation reaction [11], employed the DiPAMP diphosphine bound to a cationic Rh center.

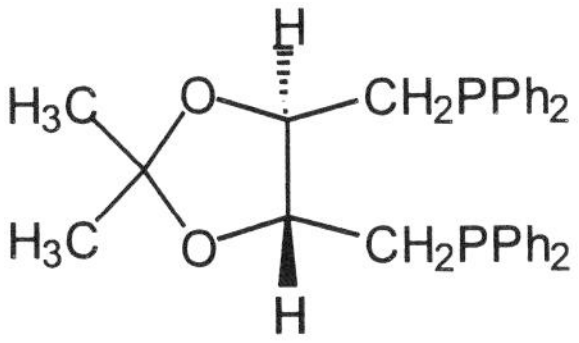

1

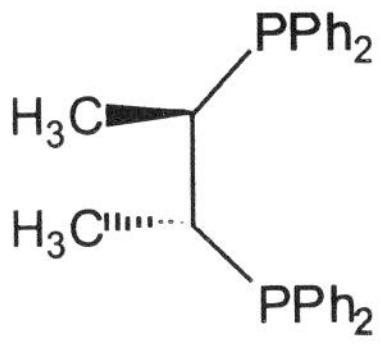

2

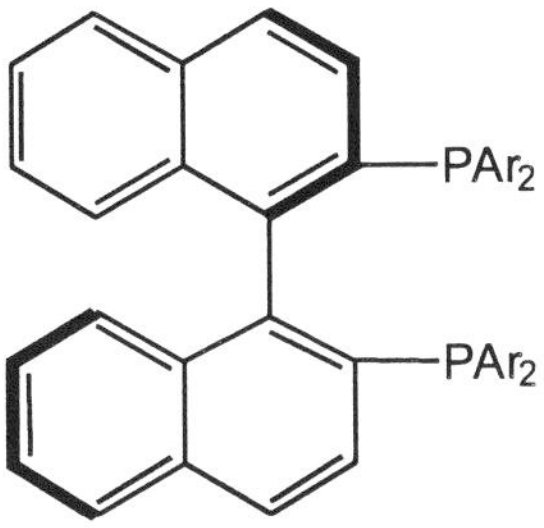

3

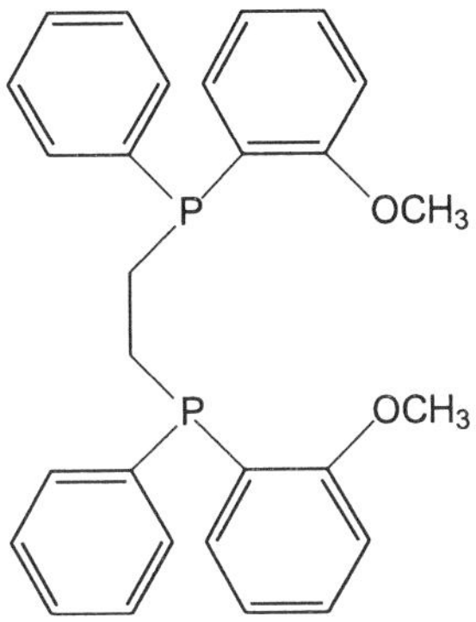

4

A key feature of the mechanism of Wilkinson's catalyst is that catalysis begins with reaction of the solvated catalyst, $RhCl(PPh_3)_2S$ (S=solvent), and H_2 to form a solvated dihydride $Rh(H)_2Cl(PPh_3)_2S$ [1]. In a subsequent step the alkene binds to the catalyst and then is transformed into product via migratory insertion and reductive elimination steps. Schrock and Osborn investigated solvated cationic complexes $[M(PR_3)_2S_2]^+$ (M=Rh, Ir and S= solvent) that are closely related to Wilkinson's catalyst. Similarly to Wilkinson's catalyst, the mechanistic sequence proposed by Schrock and Osborn features initial reaction of the catalyst with H_2 followed by reaction of the dihydride with alkene for the case of *mono*phosphine-ligated rhodium and iridium catalysts [12-17]. Such mechanisms commonly are characterized as following the "hydride pathway". In contrast, Halpern et al. [18] and Brown and Chaloner [19, 20] determined that solvated chelating-*di*phosphine catalysts do not form stable dihydrides. Instead, hydrogenation occurs by the sequence: reversible alkene binding followed by addition of dihydrogen. This sequence is commonly termed the "alkene pathway".

When the diphosphine is chiral, binding of a prochiral alkene creates diastereomeric catalyst-alkene adducts. (Diastereomers result because binding of a prochiral alkene to a metal center generates a stereogenic center at the site of unsaturation.) Through a powerful combination of ^{31}P and ^{13}C NMR methods, Brown and Chaloner first demonstrated the presence of two diastereomeric catalyst-enamide adducts with bidentate coordination of the substrate to the metal (Figure 1) [19].

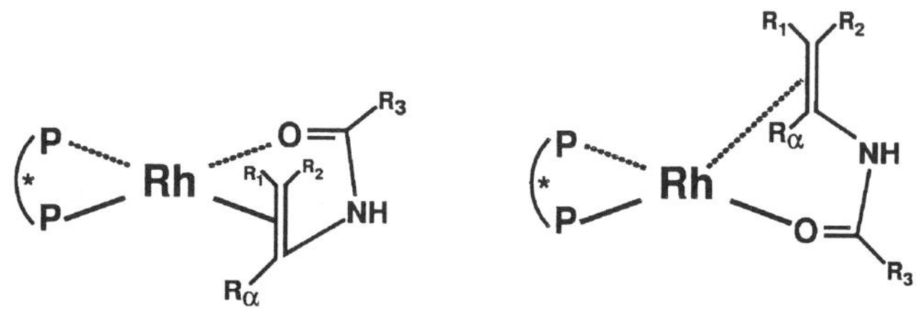

Figure 1. Diastereomeric catalyst-substrate adducts produced upon addition of olefin to solvated catalyst with chelating diphosphine ligand.

Originally, Brown and Chaloner proposed that catalytic enantioselectivity originates in the relative stabilities of the diastereomeric adducts. Superficially, such a mechanistic motif resembles the lock-and-key model, a common textbook description for the origins of selectivity in enzyme catalyzed reactions. (Here we say superficially, because proper application of the lock-and-key model must invoke a lock-and-key fit of substrate to catalyst at the selectivity-determining transition state rather than an intermediate state). However, Halpern's and Brown's groups soon thereafter showed that, in fact, the major product came from the very rapid reaction of the *less* stable diasteromer with H_2 [21, 22], a feature that we have dubbed "anti-lock-and-key" behavior. Halpern and Brown were able to identify key features of the catalytic cycle for both achiral and chiral diphosphines [18-20, 23-25]. The mechanism that arose from this work is shown in Figure 2, which is taken from the detailed kinetic study of the hydrogenation of methyl-(Z)-α-acetamidocinnamate by Rh(DiPAMP) performed by Landis and Halpern [26]. In this study the enantioselectivity of the reaction was affected by the competition between k_2 and the interconversion of the diasteromers 2^{maj} and 2^{min}. One consequence of this competition is the decrease in enantioselectivity with increasing H_2 pressure, since an enhanced rate of oxidative addition will decrease the steady-state concentration of the less stable, but more productive, "minor diastereomer".

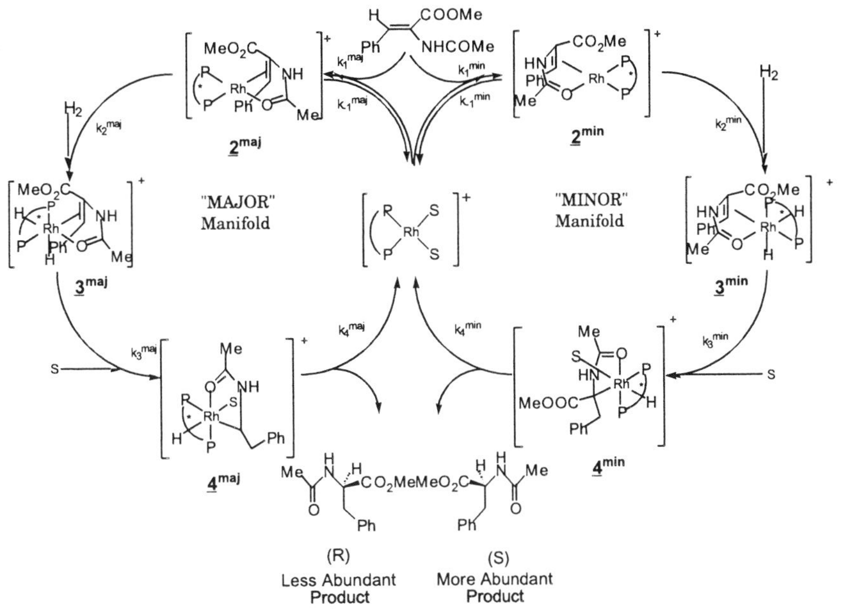

Figure 2. Mechanism of catalytic asymmetric hydrogenation. Adapted from Ref. [26].

Although chelating aryl diphosphine ligands are effective for hydrogenating dehydroamino acids with high stereoselectivity, minor changes in the composition of the substrate can drastically lower the enantiomeric excess. The best results are obtained for the following structures:

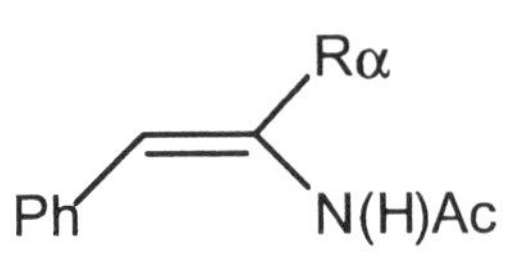

where R_α is an electron withdrawing group such as CO_2H or Ph, and the phenyl group at the β position is (Z) to the amide. Selectivities for this class of substrates routinely exceeded 95% e.e. The enantioselectivities commonly drop to below 55% for a β-disubstituted compound, and below 25% with the phenyl group in the (E) configuration [27]. Replacing the electron withdrawing R group with an electron donating group such as $-CH_3$ causes the % e.e. to drop to 51% [10].

Mark Burk and coworkers at DuPont generated a much more versatile class of chelating diphosphine ligands based on chiral phospholane ring structures (R-BPE (**5**) and R-DuPHOS (**6**), R = Me, Et, or *i*Pr) [28-30]. Enantioselectivities exceed 95% e.e. for β-disubstituted [31] and β-unsubstituted enamides [29], (Z) and (E) substrate configurations [32], and conjugated dienes (with high regioselectivity) [33], among others [34, 35].

5   **6**

The success of Burk's alkyl diphosphines spurred development of a number of other ligands with electron rich phosphines, such as Zhang's PennPHOS (**7**) [36-38], Marinetti's *i*Pr-CnrPHOS (**8**) [39], and Imamoto's BisP* ligands (**9**) [40].

7   **8**

9

In 1993 Burk, Brown, and coworkers confirmed that DuPHOS complexes exhibit the same "anti-lock-and-key" mechanistic motif as seen for aryl phosphine ligated catalysts [41]. In 1998 by Burk and coworkers reported an unexpected and interesting result [34]. With substrates having R_α = aryl, selectivity of 99% e.e. for the *S* product resulted from (*S,S*)-Me-DuPHOS-Rh hydrogenations, but the *R* product was obtained with similarly high enantioselectivity when R_α = *t*-Bu or adamantyl. In other words, the simple change of an aryl substituent to a bulky alkyl completely reverses the sense of enantioselection.

Recently, Imamoto's group has published a series of thorough mechanistic investigations of their electron-rich diphosphine catalysts [42-47]. They observe many of the same features seen in the aryl diphosphine and DuPHOS systems, such as anti-lock-and-key behavior and complete reversal of enantioselectivity with an electron donating R_α substituent. However, they also see that solvated catalyst compounds *can* react with hydrogen prior to forming the catalyst-enamide adduct complexes (Figure 1). The dihydride pathway becomes more favorable as the phosphine ligand bite angle increases: solvated BisP*-Rh undergoes reversible H_2 addition, while a catalyst containing the larger bite angle *o*-xylene derived ligand (**10**) undergoes irreversible H_2 addition [48]. These results will be addressed after we discuss the results of our computational explorations.

10

A number of attempts have been made to find an accurate empirical predictor of the sense of product chirality. The twist in the diphosphine chelating ring is one example: δ twists lead to S product while λ twists lead to R product [7, 49]. However, ligands such as DuPHOS have no backbone twist. Another predictor is the twist in the coordinated diene of the catalyst precursor: catalysts with dienes twisted counterclockwise lead to R products, with clockwise leading to S [50]. This works well for most ligands, but fails for BisP*, which exhibits a small clockwise twist but produces R product

[45]. Knowles suggested the use of quadrant diagrams (Figure 3) to predict the stability of the catalyst-enamide adducts [51]. Quadrant diagrams provide an easily processed graphic representation but can be misleading. Indeed, Knowles' original success with the quadrant model rested on two incorrect assumptions. First, application of the quadrant model led to an ordering of diastereomer stabilities that was later shown to be inverted. A second incorrect assumption, that the more stable diastereomer gave rise to most of the product, fortuitously corrected the first assumption. Thus, in this unusual case two wrongs actually do make a right! To date, most simple qualitative models do not provide a thorough understanding of the factors leading to enantioselectivity, nor an explanation for the anti-lock-and-key or enantioreversal behaviors. As stated by John Brown in a recent review:

> "The problems which remain in understanding and interpreting rhodium asymmetric hydrogenation arise from a persistent lack of information on the presumed rhodium dihydride; without which the pathway between the enamide complex and the turnover limiting TS for H_2 addition (i.e. the step in which the enantioselectivity of the reaction is set) remains opaque, and hence the overall understanding is elusive." [52]

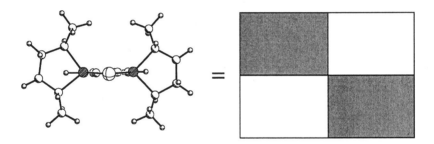

Figure 3. (R,R)-Me-DuPHOS and its quadrant representation, with shaded squares representing those areas hindered by the presence of a methyl group protruding towards the metal.

Computation allows one to circumvent nature's reluctance to offer the dihydride to direct detection. The first papers using molecular mechanics to study asymmetric hydrogenation appeared in the late 80's [53-55]. However, molecular mechanics is not the ideal technique for any reaction that involves bond-breaking or bond-forming, such as all catalytic reactions, and only a limited amount of reliable information was obtained from these early studies. An MP2/QC*IS*D(T) study of $(PH_3)_2Rh(olefin)$ structures was published in

1997 by Kless et al., but they did not compute reaction pathways [56]. It was around this time that we began our density functional theory work on asymmetric hydrogenation.

3. DENSITY FUNCTIONAL STUDIES OF RH-(PH$_3$)$_2$ MODEL SYSTEMS

As a computationally tractable model system, we chose [Rh(PH$_3$)$_2$(α-formamidoacrylonitrile)]$^+$ (**11**), shown in Figure 4. This model contains all the critical structural and electronic features of the experimental system, such as an electron withdrawing group on the α-position of the enamide. We used B3LYP for all of our calculations [57, 58], the LANL2DZ basis set ("Basis I") for geometry optimizations [59-61], and an expanded basis set from the Stuttgart group ("Basis II") for single point energies and selected reoptimizations [62, 63]. Additional details on the methods can be found in our previously published work [64].

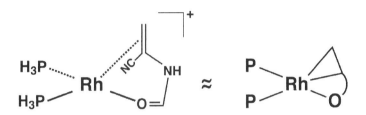

Figure 4. Model system **11** and the abbreviated representation used in this chapter.

Starting with **11**, we followed a multi-step reaction path that roughly parallels the mechanism shown in Figure 2:

a) Formation of a weakly-bound, ion-induced dipole complex (labeled **IID** in this chapter and our later papers, **SQPL** in our earlier papers),

b) Further addition of hydrogen to form a five-coordinate molecular hydrogen intermediate (**MOLH$_2$**),

c) Oxidative addition of hydrogen generating the six-coordinate Rh(III) dihydride (**DIHY**) species,

d) Migratory insertion of carbon into a Rh-H bond, forming either an α- or β-alkyl hydride (**ALHY-α/β**), depending on which carbon remained bound to rhodium,

e) Reductive elimination of C-H from the alkyl hydride to generate the alkane, still coordinated to the metal (**PROD**, equivalent to **4** in Figure 2).

The biggest divergence between our mechanism and the one in Figure 2 is the intermediacy of molecular hydrogen complexes, which are often found in other catalytic cycles [65], and had been predicted to exist by previous theoretical studies [66]. The formation of stable molecular hydrogen intermediates will have important consequences for the mechanism, as we shall demonstrate.

Since our first model system is achiral, we did not need to consider different diastereomeric manifolds. However, we did need to follow four different reaction pathways corresponding to the four possible *cis*-dihydride isomers (Figure 5). Intermediates with *trans* phosphorus orientations were not considered because the catalysts of interest have chelating diphosphine ligands.

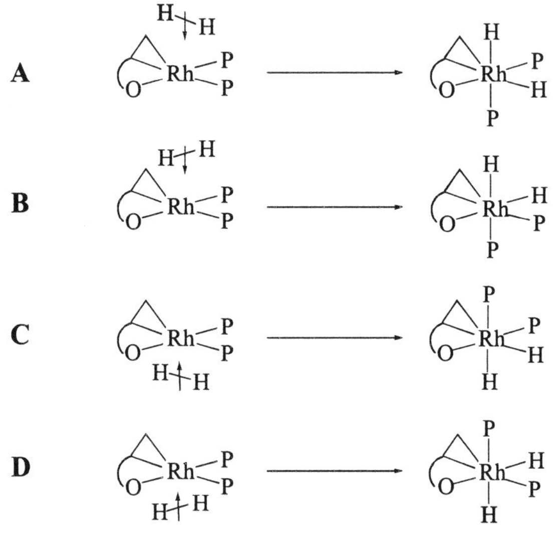

Figure 5. Isomers of dihydrides that may be formed from the addition of H_2 to **11**.

Our calculated potential energy surface is shown in Figure 6. The energies in Figure 6 are purely gas phase electronic energies; they do not take zero-point energies, solvation effects, nor the loss of H_2 entropy into account. There is little difference in the energies of the two ion-induced dipole complexes, but the four pathways diverge dramatically upon the continued approach of H_2. Pathways B and D place H_2 at the axial position of the trigonal bipyramidal intermediate, *trans* to oxygen. Hoffmann [67] and Crabtree [68] showed that in d^8 trigonal bipyramidal complexes, σ donors prefer to be axial and π acceptors prefer to be equatorial, which explains the thermodynamic instability of **MOLH$_2$-B/D** relative to the other two isomers. The large kinetic barrier for formation of **MOLH$_2$-B/D** can be rationalized by a least-nuclear-motion argument [69]: transforming the square planar structure into a trigonal bipyramid with H_2 in the axial position requires a tremendous amount of nuclear and electronic reorganization. Because of this large energetic penalty we eliminated B and D catalytic pathways on kinetic grounds, leaving **MOLH₂-A** and **MOLH₂-C** as viable intermediates.

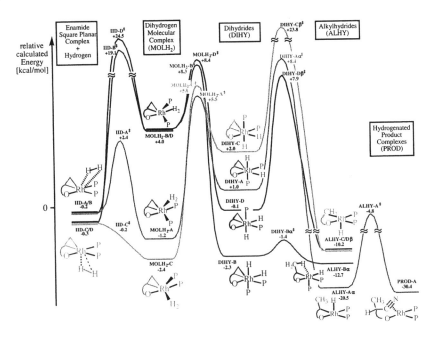

Figure 6. Potential energy surface for the reaction of **11** with H_2.

After oxidative addition to form the dihydride, however, the relative stabilities of the pathways invert. **DIHY-A** and **DIHY-C** are the thermodynamically less stable intermediates due to the *trans* influence. The *trans* influence of the double bond is greater than that of the amide oxygen, which can be seen from the two Rh-P bond lengths in **11** (Rh-P_{trans}O is ~ 0.1Å shorter than Rh-P_{trans}C). The very strong σ-donating hydride ligand prefers to be *trans* to oxygen, as it is in **DIHY-B** and **DIHY-D**. However, even though these two dihydrides are thermodynamically more stable, they are still kinetically inaccessible due to the barriers to formation of the corresponding molecular H_2 complexes. We searched for isomerization pathways between **DIHY-A** and **DIHY-B**, but found that the energetics were not competitive with the migratory insertion step. Isomerization pathways between molecular H_2 isomers were also identified, but were too high in energy as well.

There are two fundamentally different migratory insertion steps: migration of the β-carbon into a Rh-H bond to form an α-alkyl hydride (pathways A and B), and migration of the α-carbon to form a β-alkyl hydride (pathways C and D). With the presence of an electron withdrawing α-substituent, such as the nitrile in **11**, the α-alkyl hydride is more stable, resulting in a 7.1 kcal/mol lower migratory insertion barrier for pathway B than for pathway C, as per the Hammond postulate. The significant barrier along pathway A comes from a transition state structure that has the remaining hydride ligand trans to an alkyl group. Pathway C, which forms a β-alkyl hydride and contains the *trans* H-C$_{alkyl}$ structure at the transition state, exhibits a 21.8 kcal/mol barrier, by far the largest of the four. The consequence is that pathway C, which possesses a kinetically favorable dihydride, is not a viable catalytic pathway.

The pathway with the overall lowest barrier is A, which also possesses the thermodynamically most stable alkyl hydride. Because the turnover-limiting and enantiodetermining steps are known to exist prior to the alkyl hydrides, we calculated the reductive elimination only along pathway A. The relatively large barrier for reductive elimination (15.7 kcal/mol) agrees with the experimental observation that the alkyl hydrides can be isolated in catalytic hydrogenation reactions [21, 24]. The relative barrier heights along pathway A support a mechanism that involves reversible dihydride formation and turnover-limiting migratory insertion. This is contrary to the commonly accepted mechanism of catalytic asymmetric hydrogenation, which posits irreversible oxidative addition [70], but in agreement with recent experimental results that show irreversible and stereodetermining migratory insertion [45].

4.   ONIOM STUDIES OF [Rh((*R,R*)-Me-DUPHOS)(α-FORMAMIDOACRYLONITRILE)]$^+$

Addressing the origins of enantioselectivity requires a chiral model. We chose [Rh((*R,R*)-Me-DuPHOS)]$^+$ (**12**) for the model system because of its relative simplicity and high selectivity. As in the previous study, we used the simplified substrate, α-formamidoacrylonitrile (**13**). Since the double bond in **13** is prochiral, there are two different catalyst-enamide diastereomers as shown in Figure 1.

12   13

The adduct with the *re* face bound to Rhodium, leading to the *S* enantiomer, is more stable than the Pro-*R* *si*-coordinated adduct, and so would be the predominant diastereomer in solution. In our original papers on this work [71, 72], we labeled the more stable Pro-*S* diastereomer **MAJ** and the Pro-*R* diastereomer **MIN** in analogy to the nomenclature from Figure 2. For consistency with our later work, however, we use the designations **PRO-*R*** and **PRO-*S*** in this chapter.

The computational effort expended in this work is considerably greater than for the PH$_3$ model system. Not only is the ligand larger, but each diastereomer can react along the four isomeric hydrogen addition pathways seen in Figure 5, giving a total of eight multi-step reaction pathways to be followed. Performing full DFT calculations along all of these pathways was impractical at the time due to the computer resources required, so we used the ONIOM hybrid method. The ONIOM method was developed by Morokuma, Maseras, and coworkers in the mid-90's as a generalized QM/MM scheme, which allows different levels of theory to be applied to different portions of the molecule [73, 74]. Additional details on this method are available elsewhere in this volume. We used a three-layer partitioning scheme for the DuPHOS ligand, as seen in Figure 7. Our core layer contains the region of the system that actively participates in the catalysis: rhodium, a bridging bis(phospholano)ethylene ligand, the model enamide, and dihydrogen (not shown in Figure 7). We applied the same levels of theory we used for the Rh(PH$_3$)$_2$ system to this layer. The intermediate layer includes the portion of the molecule that is the most electronically relevant but outside the core: the 5-membered phospholane rings. To include these electronic effects in a time-efficient manner, we treated this level with simple Hartree-Fock theory and the LANL2MB basis set [59, 60, 75]. The remainder of the molecule (the methyl groups on the phospholane rings and the complete phenyl ring of the bridge) was assumed to have only a steric effect, and was treated with Rappé's Universal Force Field (UFF) [76].

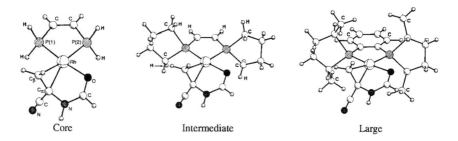

| Core | Intermediate | Large |

Figure 7. ONIOM partitioning scheme, shown here for **PRO-*R***.

All calculations were performed using Gaussian98 [77], which required making some additions and corrections to the UFF force field because Gaussian98 does not have atom types for square planar rhodium or coordinated double bonds. Details about these changes can be found in the supporting information of our original paper [72].

The free energy surface for hydrogenation along both diastereomeric manifolds is shown in Figure 8. The effect of calculating the free energy of the reaction is seen in the much larger barriers to formation of the ion-induced dipole complexes. These barriers arise from the loss of entropy upon addition of H_2 to the complex. Aside from this, however, the surfaces in Figure 8 are nearly identical to the surface obtained for our achiral system (Figure 6). Once again, there are very large barriers to hydrogen addition along pathways B and D, and pathway C experiences a large barrier for the migratory insertion step, making A the most reactive pathway along both manifolds. The relative energetics of intermediates and transition states are nearly identical to the achiral system, with one important exception. The migratory insertion barriers are noticeably smaller, particularly in relation to the oxidative addition transition states. In this more electron rich system, it appears that oxidative addition is turnover-limiting and enantiodetermining.

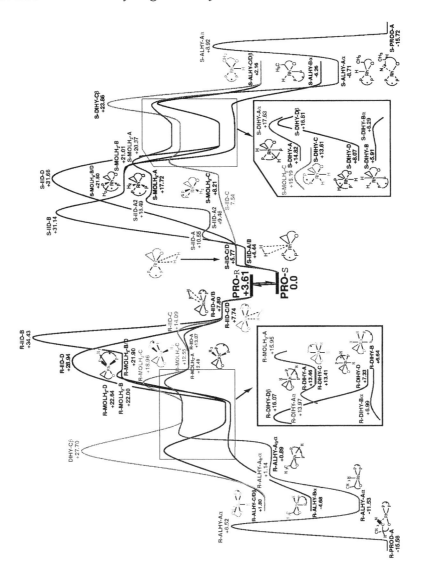

Figure 8. Free energy surface (in kcal/mol) for reaction of catalyst-enamide diasteromers with H_2

Our calculations reproduce the anti-lock-and-key behavior seen experimentally. **Pro-S** is 3.6 kcal/mol more stable than **Pro-R** but the overall barrier along pathway A on the S manifold is 20.4 kcal/mol, over 4.4 kcal/mol *greater* than the overall barrier along pathway A on the R manifold. As is observed experimentally, our calculations predict the dominant product of (R,R)-Me-DuPHOS-Rh hydrogenations to be the R enantiomer. Computed relative energies also agree reasonably with experiment. The difference in diastereomer stability of 3.6 kcal/mol corresponds to a solution concentration of **Pro-R** of less than 0.2%, which would be a low enough concentration to account for the inability of a ^{31}P NMR study of a similar DuPHOS ligated catalyst to detect any minor diastereomer [41]. The 4.4 kcal/mol difference in overall barrier heights translates to an enantiomeric excess of 99.9%. DuPHOS hydrogenations of similar substrates regularly show enantioselectivities greater than 99% e.e. [30].

The structures of the two catalyst-enamide adducts (Figure 9) do not suggest an immediate reason for their relative stabilities. In **Pro-R**, the double bond is forced downwards by one of the DuPHOS methyl groups so that the β-carbon lies in the P-Rh-P plane, while the situation is reversed with **Pro-S**, which has the α-carbon in the plane. Removing the methyl groups from DuPHOS and reoptimizing the geometries confirms there is an inherent preference for the α-carbon to be in the plane; the structure of **Pro-S** remains the same while **Pro-R** rearranges considerably so that the double bond is largely above the plane. Reasons for this behavior are discussed later in the chapter.

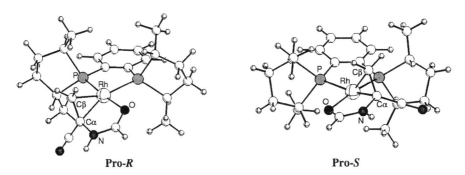

Figure 9. ONIOM optimized geometries of the two catalyst-enamide adduct diastereomers.

The difference between the structures and reactivity of the diastereomers can be visualized using Knowles' quadrant diagrams. Overlaying the structures of the two diastereomers on the quadrant diagram for DuPHOS (Figure 3) clearly reveals that displacement of the β-carbon of **Pro-R** into the plane lessens steric interactions of the β-methylene with the hindered quadrant (Figure 10).

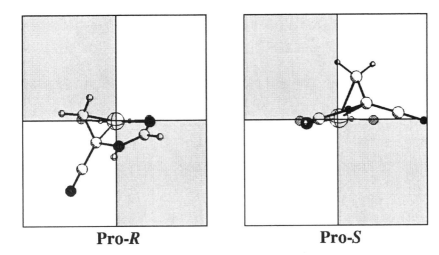

Pro-*R* **Pro-*S***

Figure 10. Orientations of substrate in catalyst-enamide diastereomers overlaid onto quadrant diagram representing DuPHOS ligand (see Figure 3). The phosphorus atoms and bridging carbons of the DuPHOS ligand are still shown, but some hydrogens have been removed for clarity.

Pathway A, corresponding to bringing hydrogen in from the top along the P-Rh-C bond axis, has already been shown to be the most reactive pathway for both diastereomers. Figure 11 shows the orientations of the enamide for both diastereomers as hydrogen adds from the top. For **Pro-*R***, very little motion has to occur in order to coordinate hydrogen, and what little motion does occur moves the double bond further into an unhindered quadrant. For **Pro-*S***, on the other hand, the enamide must distort considerably to make room for H$_2$, losing the stability of having the α-carbon in the plane, and forcing the double bond and nitrile to swing into a quadrant hindered by one of the DuPHOS methyl groups. Figure 8 confirms that the major difference between the two manifolds occurs in this step. To form **S-MOLH$_2$-A** requires 18.5 kcal/mol from **Pro-*S***, but forming **R-MOLH$_2$-A** from **Pro-*R*** only requires 9.5 kcal/mol. This more than compensates for the difference in diastereomer stability, and accounts for the enhanced reactivity of **Pro-*R***.

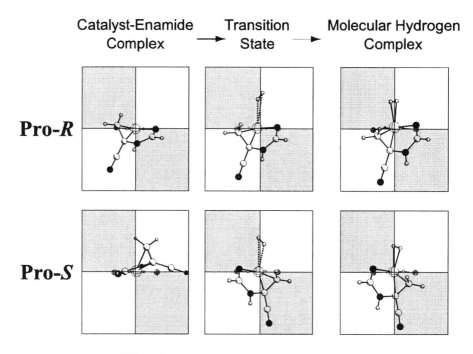

Figure 11. Addition of H_2 along pathway A of both diastereomeric manifolds.

5.   ONIOM STUDIES OF [RH((*R,R*)-ME-DUPHOS)(N-(1-*TERT*-BUTYL-VINYL)-FORMAMIDE)]⁺

Having reproduced and provided a simple rationalization for the anti-lock-and-key motif, we turned to the dramatic effect of the α-substituent on reaction enantioselectivity. Upon replacing the electron-withdrawing group on the α-carbon with an electron donating group, such as *tert*-butyl or adamantyl, the opposite enantiomer of the product is observed experimentally with 99% e.e. [34]. We replaced the nitrile group on **13** with a *tert*-Butyl group to generate N-(1-*tert*-Butyl-vinyl)-formamide (**14**) [79].

14

The computations for **14** were carried out as before. We partitioned the *t*-Bu group between two ONIOM layers, placing the full group in the Intermediate (HF) layer and representing it with a methyl in the Core (B3LYP) layer. For a few structures, constraints had to be used to ensure geometry convergence, so these energies can be viewed as upper bounds to the true ONIOM energies, but in all cases transition states had one negative vibrational frequency, and intermediates had none.

The catalyst diastereomers **Pro-'R** and **Pro-'S** have very similar structures to their nitrile counterparts, except for the double bond in **Pro-'R**, which is much less perpendicular to the P-Rh-P plane than it was in **Pro-R** (Figure 12). The relative energies of these two structures markedly differ, however. The calculated free energy difference between the **Pro-'R** and **Pro-'S** is essentially zero, and reoptimizing the geometries after removing the DuPHOS methyl groups does not result in **Pro-'R** taking the **Pro-'S** structure, as it did for the nitrile substituted enamide. Instead, the orientation of the double bond does not change.

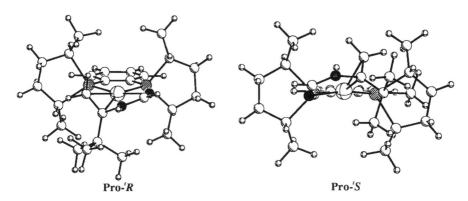

Pro-'R   Pro-'S

Figure 12. ONIOM optimized geometries of diastereomers for *tert*-Butyl substituted enamides.

The free energy surfaces for reactions of these diastereomers are shown in Figure 13. Because we knew that the barriers for addition of H_2 along pathways B and D could only increase with the bulkier system, we calculated structures along pathways A and C, only.

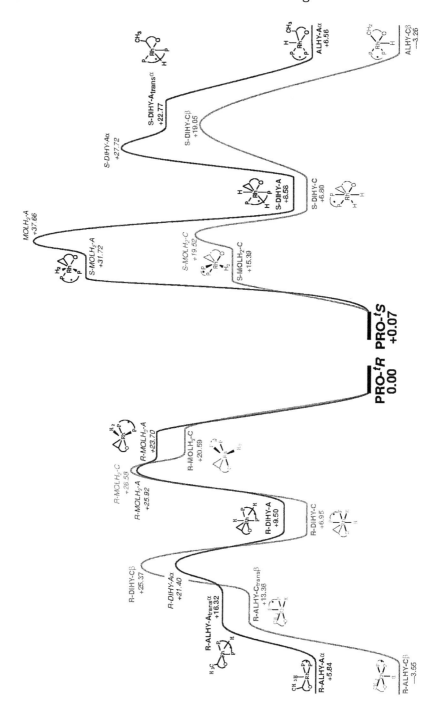

Figure 13. Free energy (kcal/mol) surface for the reaction of **Pro-'R** and **Pro-'S** with H₂. Structures with constraints on their geometry are italicized.

The pathway with the lowest overall barrier is C on the **Pro-^{t}S** manifold, meaning that for this system, we would predict the *S* enantiomer to be formed, in accord with experiment. Our calculated energy difference between the two manifolds is almost 6.5 kcal/mol, which is more than large enough to account for the observed enantiomeric excess of > 99% [78].

For the nitrile-substituted enamide (**11**), the β-alkyl hydride is considerably less stable than the α-alkyl hydride (by >13 kcal/mol), which results in large migratory insertion barriers along β-alkyl-hydride-generating pathway C. In the *t*-Bu substituted enamide, **14**, the β-alkyl hydride is more stable than the α-alkyl hydride by >9 kcal/mol, making the migratory insertion barriers for pathway A higher than for C. This dramatic shift can be attributed to the destabilizing effect that an electron donating group has on a metal-carbon bond, which strongly affects the stability of the α-alkyl hydride.

Pathway C involves addition of H₂ from the "bottom side" of the catalyst-diastereomer complex. The situation is similar to Figure 11, but inverted. Adding hydrogen from the bottom to **Pro-S** or **Pro-^{t}S** pushes the double bond further up into an unhindered quadrant, which causes little additional steric penalty. Adding hydrogen to the bottom of **Pro-R** or **Pro-^{t}R** pushes the double bond into a hindered quadrant, with a consequently higher barrier. The situation is summarized in Figure 14 for substrate **14**, but the conclusions restricting double bond movement are valid for any substituent.

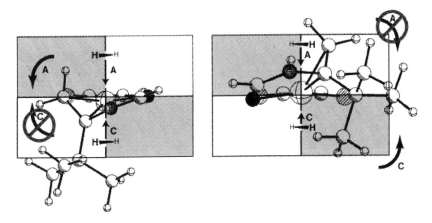

Figure 14. Addition of dihydrogen to both diastereomers, showing the direction of motion of the double bond upon addition via each pathway.

We can summarize our conclusions as follows:
- The electronic characteristics of the carbon substituents on the enamide determine the preference for the alkyl hydride:

    Electron-donating → β-alkyl hydride

Electron-withdrawing → α-alkyl hydride
- The direction of hydrogen addition determines which alkyl hydride will be formed:

From top (Pathways A/B) → α-alkyl hydride
From bottom (Pathways C/D) → β-alkyl hydride

- Hydrogen approach can only occur when the double bond is being pushed into an unhindered quadrant:

From top (Pathways A/B) → Pro-*R* diastereomer
From bottom (Pathways C/D) → Pro-*S* diastereomer

So for an electron withdrawing substituent, an α-alkyl hydride is favored, which can only form when hydrogen adds from the top, and addition from the top occurs more readily for the Pro-*R* diastereomer, producing *R* product. For an electron-donating substituent, the argument is reversed, leading to *S* product.

6.   EFFECT OF THE α-SUBSTITUENT ON CATALYST-ENAMIDE ADDUCT STABILITIES

Although understanding the effect of the α-substituent on relative alkyl hydride stabilities is straightforward, understanding the effect on the catalyst-enamide diastereomers is not. To clarify this matter we performed a series of calculations on a variety of substituted enamides [78]. To eliminate the effect of the amide oxygen, we examined frontier orbitals of the $[Rh(PH_3)_2(formamide)]^+$ fragment (**15**) and those of model enamides (**16**).

15   **16-R$_\alpha$**

The bonding between these two fragments can be understood using the Dewar-Chatt-Duncanson model of donation and backdonation [79, 80]. The frontier orbitals responsible for these interactions between **15** and **16-R$_\alpha$** are drawn to scale in Figure 15.

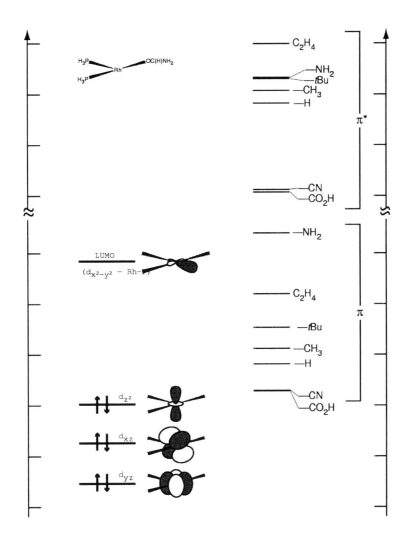

Figure 15. Canonical molecular orbital energy levels for [Rh(PH$_3$)$_2$(formamide)]$^+$, showing the filled and unfilled orbitals with the proper symmetry to interact with C–C π and π· orbitals, seen on the right. Scale markings are in eV. All calculations were done at the B3LYP/LANL2DZ level.

In accord with the Dewar-Chatt-Duncanson model, we find that the dominant interaction is donation from the C C π bond into the rhodium LUMO. This interaction is enhanced when the double bond lies in the rhodium-diphosphine plane and with electron donating substituents which raise the energy of π to more closely match the LUMO Charge Decomposition Analysis (CDA) [81] shows that the amount of donation is maximized with **R–*t*Bu** and **R–CH₃**, but the reduction in donation ability is fairly minor as one moves to electron withdrawing substituents. The change in the amount of backdonation, however, can be extremely dramatic. Backdonation is enhanced by electron withdrawing substituents which polarize and lower the energy of $\pi^\cdot$ and also by the ability of the $\pi^\cdot$ orbital to geometrically overlap with the highest energy metal lone pair, d_{z^2}. Only C=C double bonds whose centroid is displaced above the P-Rh-P plane effect net overlap of $\pi^\cdot$ with d_{z^2} (Figure 16). Maximum overlap results when the polarization of the $\pi^\cdot$ orbital matches the spatial distribution of the d_{z^2}; as shown in Figure 16 this occurs when the β-carbon is pseudo-axial. The major diastereomer **Pro-*S*** adopts this conformation. If an electron-withdrawing group is present, but the β-carbon of the double bond is sterically prevented from rising above the plane to accept electrons from d_{z^2}, a less stabilizing interaction results. Such is the case for the minor diastereomer **Pro-*R***. For an electron donating substituent, as in the case of *t*-Bu, the potential backbonding interaction is much smaller, decreasing the preference for the β-carbon to be above the plane. These arguments agree with those made by Hoffmann and coworkers in their studies on the bonding of substituted ethylenes to metals [82, 83].

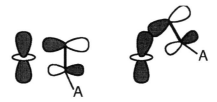

Figure 16. Interaction between the C=C LUMO and d_{z^2} with an electron acceptor (A) on C_α. The relative sizes of the lobes on $\pi^\cdot$ indicate the polarization due to A

The Valence Bond perspective on the substrate-catalyst interaction emphasizes the resonance between two 16 electron configurations: one corresponding to a simple Lewis donor-acceptor interaction with the alkene π electrons acting as donor and the metal as an acceptor and one corresponding to a metallocyclopropane configuration (Figure 17). As we previously have shown, the idealized geometry of the 16 electron metallocyclopropane configuration is a monovacant octahedron (or square pyramid with 90° bond angle between axial and basal bonds) [84]. Therefore, mixing of metallocyclopropane character with the square planar donor-acceptor will tend to move the midpoint of the alkene out of the P-Rh-P plane. Whereas the axial Rh-C bond of the metallocyclopropane conforms to a standard two center-two electron bond, the bonds of the basal plane are dominated by three center-four electron interactions among *trans* ligand pairs. Such interactions create substantial charge accumulation on the basal ligands. When the alkene has an electron withdrawing substituent, maximum stabilization occurs with that C placed close to the basal plane. For electron donating substituents, the situation is reversed.

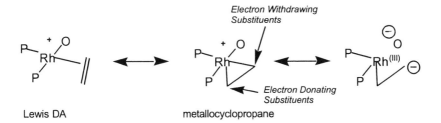

Figure 17. Alternative valence bond pictures of the catalyst-enamide adduct. Only one of the four three center-four electron resonance structures is shown (far right). Electron withdrawing substituents prefer the basal position of the square pyramidal metallocyclopropane structure, while electron donating substituents prefer the axial position.

7.   RECENT EXPERIMENTAL RESULTS

The discovery by the Imamoto and Brown groups that hydride pathways may be feasible for catalytic asymmetric hydrogenation may limit the generality of our conclusions [42-47, 85]. The hydrogen addition portion of our catalytic cycle is, of course, not applicable to catalysts where the enamide adds to a dihydride solvate. Since our argument about the stereoselectivity of Rh-DuPHOS catalysts is based on the hydrogen addition, our model does not explain the high enantioselectivities observed by BisP* or PHANEPHOS ligands, *if* these catalysts do operate primarily via a hydride pathway. At this point, the experimental data show that the hydride pathway is possible for the BisP* and PHANEPHOS ligands, but do not require that it carries a significant amount of the catalytic flux.

It is interesting to apply our computed free energy surface to a reaction occurring via the hydride pathway. To do this we simply assume that binding of the substrate to BisP*-dihydride-disolvate is equivalent to dropping onto the **DIHY** portion of the computed surface. By experiment, an equilibrium distribution of BisP*-substrate-dihydride isomers (analogues of the **DIHY** complexes) is obtained upon reaction of the substrate with the solvate dihydrides [45], meaning that both the thermodynamic stabilities and the reactivities of the **DIHY** isomers contribute to the observed enantioselectivity. According to the computed free energy surface the lowest energy and fastest reacting dihydrides are the **R-** and **S-DIHY-B** isomers.

We calculate only a 0.7 kcal/mol difference between **R-DIHY-B** and **S-DIHY-B**, with nearly identical migratory insertion barriers, so we would predict only modest enantioselectivity if a DuPHOS-ligated catalyst reacted along the hydride route. However, we are not aware of any evidence that a solvated Rh-DuPHOS catalyst reacts with hydrogen to form dihydrides.

Aside from the question of selectivity along the hydride pathway, many of the new experimental results are in accord with reported results for the (*R,R*)-Me-DuPHOS-Rh catalytst and preliminary computations with (*S,S*)-BisP*-Rh. The steric profile of (*S,S*)-BisP*-Rh is the same as (*R,R*)-Me-DuPHOS, with the quadrant diagram shown in Figure 3. Preliminary computations reveal that when the olefin route is followed for BisP*-Rh and the substrate has an electron withdrawing group in the alpha position, the more stable catalyst-enamide diastereomer has the same relative orientation on the quadrant diagram as **Pro-S**. But as in DuPHOS-ligated catalysts, the **Pro-R** diastereomer is more reactive. When the electron withdrawing group is replaced with *tert*-Butyl, preliminary computations demonstrate little energetic differentiation between catalyst-enamide diastereomers, but the **Pro-S** diastereomer is substantially more reactive and the *S* product is produced. More importantly, β-alkyl hydrides are observed with the *t*-Bu substituted enamide, while α-alkyl hydrides are formed with the electron-withdrawing substituted enamides, regardless of whether the reaction proceeded via the olefin or hydride route [44]. Furthermore, the experimentally observed structures of several low-energy intermediates are similar to our calculated structures. The most stable experimental dihydride has the structure of the most stable dihydride (**DIHY-B**) obtained by computation. Migratory insertion from **DIHY-B** is computed to initially form a structure identical to our **ALHY-B** which is thermodynamically poised to isomerize to **ALHY-A** with a small barrier. Experimentally, the observed alkyl hydride has the coordination geometry of **ALHY-A** [44].

8.   CONCLUSIONS

We see the development of an interesting pas de deux involving experiment and computation in catalysis. With the support provided by computations, experiments can jump higher than ever before. As a result, interpretations of challenging experimental data, such as Brown and Bargon's identification of "agostic di-hydrides" (which we interpret as alkyl hydrides with a strong agostic interaction) become tractable. However, computations can provide more than just anonymous support for experiment: in some instances computations lead to genuinely new insights. For example, only computations have enabled robust models for understanding the "anti lock-and-key" motif and the origin of enantioreversal upon changing enamide α-substituents. In our opinions, these computational models carry the same attributes: simplicity, beauty, and logic, as the best models derived from experiment. Like any good research, success creates new questions, thus placing further pressure on both experiment and computations. For example, testing the validity of some of our computational conclusions motivates us to perform further experimental work with the DuPHOS ligand. Similarly, the unexpected and fascinating dihydride chemistry observed with the BisP* and PHANEPHOS ligands calls for further, and more sophisticated, computation.

REFERENCES

1   Osborn, J. A.; Jardine, F. H.; Young, J. F.; Wilkinson, G. *J. Chem. Soc. A* **1966**, 1711-1730.
2   Horner, L.; Siegel, H.; Büthe, H. *Angew. Chem. Int. Ed. Engl.* **1968**, *7*, 942.
3   Knowles, W. S.; Sabacky, M. J. *Chem. Commun.* **1968**, 1445-1446.
4   Knowles, W. S.; Sabacky, M. J.; Vineyard, B. D. *J. Chem. Soc. Chem. Commun.* **1972**, 10-11.
5   Dang, T. P.; Kagan, H. B. *J. Am. Chem. Soc.* **1972**, *94*, 6429-6433.
6   Kagan, H. B.; Dang, T.-P. *J. Am. Chem. Soc.* **1972**, *94*, 6429-6433.
7   Fryzuk, M. D.; Bosnich, B. *J. Am. Chem. Soc.* **1977**, *99*, 6262-6267.
8   Miyashita, A.; Yasuda, A.; Takaya, H.; Toriumi, K.; Ito, T.; Souchi, T.; Noyori, R. *J. Am. Chem. Soc.* **1980**, *102*, 7932-7934.
9   Knowles, W. S.; Sabacky, M. J.; Vineyard, B. D.; Weinkauff, D. J. *J. Am. Chem. Soc.* **1975**, *97*, 2567-2568.
10   Vineyard, B. D.; Knowles, W. S.; Sabacky, M. J.; Bachman, G. L.; Weinkauff, D. J. *J. Am. Chem. Soc.* **1977**, *99*, 5946-5952.
11   Knowles, W. S. *J. Chem. Ed.* **1986**, *63*, 222-225.
12   Shapley, J. R.; Schrock, R. R.; Osborn, J. A. *J. Am. Chem. Soc.* **1969**, *91*, 2816-2817.
13   Schrock, R. R.; Osborn, J. A. *Inorg. Chem.* **1970**, *9*, 2339-2343.
14   Schrock, R. R.; Osborn, J. A. *J. Am. Chem. Soc.* **1971**, *93*, 2397-2402.
15   Schrock, R. R.; Osborn, J. A. *J. Am. Chem. Soc.* **1976**, *98*, 2134-2143.
16   Schrock, R. R.; Osborn, J. A. *J. Am. Chem. Soc.* **1976**, *98*, 2143-2147.
17   Schrock, R. R.; Osborn, J. A. *J. Am. Chem. Soc.* **1976**, *98*, 4450-4455.
18   Halpern, J.; Riley, D. P.; Chan, A. S. C.; Pluth, J. J. *J. Am. Chem. Soc.* **1977**, *99*, 8055-8057.
19   Brown, J. M.; Chaloner, P. A. *Tet. Lett.* **1978**, *21*, 1877-1880.
20   Brown, J. M.; Chaloner, P. A. *J.C.S. Chem. Commun.* **1978**, 321-322.
21   Brown, J. M.; Chaloner, P. A. *J. Chem. Soc. Chem. Commun.* **1980**, 344-346.
22   Chan, A. S. C.; Pluth, J. J.; Halpern, J. *J. Am. Chem. Soc.* **1980**, *102*, 5952-5954.
23   Chan, A. S. C.; Pluth, J. J.; Halpern, J. *Inorg. Chim. Acta.* **1979**, *37*, L477.
24   Chan, A. S. C.; Halpern, J. *J. Am. Chem. Soc.* **1980**, *102*, 838-840.
25   Brown, J. M.; Parker, D. *Organometallics* **1982**, *1*, 950-956.
26   Landis, C.; Halpern, J. *J. Am. Chem. Soc.* **1987**, *109*, 1746-1754.

27 Koenig, K. E. In *Asymmetric Synthesis*; Morrison, J. D., Ed.; Academic: Orlando, 1985; Vol. 5, pp 71-101.

28 Burk, M. J.; Feaster, J. E.; Harlow, R. L. *Organometallics* **1990**, *9*, 2653-2655.

29 Burk, M. J. *J. Am. Chem. Soc.* **1991**, *113*, 8518-8519.

30 Burk, M. J.; Feaster, J. E.; Nugent, W. A.; Harlow, R. L. *J. Am. Chem. Soc.* **1993**, *115*, 10125-10138.

31 Burk, M. J.; Gross, M. F.; Martinez, J. P. *J. Am. Chem. Soc.* **1995**, *117*, 9375-9376.

32 Burk, M. J.; Wang, Y. M.; Lee, J. R. *J. Am. Chem. Soc.* **1996**, *118*, 5142-5143.

33 Burk, M. J.; Allen, J. G.; Kiesman, W. F. *J. Am. Chem. Soc.* **1998**, *120*, 657-663.

34 Burk, M. J.; Casy, G.; Johnson, N. B. *J. Org. Chem.* **1998**, *63*, 6084-6085.

35 Burk, M. J. *Acc. Chem. Res.* **2000**, *33*, 363-372.

36 Jiang, Q.; Jiang, Y.; Xiao, D.; Cao, P.; Zhang, X. *Angew. Chem. Int. Ed.* **1998**, *37*, 1100-1103.

37 Jiang, Q.; Xiao, D.; Cao, P.; Zhang, X. *Angew. Chem. Int. Ed.* **1999**, *38*, 516-518.

38 Zhang, Z.; Zhu, G.; Jiang, Q.; Xiao, D.; Zhang, X. *J. Org. Chem.* **1999**, *64*, 1774-1775.

39 Marinetti, A.; Jus, S.; Genêt, J.-P. *Tet. Lett.* **1999**, *40*, 8365-8368.

40 Imamoto, T.; Watanabe, J.; Wada, Y.; Masuda, H.; Yamada, H.; Tsuruta, H.; Matsukawa, S.; Yamaguchi, K. *J. Am. Chem. Soc.* **1998**, *120*, 1635-1636.

41 Armstrong, S. K.; Brown, J. M.; Burk, M. J. *Tet. Lett.* **1993**, *34*, 879-882.

42 Gridnev, I. D.; Higashi, N.; Asakura, K.; Imamoto, T. *J. Am. Chem. Soc.* **2000**, *122*, 7183-7194.

43 Yasutake, M.; Gridnev, I. D.; Higashi, N.; Imamoto, T. *Organic Letters* **2001**, *3*, 1701-1704.

44 Gridnev, I. D.; Yasutake, M.; Higashi, N.; Imamoto, T. *J. Am. Chem. Soc.* **2001**, *123*, 5268-5276.

45 Gridnev, I. D.; Yamanoi, Y.; Higashi, N.; Tsuruta, H.; Yasutake, M.; Imamoto, T. *Adv. Synth. Catal.* **2001**, *343*, 118-136.

46 Gridnev, I. D.; Higashi, N.; Imamoto, T. *J. Am. Chem. Soc.* **2001**, *123*, 4631-4632.

47 Gridnev, I. D.; Higashi, N.; Imamoto, T. *J. Am. Chem. Soc.* **2000**, *122*, 10486-10487.

48 Gridnev, I. D.; Higashi, N.; Imamoto, T. *Organometallics* **2001**, *20*, 4542-4553.

49 Samuel, O.; Couffignal, R.; Lauer, M.; Zhang, S. Y.; Kagan, H. B. *Nouv. J. Chim.* **1981**, *5*, 15.

50 Kyba, E. P.; Davis, R. E.; Juri, P. N.; Shirley, K. R. *Inorg. Chem.* **1981**, *20*, 3616-3623.

51 Knowles, W. S. *Acc. Chem. Res.* **1983**, *16*, 106-112.

52 Brown, J. M. In *Comprehensive Asymmetric Catalysis*; Jacobsen, E. N., Pfaltz, A., Yamamoto, H., Eds.; Springer: Berlin, pp 121-182.

53 Brown, J. M.; Evans, P. L. *Tetrahedron* **1988**, *44*, 4905-4916.

54 Bogdan, P. L.; Irwin, J. J.; Bosnich, B. *Organometallics* **1989**, *8*, 1450-1453.

55 Giovanetti, J. S.; Kelly, C. M.; Landis, C. R. *J. Am. Chem. Soc.* **1993**, *115*, 4040-4057.

56 Kless, A.; Börner, A.; Heller, D.; Selke, R. *Organometallics* **1997**, *16*, 2096-2100.

57 Becke, A. D. *Phys. Rev. A* **1988**, *38*, 3098-3100.

58 Lee, C.; Yang, W.; Parr, R. G. *Phys. Rev. B* **1988**, *37*, 785-789.

59 Hay, P. J.; Wadt, W. R. *J. Chem. Phys.* **1985**, *82*, 285-298.

60 Hay, P. J.; Wadt, W. R. *J. Chem. Phys.* **1985**, *82*, 299-310.

61 Dunning, T. H.; Hay, P. J. In *Modern Theoretical Chemistry*; Schaefer, H. F., Ed.; Plenum: New York, 1976; Vol. 3, p 1.

62 Andrae, D.; Haeussermann, U.; Dolg, M.; Stoll, H.; Preuss, H. *Theor. Chim. Acta* **1990**, *77*, 123-141.

63 Bergner, A.; Dolg, M.; Kuenchle, W.; Stoll, H.; Preuss, H. *Mol. Phys.* **1993**, *80*, 1431-1441.

64 Landis, C. R.; Hilfenhaus, P.; Feldgus, S. *J. Am. Chem. Soc.* **1999**, *121*, 8741-8754.

65 Esteruelas, M. A.; Oro, L. A. *Chem. Rev.* **1998**, *98*, 577-588.

66 Daniel, D.; Koga, N.; Han, J.; Fu, X. Y.; Morokuma, K. *J. Am. Chem. Soc.* **1988**, *110*, 3773.

67 Rossi, A. R.; Hoffmann, R. *Inorg. Chem.* **1975**, *14*, 365-374.

68 Burk, M. J.; McGrath, M. P.; Wheeler, R.; Crabtree, R. H. *J. Am. Chem. Soc.* **1988**, *110*, 5034-5039.

69 Lowry, T. H.; Richardson, K. S. *Mechanism and Theory in Organic Chemistry*; 2nd ed.; Harper & Row: New York, 1981.

70 Halpern, J. In *Asymmetric Synthesis*; Morrison, J. D., Ed.; Academic: Orlando, 1985; Vol. 5, pp 41-69.

71 Landis, C. R.; Feldgus, S. *Angew. Chem. Int. Ed. Engl.* **2000**, *39*, 2863-2866.

72 Feldgus, S.; Landis, C. R. *J. Am. Chem. Soc.* **2000**, *122*, 12714-12727.

73 Maseras, F.; Morokuma, K. *J. Comp. Chem.* **1995**, *16*, 1170-1179.

74 Svensson, M.; Humbel, S.; Froese, R. D. J.; Matsubara, T.; Sieber, S.; Morokuma, K. *J. Phys. Chem.* **1996**, *100*, 19357-19363.

75 Hehre, W. J.; Stewart, R. F.; Pople, J. A. *J. Chem. Phys.* **1969**, *51*, 2657-2664.

76 Rappé, A. K.; Casewit, C. J.; Colwell, K. S.; Goddard, W. A. I.; Skiff, W. M. *J. Am. Chem. Soc.* **1992**, *114*, 10024-10035.

77 Frisch, M. J. et al. *Gaussian 98*, Gaussian, Inc.: Pittsburgh PA, USA, **1998.**

78 Feldgus, S.; Landis, C. R. *Organometallics* **2001**, *20*, 2374-2386.

79 Dewar, J. M. S. *Bull. Soc. Chim. Fr.* **1951**, *18*, C71-C79.

80 Chatt, J.; Duncanson, L. A. *J. Chem. Soc.* **1953**, 2939-2947.

81 Dapprich, S.; Frenking, G. *J. Phys. Chem.* **1995**, *99*, 9352-9362.

82 Albright, T. A.; Hoffmann, R.; Thibeault, J. C.; Thorn, D. L. *J. Am. Chem. Soc.* **1979**, *101*, 3801-3812.

83 Eisenstein, O.; Hoffmann, R. *J. Am. Chem. Soc.* **1981**, *103*, 4308-4320.

84 Firman, T. K.; Landis, C. R. *J. Am. Chem. Soc.* **1998**, *120*, 12650-12656.

85 Heinrich, H.; Giernoth, R.; Bargon, J.; Brown, J. M. *Chem. Commun.* **2001**, 1296-1297.

Chapter 6

Isomerization of Double and Triple C-C Bonds at a Metal Center

Eric Clot* and Odile Eisenstein*

LSDSMS (UMR 5636), Case courrier 14, Université Montpellier II, Place Eugène Bataillon, 34095 Montpellier Cedex 5, France

Abstract: The computational studies of the isomerization of alkyne into vinylidene and alkene into carbene in presence of a transition metal fragment are described. Three different routes have now been proposed in the literature.

Key words: DFT, isomerization, alkene, alkyne, carbene, vinylidene.

1. INTRODUCTION

Homogeneous catalysis with organometallic complexes is an efficient and selective way to functionalize organic molecules [1]. The properties of an organometallic complex reflect the properties of its organic and organometallic parts tuned by their mutual interaction. From an organic perspective, the metallic part can efficiently stabilize reactive organic species, difficult or impossible to isolate and characterize in the free state. This, in turn, changes the properties of the organic ligand, which can enter new reactions not feasible for the isolated species.

Of particular interest are transition metal complexes containing reactive M=C or C=C double bonds such as carbene (**1**) or vinylidene (**2**) complexes.

$$L_nM=C\underset{R^3}{\overset{CHR^1R^2}{<}} \qquad L_nM=C=C\underset{H}{\overset{R^1}{<}}$$

$$\mathbf{1} \qquad\qquad \mathbf{2}$$

The corresponding organic fragments are tautomers (equations 1 and 2) of stable molecules and do not exist with significant lifetime in the free state.

F. Maseras and A. Lledós (eds.), Computational Modeling of Homogeneous Catalysis, 137–160.

Although the transformation of a primary alkyne into a vinylidene complex, **2**, in presence of a number of transition metal systems is well reported [2, 3], only rare examples are known for the transformation of an alkene into a carbene complex [4, 5]. Given the increased role played by vinylidene and carbene complexes as key partners in metathesis reactions and related catalytic processes [6, 7], opening up new efficient and easy synthetic routes to such complexes is an important challenge.

$$\underset{R^1}{\overset{H}{\diagdown}}C=C\underset{R^3}{\overset{R^2}{\diagup}} \rightleftharpoons \quad \vert\,C\underset{R^1}{\overset{CHR^2R^3}{\diagup}} \tag{1}$$

$$H-C\equiv C-R^1 \rightleftharpoons \quad \vert\,C=C\underset{R^1}{\overset{H}{\diagup}} \tag{2}$$

These isomerization reactions are of great interest to theoreticians because the role of many factors (metal, substituents on the organic fragment, ancillary ligands) on the outcome of the reaction can be studied through computations. The purpose of this chapter is to describe the theoretical studies carried out on the isomerization of alkyne to vinylidene and alkene to carbene in the presence of transition metal fragments.

The essential features of the isomerization reactions of alkyne into vinylidene and alkene into carbene in their free state are first presented. The bonding properties of these systems are then briefly discussed. This illustrates how the transition metal fragment influences the energy of the isomerization reaction. The key role of substituents on the organic fragment and ancillary ligands on the metal become thus apparent. We then discuss the possible reactions pathways for isomerization in the case of alkyne. The key role played by a metal hydride fragment to give access to a multistep reaction with especially low activation barriers is then presented for the alkyne and the alkene.

2.   ISOLATED LIGANDS

2.1   Thermodynamics of the isomerization reaction

The thermodynamic outcome of a transformation A to B can be rationalized through an analysis of the bonding within each system. The transformation of a primary alkyne into vinylidene corresponds mostly to the loss of one of the π components of a C≡C triple bond since the C(sp)-H and

$C(sp^2)$-H energies are close. Likewise, the transformation of an alkene to a substituted carbene corresponds mostly to the loss of the π component of a C=C double bond. The loss of these bonds is clearly costly in energy but a key point to the present story is that the energy cost is different for the two systems. Loss of one of the π bonds of an alkyne corresponds to 40-50 kcal.mol^{-1}, whereas the loss of the π bond in an alkene amounts to 70-80 kcal.mol^{-1}. Therefore the isomerization of primary alkyne to vinylidene and of alkene to substituted carbene are both endothermic with the latter having the larger endothermicity.

Acetylene and ethylene are two convenient models for quantitative studies. A large number of studies have been carried out on these systems but the purpose of this section is not to give a comprehensive discussion of these transformations. For these reasons, the numerical values given here are the one used later for the organometallic systems. B3PW91/6-31(d,p) calculations [8] give values in accord with this simple description (equations 3 and 4)

$$HCCH \rightarrow C=CH_2 \qquad \Delta E = +43.2 \text{ kcal.mol}^{-1} \qquad (3)$$

$$H_2C=CH_2 \rightarrow CH(CH_3) \quad \Delta E = +79.5 \text{ kcal.mol}^{-1} \qquad (4)$$

Substituent G has no drastic influence on the energy associated with equation 3 since G is remote from the terminal end of vinylidene. This is not the case for equation 4 where a substituent G can be on the carbene carbon (equation 5).

$$HGC=CH_2 \rightarrow CG(CH_3) \quad \Delta E \qquad (5)$$

It is well known that π electron donor groups such as OR or NR$_2$ give electron to the low lying empty p orbital of the carbene (see below).

In contrast, there is little to no electron transfer of a π electron donor substituent to the π system of an alkene. Therefore a π electron donor substituent stabilizes more the carbene side of equation 5 and diminishes the endothermicity of the reaction. Calculations confirm the key role of G substituents of increased π donating ability to diminish the ΔE of equation 5 (G = F +54.2 kcal.mol^{-1}, G = OMe +41.6 kcal.mol^{-1}) [8].

Thus in the absence of any coordination to a transition metal fragment, the energy difference between the two tautomers ranges from 40 to 80 kcal.mol^{-1} depending on the nature of the π systems. The isomerization of these unsaturated species could become thermodynamically accessible if the transition metal fragment interacts more strongly with the products than with the reactants. A transition metal fragment capable of isomerizing alkyne into a vinylidene is also a good candidate for isomerizing an alkene substituted by a π electron donor group G into a stabilized carbene group, since it involves the same range of reaction energies. In contrast, isomerization of alkene deprived of a π electron donor group is energetically more difficult.

2.2   Bonding capabilities of the organic ligands

The frontier orbitals of the interacting species are responsible for the bond between the organic and inorganic fragments. The d^6 ML$_5$ fragment is conveniently used as representative of the bonding capabilities of any metallic fragment with an empty coordination site (low lying LUMO of cylindrical symmetry) and back bonding capability from a high lying doubly occupied d orbital. The alkyne acts as an electron donor towards the metal through π_1 and as an electron acceptor through π_1^* (the orthogonal π_2 and π_2^* orbitals are of lesser importance when the alkyne is acting as a two-electron donor). The vinylidene is an electron donor from its HOMO which is along the C-C direction and an electron acceptor through its empty p orbital (Figure 1) [9].

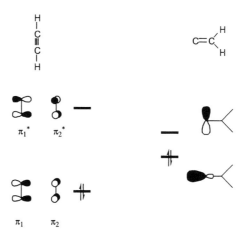

Figure 1. Comparison of the frontier orbitals for acetylene and vinylidene

The orthogonal π and π^* orbitals of vinylidene can be neglected. Vinylidene has thus a HOMO which is higher and a LUMO which is lower than that of alkyne and thus closer to the frontier orbitals of the metal fragment. As also schematically shown on Figure 1 the overlap between the frontier orbitals is clearly larger for vinylidene than alkyne. As a result, the metallic fragment makes a stronger bond to the vinylidene than to the alkyne. A similar analysis can be done for the alkene/carbene isomeric forms since the frontiers orbital of the organic species are closely related. The carbene makes stronger bond to the metal than the olefin. While numerical studies are necessary to assign the magnitude of the interactions, it is clear, from this qualitative analysis, that any metal fragment with an empty coordination site and back bonding ability stabilizes more the right side than the left sides of equation 3 and 4 resulting necessarily in a decrease of the endothermicity for the isomerization.

3.   ALKYNE TO VINYLIDENE: MECHANISTIC CONSIDERATION

Several mechanisms have been currently proposed for isomerizing primary alkyne to vinylidene in presence of transition metal fragments [2, 3].

3.1 $L_nM(HCCR)$ to $L_nM(CCHR)$

The reaction consists formally of a 1,2 hydrogen shift. Ab initio calculations have been carried out for free HCCH. The transition state resembles the vinylidene and lies 45 kcal.mol^{-1} above HCCH. A transition metal fragment could favor this path by stabilizing the vinylidene species and all structures relatively close to this structure on the potential energy surface. Alternatively, the transition metal fragment can give entry to a multistep reaction pathway which is no more a 1,2 hydrogen shift. Two paths have been considered.

a) approach of the metal to C_α while the hydrogen atom concomitantly moves towards C_β (Figure 2, pathway a). This path was first studied with Extended Hückel theory (EHT) [10]. Further studies by more quantitative methods have agreed with the basic aspects found by the EHT study [11-13].

b) oxidative addition of the 1-alkyne to the metal centre to give a hydrido-alkynyl complex which then isomerizes by a 1,3 hydrogen shift from the metal to C_β (Figure 2, pathway b). This path was studied with ab-initio methods [14, 17].

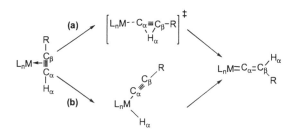

Figure 2. Proposed mechanisms for the alkyne to vinylidene isomerization coordinated to a transition metal fragment

3.1.1   1,2 intraligand H shift

Orbital interaction diagram and EHT calculations show that the 1,2 intramolecular shift of hydrogen is symmetry disfavored [10]. In presence of a transition metal fragment to which the alkyne coordinates, the activation energy is considerably lower. This has been attributed to the tendency of H to shift as a proton rather than as a hydride.

The orbital analyses by Silvestre and Hoffmann [10] have been substantiated by quantitative *ab initio* (MP2) calculations by Wakatsuki *et al.* on the transformation of $RuCl_2(PH_3)_2(HC\equiv CH)$ to $RuCl_2(PH_3)_2(C=CH_2)$ [11]. The more stable acetylene π-complex $HC\equiv CH$ transforms in a σ C-H complex by slippage of the metal along the C-C direction. The σ C-H complex undergoes a 1,2 hydrogen shift to give a vinylidene complex which is an isomer of the most stable vinylidene complex (Figure 3).

The intermediates that are thus connected in the rate determining step are of equal energy. The reaction is exothermic when the more stable alkyne and vinylidene complexes are respectively considered (Figure 3, bottom). The Ru^{II} d^6 fragment $RuCl_2(PH_3)_2$ is sufficiently both σ-acid and π-basic to reverse the thermodynamic outcome of the isomerization for the free ligand. Analysis of localized molecular orbitals (LMO) contour diagrams along the reaction coordinate demonstrates that the σ C_α-H_α bond transforms into the C_α lone pair orbital while the in plane C-C π orbital changes into the newly formed C_β-H_α σ bond. The p orbital at C_α, which has been a part of the in-plane π orbital, is left empty as the C_β-H_α σ bond is being formed, and becomes the empty p orbital of the vinylidene. This orbital is involved in back-donation from the metal. It results that the migrating hydrogen behaves as a naked proton rather than a hydride.

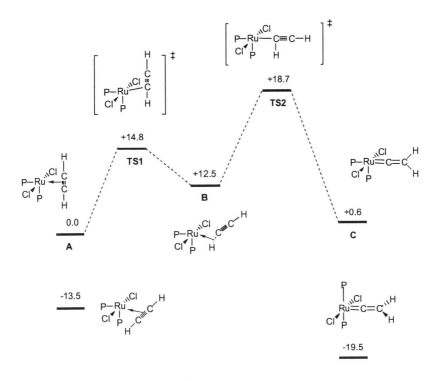

Figure 3. Energy diagram (kcal.mol⁻¹) of RuCl₂(PH₃)₂(HC=CH⌡. RuCl₂(PH₃)₂(C=CH₂), intermediates and the transition states as calculated by Wakatsuki *et al.* [11].

The isomerization, itself, originates from the σ complex (B in Figure 3). However the total activation energy depends critically on the relative energy of A and B (Figure 3). An alkyne C≡C triple bond binds more efficiently to a transition metal complex than a σ C-H bond since the π C-C orbital is a better electron-donor and the π* C-C orbital a better electron acceptor than the σ and σ* C-H orbitals, respectively. However, the difference in energy between the two isomers is relatively low for a d⁶ metal center because four-electron repulsion between an occupied metal d orbital and the other π C-C orbital destabilizes the alkyne complex. This contributes to facilitate the transformation for the Ru^{II} system studied by Wakatsuki *et al.*

For other systems the transformation is still energetically accessible but the barrier may be higher if the alkyne complex is less destabilized. In d² and d⁴ complexes, the four-electron repulsion with the alkyne can be avoided (the alkyne behaves like a 4-electron donor) and this stabilizes the alkyne complex. The intraligand 1,2 shift is then associated to a higher activation barrier. This has been illustrated in a study by Lledós *et al.* on [Cp₂Nb(HC≡CH)(L)]⁺ and [Cp₂Nb(C=CH₂)(L)]⁺ (L = CO, PH₃) where DFT calculations show that the barrier to isomerization is +29.2 kcal.mol⁻¹ (resp.

+35.6 kcal.mol^{-1}) for L = CO (resp. PH$_3$) [12]. The thermo neutral character of the transformation with a reaction energy of -0.8 kcal.mol^{-1} (resp. -2.9 kcal.mol^{-1}) for L = CO (resp. PH$_3$) highlights the additional stabilization of the alkyne complex. The two tautomers become isoenergetic and the vinylidene complex [Nb(η5-C$_5$H$_4$SiMe$_3$)$_2$(C=CHPh)(CO)]$^+$ undergoes an isomerization process to give the corresponding η^2-alkyne derivative.

The ability for a transition metal to stabilize the alkyne can even be larger as shown by Stegmann *et al.* for d^0 F$_4$W(HCCH) [13]. The acetylene complex is 10.4 kcal.mol^{-1} lower in energy than the isomeric vinylidene complex F$_4$W(CCH$_2$). The direct 1,2 hydrogen migration has a barrier of 84.8 kcal.mol^{-1} and proceeds via a transition state which has a non planar C$_2$H$_2$ moiety. In contrast to the preceding cases, no σ C-H complex is obtained as an intermediate, probably because the d^0 W^{IV} metal center cannot stabilize this intermediate through back donation.

The isomerization processes described above do not imply a change in the oxidation state of the metal but only a reorganization of the electron density within the organic ligand with the hydrogen atom migrating as a proton. This is particularly favored for d^6 metal centers not prone easily to oxidation (for instance Ru^{II} $\rightarrow$ Ru^{IV}).

In summary to this section, the metal fragment always makes the 1,2 shift easier than for the free alkyne ligand: the isomerization is much less endothermic and the activation energy is lower. This is mostly due to the efficient stabilization by the metal fragment of the incipient lone pair on the carbon that is loosing its hydrogen. Interestingly, the other π orbital, which is not in the metal-C-C plane, plays also an important role. It can destabilize the alkyne complex and thus facilitates the reaction when involved in a 4-electron destabilization with an occupied metal d orbital. It can inhibit the reaction if this 4-electron destabilization is not present. In this case the alkyne behaves like a 4-electron donor and makes especially stable complexes with the metal fragment.

3.1.2   Hydrido(alkynyl) complex as intermediate

Already 20 years ago, Antonova *et al.* proposed a different mechanism, with a more active role of the transition metal fragment [3]. The tautomerization takes place via an alkynyl(hydrido) metal intermediate, formed by oxidative addition of a coordinated terminal alkyne. Subsequent 1,3-shift of the hydride ligand from the metal to the β-carbon of the alkynyl gives the vinylidene complex (Figure 2, pathway b).

Wakatsuki *et al.* carried out theoretical calculations on the isomerization reaction of RhCl(PH$_3$)$_2$(HC≡CH) [14], as a model for RhCl(P^{i}Pr$_3$)$_2$(HC≡CR) studied by Werner and co-workers [15]. One purpose of the study was to discuss the existence of the (hydrido)alkynyl metal complex as an intermediate during the reaction.

The transformation of $RhCl(PH_3)_2(HC\equiv CH)$ to $RhCl(PH_3)_2(C=CH_2)$ has been calculated (MP2) to be exothermic by 7.8 kcal.mol^{-1}. The intraligand 1,2-hydrogen shift mechanism found in the Ru^{II} system is not relevant to the present rhodium case. Starting from a $\eta^2 C\equiv C$ complex, both systems give a metal-(η^2 C-H) species in a subsequent step. In the case of the d^6 Ru^{II} system this η^2 C-H complex is an intermediate. In contrast, the η^2 C-H coordinated state is a transition state in the d^8 Rh^I system, the oxidative addition being a very facile process.

From the hydrido(alkynyl) intermediate, the hydrogen atom can be transferred to C$_\beta$, either unimolecularly through a 1,3-shift, or bimolecularly from one hydrido(alkynyl) complex to an other one. The unimolecular hydrogen migration was calculated to have activation energy of 33.5 kcal.mol^{-1}. In contrast, the bimolecular rearrangement takes place with a very low activation barrier of 3.4 kcal.mol^{-1} (Figure 4). This barrier has been further calculated for a more realistic system with "real" phosphine P^{i}Pr$_3$ and alkyne HCCR using the IMOMM method [16]. It was concluded that the bimolecular hydrogen shift is still favored by ca. 15 kcal.mol^{-1} over the unimolecular 1,3 H-migration.

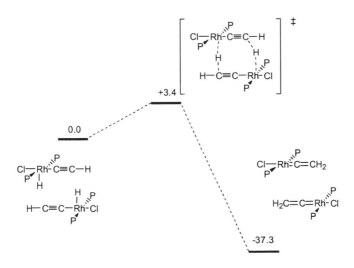

Figure 4. Energy diagram (kcal.mol^{-1}) for bimolecular rearrangement calculated by Wakatsuki *et al* [14]: $[RhCl(PH_3)_2(H)(C=CH)]_2$, $[RhCl(PH_3)_2(C=CH_2)]_2$, and transition state.

For more bulky systems the bimolecular pathway may be disfavored over the unimolecular 1,3 shift as shown by Mealli and co-workers for the metal fragment $[(pp_3)M]^+$ (M = Co, Rh; pp$_3$ = P(CH$_2$CH$_2$PPh$_2$)$_3$) [17]. The bulky tetrapodal pp$_3$ was modeled by P(CH$_2$CH$_2$PH$_2$)$_3$ and the energy barrier associated with the transformation of the hydrido(alkynyl) intermediate to vinylidene (20.6 kcal.mol^{-1}) is easier to overcome compared to that for reversion to the alkyne complex (28.6 kcal.mol^{-1}). The situation is reversed for the analogous Rh^I system, with the initial π-acetylene adduct being slightly more stable.

In conclusion to this section, the two step reaction requires the presence of a metal that can be easily oxidized (d^8 over d^6). However the 1,3 migration step does not appear to be easy in an intramolecular way. An intermolecular 1,3 shift seems feasible but is probably highly sensitive to steric effects. The pathways for the isomerization reactions with d^6 and d^8 metal complexes are significantly different. They have in common the preference for the hydrogen to move as a proton, whether the shift starts from C$_\alpha$ or from the metal center.

3.2   L$_n$$_,$M(H*)(HCCR) to L$_n$$_,$M(H)(CCH*R)

In the transformation of a 1-alkyne to a vinylidene in the coordination sphere of a transition metal, the migrating hydrogen atom plays a key role. Usually, ancillary ligands on the metal are only spectators and do contribute to small modifications of the bonding properties of the metal fragment. However, if a hydride is present as a ligand to the transition metal center, it may interfere with the alkyne to vinylidene transformation. This may open up new selective and efficient routes to vinylidene complexes.

Recently, Caulton *et al.* have studied the reaction of RuHX(H$_2$)L$_2$ (L = P^{t}Bu$_2$Me; X = Cl, I) and OsH$_3$XClL$_2$ (L = P^{i}Pr$_3$) with terminal alkynes [18, 19]. The reactions of RuHX(H$_2$)L$_2$ with PhC≡CD lead in time of mixing to RuDX(C=CHPh)L$_2$ as the only isotopomer. The results are in total disagreement with the two different mechanisms usually considered for such transformations. The 1,2 intraligand hydrogen shift would yield RuHX(C=CDPh)L$_2$ as the only isotopomer. In the case of a hydrido(alkynyl) intermediate RuHDX(C≡CPh), the subsequent 1,3 hydrogen shift from the metal to C$_\beta$ would yield a mixture of RuHX(C=CDPh)L$_2$ and RuDX(C=CHPh)L$_2$. The mechanism of the isomerization for these particular systems has to be different.

Experimentally, two equivalents of the alkyne for one equivalent of the metal species are necessary to achieve reaction. One equivalent serves to abstract H$_2$ from the starting materials, RuH(H$_2$)XL$_2$ or OsH$_3$XL$_2$, hence generating *in situ* the very reactive 14-electron compound MHXL$_2$ (M = Ru, Os). It is assumed that the reaction starts from the 14-electron complex modeled as MHCl(PH$_3$)$_2$ (M = Ru, Os) and the acetylene is used as a model for phenyl acetylene. DFT (B3LYP) calculations were carried out to elucidate the reaction mechanism and to account for the results of the labeling experiment [19].

3.2.1 Mechanism for M = Ru

The four-coordinate model complex RuHCl(PH$_3$)$_2$ is not planar but has a saw-horse geometry with trans phosphines and H-Ru-Cl = 101.3°. This angle illustrates that a d^6 tetra coordinated complex prefers to be a piece of an octahedron with two empty coordination sites in order to keep the six electrons of the metal in nonbonding orbitals (essentially similar to the t$_{2g}$ set of an octahedron).

The acetylene coordinates trans to the least σ electron donor group, chlorine. Coordination of the C-H bond is a less favorable alternative to coordination of the π system. The σ C-H complex is 17.1 kcal.mol^{-1} less stable than the π-alkyne complex (Figure 5). From this σ C-H intermediate the 1,2 shift is possible with a relatively small activation barrier (+15.5 kcal.mol^{-1}) to yield the vinylidene complex. However this mechanism is in contradiction with the labeling experiment.

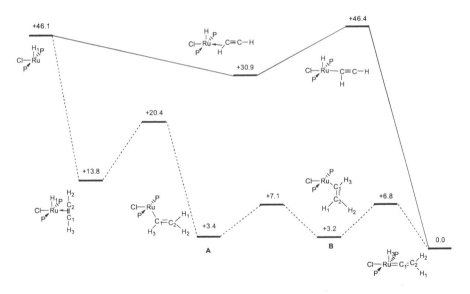

Figure 5. Energy diagram (kcal.mol^{-1}) for the formation of RuHCl(C=CH$_2$)(PH$_3$)$_2$ from RuHCl(PH$_3$)$_2$ and C$_2$H$_2$ as calculated by Oliván *et al.* [19].

From the energetically preferred π-alkyne complex there is an alternative pathway involving the hydride ligand (Figure 5). The first step is an easy (ΔE$^{\#}$ = 6.6 kcal.mol^{-1}) migratory insertion of the C≡C triple bond into the *cis* Ru-H bond to yield a σ-vinyl complex, **A**, 10.4 kcal.mol^{-1} below the π-alkyne complex. This 14-electron σ-vinyl complex has also a saw-horse geometry adapted to four-coordinate Ru^{II} systems. Compared to the 16-electron acetylene complex, the lower energy of **A** is a consequence of the four-electron repulsion in the alkyne complex, which destabilizes the latter.

The structure of the σ-vinyl intermediate **A** is not adapted to a α-hydride migration since the C_1-H_3 bond is not directed toward the empty site of the metal. To achieve this requirement, the whole vinyl group easily rotates around the Ru-C_1 single bond with a transition state lying only 3.7 kcal.mol^{-1} above **A**. Not surprisingly the second σ-vinyl complex **B** is isoenergetic with **A** and has the proper geometry to undergo a α-hydride migration, which indeed occurs easily since the associated transition state is 3.6 kcal.mol^{-1} above **B**.

The vinylidene complex is thus obtained as the thermodynamic product in three steps, each associated with a low activation barrier. This agrees well with the fact that no intermediates could be observed during the reaction. This new 3-step mechanism is in agreement with the labeling experiment where the hydride initially on Ru terminates on the vinylidene ligand. The high specificity of the reaction suggests that the two other routes cannot compete with this sequence of insertion-rearrangement-migration reaction. This is supported by the fact that the intermediate and transition state for the 1,2 shift are significantly higher in energy than any of the transition states shown in Figure 5. Likewise the mechanism starting with the oxidative addition to the C-H bond of the alkyne was eliminated by finding that $Ru(H)_2Cl(CCH)(PH_3)_2$ is also at very high energy. The key conclusion is that a new and efficient route for the 1-alkyne-vinylidene isomerization is made available by the presence of a hydride on the transition metal center. Hydride does not remain as a spectator ligand in the formation of hydrido-vinylidene complexes.

3.2.2   Mechanism for M = Os

The reaction path was calculated for the same system in which Ru was replaced by Os with the purpose to discuss the difference induced by the change of a 4d to a 5d metal [19]. The binding dissociation energy of the alkyne complex is larger with Os (46.6 kcal.mol^{-1}) than with Ru (32.3 kcal.mol^{-1}), reflecting the stronger metal-ligand interaction of a 5d metal. The increase of metal-ligand bond is even more in favor of the vinylidene complex: the energy of reaction is 23.7 kcal.mol^{-1} with Os to compare with 13.8 kcal.mol^{-1} with Ru. The main differences between Ru and Os are to be found in the structure of vinyl intermediate. There is a unique η^2-vinyl complex in the case of Os and from this vinyl complex one goes in one step to either the alkyne or the vinylidene complexes.

In the η^2-vinyl complex, the C=C is long (1.411Å), the Os-C_α very short (1.876Å) and the Os-C_β rather long (2.236 Å). No σ-bonded vinyl Os complex on one hand and no π-bonded vinyl Ru complex on the other hand could be located as minima on the potential energy surfaces. This illustrates that, among several isomeric structure, Ru favors the structure with the maximum bonding within the organic ligand (full σ and π bond in the vinyl ligand) whereas Os favors the structure with the maximum number of metal-ligand bonds (Os-C_α and Os-C_β).

4. ALKENE TO CARBENE COMPLEXES: HOW TO FAVOR PRODUCT FORMATION

The alkene to carbene isomerization can be viewed as a 1,2 hydrogen shift similar to the alkyne to vinylidene tautomerization (equation 3). However, until recently, very few examples of the transformation have been observed in organometallic chemistry. Schrock *et al.* [4] have reported that $[N_3N]Ta(H_2C=CH_2)$ ($[N_3N]^{3-} = [(Me_3SiNCH_2CH_2)_3N]^{3-}$) react with ethylene in presence of $PhPH_2$ to give $[N_3N]Ta(CHMe)$ but the mechanism has remained unclear. In 1998, Caulton and co-workers reported an unusual direct synthesis of coordinated carbenes from olefins [20]. Dehydrohalogenation of $Ru(H)_2Cl_2L_2$ (L = P^iPr_3) with 1 equivalent of lithium 2,2,6,6-tetramethylpiperide in benzene gives $RuHClL_2$ a 14-electron d^6 moiety [21]. Reaction of this 14-electron complex with ethyl vinyl ether $H_2C=C(H)(OEt)$ at 25 °C gives immediate formation of $RuHClL_2[C(Me)(OEt)]$.

DFT calculations confirmed the similarities with the alkyne/vinylidene transformation but have revealed that additional parameters were essential to achieve the isomerization [8, 20-23]. The hydride ligand on the 14-electron fragment $RuHClL_2$ opens up a pathway for the transformation similar to that obtained for the acetylene to vinylidene isomerization. However, thermodynamics is not in favor of the carbene isomer for unsubstituted olefins and the tautomerization is observed only when a π electron donor group is present on the alkene. Finally the nature of the X ligand on the $RuHXL_2^{+q}$ (X = Cl, q=0; X = CO, q=1) 14-electron complex alters the relative energy of the various intermediates and enables to stop the reaction on route to carbene.

4.1   Mechanism of isomerization: the case of C_2H_4

The mechanism for the ethylene to methylcarbene reaction has been calculated at the DFT (B3PW91) level with the model system $RuHCl(PH_3)_2$ [8, 21]. As in the case of the acetylene to vinylidene reaction, the starting complex was assumed to be the 14-electron complex $RuHCl(PH_3)_2$ generated *in situ*. The reaction path is very similar to that obtained with C_2H_2. They differ mainly in the overall direction of the energy pattern: downhill for acetylene and uphill for ethylene.

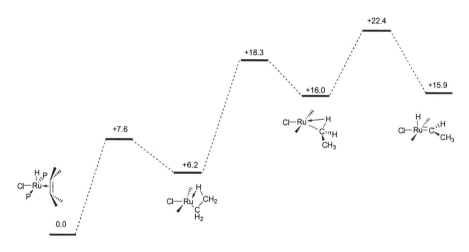

Figure 6. Energy diagram (kcal.mol^{-1}) for the formation of $RuHCl(C(H)(CH_3))(PH_3)_2$ from $RuHCl(PH_3)_2$ and C_2H_4 as calculated by Gérard *et al.* [21].

The first step is coordination of the ethylene through its π orbital. The ethylene is trans to Cl with the C=C bond in the Cl-Ru-H plane. Facile migratory insertion ($\Delta E^\# = 7.6$ kcal.mol^{-1}) of the coordinated ethylene in the Ru-H bond leads to an alkyl intermediate 6.2 kcal.mol^{-1} less stable than the π ethylene complex. The alkyl intermediate has a strong β C-H agostic interaction as illustrated by the unusually long agostic C-H bond (1.221 Å) which helps to stabilize the unsaturation in the formally 14-electron alkyl intermediate.

Disruption of the C-H β-agostic interaction and rotation around the Ru-C bond leads to a less stable alkyl intermediate, 9.8 kcal.mol^{-1} above the β-agostic alkyl complex. This new alkyl complex has a C-H α-agostic interaction weaker than the β-agostic one, as illustrated by a shorter C-H bond (1.126 Å vs. 1.221 Å). The transition state between the two alkyl intermediates can be considered as a true 14-electron complex devoid of any stabilizing interaction. This gives an estimate of stabilization associated with the agostic C-H interaction. The β-agostic interaction lowers the transition state (TS) by ca. 12 kcal.mol^{-1}, whereas the α-agostic lowers it by only 2.5 kcal.mol^{-1}. The last step of the isomerization is a α-hydrogen migration from the α-agostic alkyl complex to yield with a low energy barrier ($\Delta E^{\#} = 6.4$ kcal.mol^{-1}) the isoenergetic methylcarbene product.

The hydride initially on the metal goes to the organic ligand as in the case for the acetylene to vinylidene reaction. The isotope labeling experiment, not as selective as in the alkyne system, supports nevertheless the calculated pathway. Reaction of RuHClL$_2$ (L = P^{i}Pr$_3$) with excess C$_2$D$_4$ in C$_6$H$_6$ for 1 hour at 20 °C shows (^{2}H NMR) deuteration of the H on Ru and also the methyl groups of coordinated phosphine. No intermediate could be detected. This is indicative of reversible insertion of C$_2$D$_4$ into the Ru-H bond and it also suggests that the 16-electron olefin hydride species is more stable than any RuCl(C$_2$H$_5$)L$_2$ complex. Insertion of ethylene into Ru-H is also evidenced by the formation of ethane, detected by ^{1}H NMR.

Other possible mechanisms, corresponding to those discussed for the alkyne, have been considered. The oxidative addition pathway is excluded because all vinyl intermediates (see below) are found at high energy with respect to the π olefin complex.

Two σ C-H ethylene complexes have been located on the potential energy surface at a high energy with respect to the π olefin complex. From any of these two σ C-H complexes, the carbene product is reached in one step with a high activation barrier (+32.8 and +28.4 kcal.mol^{-1}, respectively). The 1,2 shift is thus a very unlikely route and it is not surprising that no isomerization was observed for transition metal fragment deprived of a hydride ligand. Thus the reaction observed by Schrock [4] suggests that the H of the PhPH$_2$ phosphine plays a key role in the reaction since there is no H initially on the Ta complex.

+23.7   +29.7

4.2 Thermodynamic influence of the substituent on the alkene

The main difference between the alkyne and olefin isomerization processes lies in the thermodynamic pattern: the reaction is calculated to be exothermic for acetylene to vinylidene (-13.8 kcal.mol^{-1}) but endothermic for ethylene to methylcarbene (+15.9 kcal.mol^{-1}) [8, 20, 21]. Comparison with the energies of reaction for the free ligand shows the stabilization influence of RuHCl(PH$_3$)$_2$, in favor of the vinylidene and carbene ligands, due to more efficient σ-donation and π back-donation, when compared to the alkyne and alkene ligands, respectively. The stabilization is slightly less for the acetylene/vinylidene (43.2 + 13.8 = 57 kcal.mol^{-1}) than for the ethylene/methylcarbene (79.5 – 15.9 = 63.6 kcal mol^{-1}). This strong binding energy for the vinylidene is sufficient to reverse the thermodynamics of the alkyne to vinylidene isomerization (equation 1) making the formation of vinylidene complexes an accessible thermodynamic target. In the case of the alkene to carbene tautomerization, the stabilizing energy for the carbene side is not sufficient to favor the transformation. It is necessary to introduce additional stabilizing factors for the carbene side. Substitution of the alkene by a π electron donor group offers this possibility. Thus the energy of reaction for the isomerization of H$_2$C=C(H)G into CG(CH$_3$) is equal to 54.2 kcal.mol^{-1} for G = F, 41.6 kcal.mol^{-1} for G = OMe and decreases with the π donating ability of G [8]. It is remarkable that the difference in energy between the two isomers H$_2$C=C(H)G and CG(CH$_3$) is similar to that between acetylene and C=CH$_2$.

The reaction path was not recalculated for the methyl vinyl ether, H$_2$C=C(H)(OCH$_3$) which was chosen as a model for the experimental H$_2$C=C(H)(OEt). It is assumed that the path is not modified; this hypothesis has been validated with H$_2$C=C(H)F [8]. The calculations are thus limited to the products and key intermediates.

Substitution has no significant effect on the geometry of the π complex and RuHCl(PH$_3$)$_2$(CH$_2$=CH$_2$) and RuHCl(PH$_3$)$_2$(H$_2$C=CH(OMe)) have very similar shape. Of interest the OMe group cannot reach the second empty coordination site (trans to H) of Ru. Remarkably, the Ru-alkene bond dissociation energy is also not affected by the presence of OCH$_3$ (less than 3 kcal.mol^{-1} difference in binding dissociation energy). The methyl vinyl ether prefers to coordinate the metal in the mirror plane of the molecule, as ethylene does(Figure 7). Two isomers **I** and **II** (Figure 7) are found to be minima on the potential energy surface with a preference of 0.9 kcal.mol^{-1} for **II**.

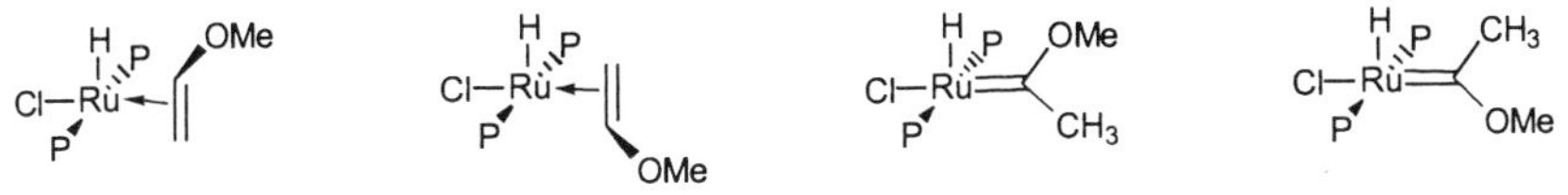

Figure 7. π olefin complexes and corresponding carbene isomers for methyl vinyl ether.

Two conformations for the carbene complex were located separated by only 0.25 kcal.mol^{-1}. The preferred orientation for $C(CH_3)(OCH_3)$ is rotated by 90° with respect to that in the $C(H)(CH_3)$ complex (Figures 6 and 7). The π electron donor group OCH_3 has decreased the reaction energy. In the case of methyl vinyl ether, the isomerization is calculated to be essentially thermo neutral ($\Delta E = 1.6$ kcal.mol^{-1}). Why is it only thermo neutral and not exothermic like in the vinylidene case? Comparison of the difference in energy between the unsubstituted or substituted alkene complexes and the carbene complexes with the corresponding values in the absence of $RuHCl(PH_3)_2$, represented as [Ru], give a clue on the individual and combined stabilizing effects of [Ru] and OCH_3. These two species act as π electron donor to the same empty p orbital of the carbene carbon, [Ru] with a d occupied orbital, OCH_3 with its p lone pair. This results in a competition which can be summarized in the following manner: the combined stabilizing influence of the two groups is *more* than the stabilizing influence of each group taken individually but *less* than the sum of the same stabilizations. This is illustrated by the following set of isodesmic reactions.

$$C_2H_4 \ + \ C(OCH_3)(CH_3) \ \rightarrow \ H_2C=CH(OCH_3) \ + \ C(H)(CH_3)$$
$$\Delta E = 37.7 \text{ kcal.mol}^{-1} \tag{6}$$

$$C_2H_4 \ + \ [Ru]C(H)(CH_3) \ \rightarrow \ [Ru](C_2H_4) \ + \ C(H)(CH_3)$$
$$\Delta E = 63.2 \text{ kcal.mol}^{-1} \tag{7}$$

$$H_2C=CH(OCH_3) \ + \ [Ru](C(OCH_3)(CH_3)) \ \rightarrow \ [Ru](H_2C=CH(OCH_3)) \ +$$
$$C(OCH_3)(CH_3) \qquad \Delta E = 39.8 \text{ kcal.mol}^{-1} \tag{8}$$

$[Ru](C_2H_4) + H_2C=CH(OCH_3) \rightarrow [Ru](H_2C=CH(OCH_3)) + C_2H_4$
$\Delta E = 2.6 \text{ kcal.mol}^{-1}$ (9)

$[Ru]C(H)(CH_3) + C(OCH_3)(CH_3) \rightarrow C(H)(CH_3) + [Ru](C(OCH_3)(CH_3))$
$\Delta E = 25.8 \text{ kcal.mol}^{-1}$ (10)

Equations 6 and 7 show that OCH_3 and [Ru] independently stabilize the carbene species by large amounts. Combining the two effects does not lead to a stabilization that is the sum of the individual stabilizations. However, the stabilization by [Ru] is larger for unsubstituted carbene (equation 8 vs. equation 7) due to the electronic influence of the methoxy on the olefin and/or on the carbene. There is no particular difference in binding energy of the alkene (equation 9), therefore [Ru] stabilizes the non-substituted carbene significantly better (equation 10) than the $C(CH_3)(OCH_3)$ molecule.

These equations show why the combined influence of [Ru] and the methoxy group is not as large as their individual influence could have wrongly suggested. This competition also suggests that one can change the metal fragment or the substituent on the alkene without greatly influencing the overall thermodynamic of the reaction. This has been verified numerically as shown in the following sets of equations [8].

$RuHCl(PH_3)_2(H_2C=CHF) \qquad\qquad \rightarrow \qquad\qquad RuHCl(PH_3)_2(CF(CH_3))$
$\Delta E = 3.6 \text{ kcal.mol}^{-1}$ (11)

$RuHCl(PH_3)_2(H_2C=CH(OCH_3)) \rightarrow RuHCl(PH_3)_2(C(OCH_3)(CH_3))$
$\Delta E = 1.6 \text{ kcal.mol}^{-1}$ (12)

$RuH(CO)(PH_3)_2(H_2C=CH(OMe))^+ \rightarrow RuH(CO)(PH_3)_2(C(OCH_3)Me)^+$
$\Delta E = 2.6 \text{ kcal.mol}^{-1}$ (13)

$OsHCl(PH_3)_2(H_2C=CH(OCH_3)) \qquad \rightarrow \qquad OsHCl(PH_3)_2(C(OCH_3)(CH_3))$
$\Delta E = -1.3 \text{ kcal.mol}^{-1}$ (14)

In these equations, the substituents are changed from weak (F) to strong (OMe) π electron donor. Likewise the ability of the metal fragment to back donate into the empty p orbital of the carbene is varied through ligand (π acid CO vs. π donor Cl with associated charge adjustment) or through metal (the 5d orbital of Os are more efficient in back donation than the 4d of Ru).

The overall energy of reaction differs by less than 4 kcal.mol^{-1} which is remarkably small especially in comparison to the energy of reaction for the unsubstituted ethylene. The reaction should be possible for all these systems. We will see below how this has been verified.

The energy of reaction for these isomerization processes illustrates how some of the properties of the entire chemical system are influenced by the properties of its parts. It is commonly accepted that alkyne and alkene have analogous bonding properties. The same is true for the bonding properties of vinylidene and carbene. They are in fact isolobal. This is however not sufficient to lead to analogous behavior in this isomerization process for alkyne and olefin. The difference in the energy pattern between the unsaturated ligand and its corresponding unstable isomer are too important. Despite the powerful ability of the metal fragment to stabilize unstable ligands, the very large difference in energy in the case of unsubstituted olefin cannot be compensated. Introducing substituents allow the overall isomerization process to be thermodynamically feasible. The competing stabilizing influence of the substituent G on the olefin and the metal fragment [M] leads to the unexpected result of equivalent energy of reaction for a wide variety of G and metal fragment.

4.3 Influence of ancillary ligand: $RuHXL_2^{+q}$ (X = Cl, q = 0; X = CO, q = 1)

The favorable relative energy of the carbene complex with respect to the π-olefin adduct in the reaction of a vinyl ether with $RuHCl(PH_3)_2$ is a consequence of the combined electron donation of the methoxy group OCH_3 and of the $RuHCl(PH_3)_2$ fragment [8, 22-24]. In presence of a π electron donor group like Cl, the 14-electron fragment is sufficiently π basic to stabilize the carbene π acid ligand through a push-pull interaction. However, electron donation from the methoxy group is required to observe experimentally the carbene complex. In the reaction of $RuH(CO)(P^tBu_2Me)_2^+$ with a vinyl ether, a substituted alkyl complex is obtained as shown by Caulton *et al.* (equation 15) [22]. The solid state structure shows that the alkoxy group on C_β makes a donor bond to Ru trans to CO.

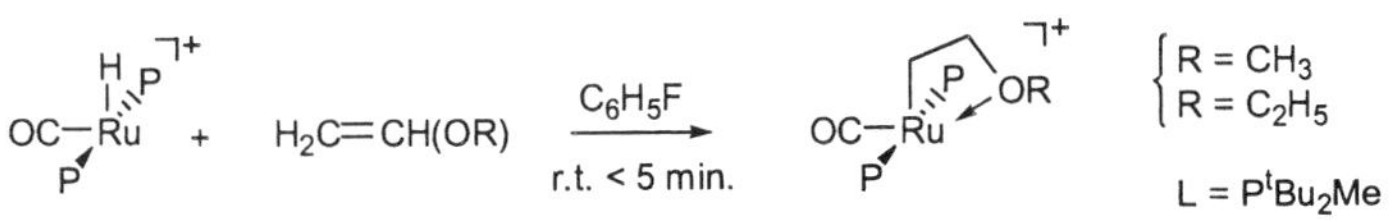

$$OC-\underset{P}{\overset{\overset{H}{\underset{|}{P}}}{Ru}}^{+} \quad + \quad H_2C=CH(OR) \quad \xrightarrow[\text{r.t. < 5 min.}]{C_6H_5F} \quad OC-\underset{P}{\overset{\overset{P}{\underset{|}{}}}{Ru}}\overset{OR}{\diagup}^{+} \qquad \left\{ \begin{array}{l} R = CH_3 \\ R = C_2H_5 \end{array} \right.$$

$$L = P^tBu_2Me \qquad (15)$$

A first explanation for this result is that the carbene complex is destabilized with respect to alternative isomers because of the poor π-basicity of the 14-electron fragment $RuH(CO)(PR_3)_2^+$. DFT calculations do not support this analysis. As already mentioned above, calculations were carried   out   on $RuHX(PH_3)_2(H_2C=CH(OCH_3))^{+q}$ and $RuHX(PH_3)_2(C(CH_3)(OCH_3))^{+q}$ (X = Cl, q = 0; X = CO, q = 1). These calculations show that the significant binding energy of the π-olefin adduct to $RuHX(PH_3)_2^{+q}$ (X = Cl, q = 0; X = CO, q = 1) is remarkably insensitive to the nature of X (Cl, 35.7 kcal.mol^{-1}; CO, 33.1 kcal.mol^{-1}). The difference of energy between the olefin complex and the carbene isomer is also mostly insensitive to the nature of X: +1.6 kcal.mol^{-1} for Cl and +2.2 kcal.mol^{-1} for CO. This is due to the competitive electron donation of the alkoxy group and of the metal fragment in the empty p orbital of the carbene. Less electron transfer from a less π basic metal (CO in place of Cl) is compensated by increased electron donation from OCH_3. The calculations show that the total effect is almost constant. Therefore, replacing Cl by CO does not change in any significant manner the thermodynamics of the transformation of the vinyl ether into alkoxycarbene complexes. From these results, isomerization of $H_2C=CH(OMe)$ should be obtained with both $RuHCl(PH_3)_2$ and $RuH(CO)(PH_3)_2^+$. However this is not the case.

The result of equation 15 can be explained only if alkyl intermediates are particularly stabilized in the case of $RuH(CO)(PH_3)_2^+$. Several 14-electron alkyl complexes were located as minima on the potential energy surface, differing by the position of the OCH_3 group in the ethyl chain and the nature of the group which interacts with the Ru vacant site trans to CO. The 14 electron $RuH(CO)(PH_3)_2(ethyl)^+$ system is electron deficient and needs additional electron donation from the available groups. On the ethyl chain, a C-H bond and the OCH_3 group are both candidates for this interaction. This yields several isomers depending on the position of the methoxy group (on C_α or on C_β of the coordinated ethyl chain) and of the nature of the interaction with Ru (C-H α–agostic, β-agostic, or CH_3O -> Ru dative bond).

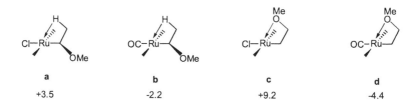

Figure 8. Selected 14-electron alkyl intermediates and their relative energy (kcal.mol^{-1}) with respect to the corresponding π-olefin complex

Figure 8 represents the most stable isomer for each substitution case, OCH_3 on C_α or C_β, with the energy relative to the most stable π-olefin complex. For the Cl ligand, the alkyl intermediates, **a** and **c**, are *above* the π-olefin complex. It is therefore understandable that the only isolated product is the carbene complex, calculated to be isoenergetic with the starting olefin complex. For the CO ligand, the alkyl complexes, **b** and **d**, are more stable than the olefin complexes. This is associated with the diminished ability of $RuH(CO)(PH_3)_2^+$ to back donate in the olefin π* orbital. The most stable isomer, **d**, corresponds to the experimentally observed system. It is strongly stabilized by the Ru-O interaction, which has a dominating electrostatic component associated with the total positive charge of the complex. Remarkably the ethyl complex with OMe on C_α and a β agostic C-H interaction, **b**, is only marginally above **d**. Its stability originates from the preference for the coordinated alkyl chain to be α substituted by an electron withdrawing group. The potential energy surface for the isomerization of ethylene into the carbene complex shows that the first intermediate is an alkyl complex with a β C-H agostic bond (Figure 6). Variable temperature NMR has confirmed that **b** is a kinetic product seen only at low temperature. The thermodynamic product is **d**. No carbene product is observed.

The calculations have pointed out that **d** derives a great part of its stability from the Ru-O interaction. Preventing this interaction should disfavor **d** as the product of reaction. What would then happen? Calculations have shown that the energy of reaction of a carbene complex from a vinyl ether is similar with $RuH(CO)(PH_3)_2^+$ and $RuHCl(PH_3)_2$. The 2,3 dihydrofurane was reacted with $RuH(CO)(P^tBu_2Me)_2^+$ [24]. The presence of the ring makes the Ru-O interaction impossible and the only observed product is the carbene complex shown in equation 16. The reaction of 2,3 dihydrofurane with $RuHCl(P^iPr_3)_2$ also gives the corresponding carbene complex [21].

$$(16)$$

It thus appears that the products of the reaction vary with changes in the two reactants. This is not due to competing mechanisms but to variation in the relative energies of reactants, products and intermediates. All reactions show that alkene complex has first inserted into the Ru-H bond to form an alkyl intermediate. Thus the ability for the metal fragment to coordinate *cis* to the metal hydride bond is essential to the story. Depending on the nature of the alkene, different scenarios occur from this intermediate. In the case of unsubstituted alkene, the only observed product is the alkene complex but D scrambling when using C_2D_4 supports the insertion into the Ru-H bond as an occurring step. With OR substituted alkene, the reaction leads to products other than the alkene complex. The alkyl complex is seen as a kinetic intermediate and a product when the alkene is poorly bonded to the metal fragment (transition metal fragment with poor back donating ability). Trapping the 14 electron alkyl complex cannot be achieved without additional strong electrostatic and donor-acceptor interaction between the metal and a substituent on the alkyl chain. When this is not achieved, the carbene complex is obtained. A large variety of substituted alkenes have been tested and the carbene complex is obtained each time the substituent on the alkene is sufficiently π donor [25-28]. The computational studies have shown how the combined and competing donation of the metal fragment and the substituent on the alkene results in a apparent lack of sensitivity to the substitution pattern.

5.   CONCLUSION

In this chapter, we have illustrated how computational studies have been used for understanding several related reactions which could not be easily studied by experimental means. The computational studies show how the metal fragment can be either a simple "help" or "spectator" for making feasible an isomerization not occurring in the free ligand (1,2 shift) . It can be also more active by cleaving bonds in the organic ligand or even by exchanging atoms (H) with the unsaturated ligand. In all studies, important approximations have been used especially in the modeling of large ligands such as phosphines. This should be kept in mind and all numerical results should be taken with a grain of salt. However permanent confrontation to the experimental results is a safeguard against unrealistic proposals.

ACKNOWLEDGEMENTS

A large part of the work described in this chapter originates from a long standing collaboration between the group of Professor Kenneth G. Caulton (Indiana University) and our group. We are thus grateful to the long list of students and post-doctoral fellows in Caulton's group who have shared their results with us over many years and even more grateful to Ken Caulton for being so deeply convinced of the synergy that could result from this collaboration and for the innumerous exciting discussions of all points. We are grateful to those at Indiana who devoted research times to verify predictions from calculations. All calculations on the alkene systems represent the PhD work of Hélène Gérard (Montpellier January 2000). We are extremely grateful to her dedication, hard work and creativity over the years in Montpellier. We would like to thank deeply Professor E.R. Davidson (Indiana University) for his numerous insightful comments and discussions. Finally we are grateful to the Indiana Computing center for a very generous donation in computing time. We would like to thanks the University Montpellier 2, CNRS and Region Languedoc-Roussillon for generous financial support.

REFERENCES

1. Crabtree, R. H. *The organometallic chemistry of the transition metals*, third edition, John Wiley & Sons (2001)
2. Bruce, M. I. *Chem. Rev.* **1991**, *91*, 197.
3. Antonova, A. B.; Johansson, A. A. *Russ. Chem. Rev.* **1989**, *58*, 693.
4. Freundlich, J. S.; Schrock, R. R.; Cummins, C. C.; Davis, W. M. *J. Am. Chem. Soc.* **1994**, *116*, 6476.
5. Ford, F.; Yuzawa, T.; Platz, M. S.; Matzinger, S.; Fülscher, M. *J. Am. Chem. Soc.* **1998**, *120*, 4430.
6. Trnka, T. M.; Grubbs, R. H. *Acc. Chem. Res.* **2001**, *34*, 18.
7. Bruneau, C.; Dixneuf, P. H. *Acc. Chem. Res.* **1999**, *32*, 311.
8. Gérard, H. *Ph D Thesis*, 2001, Université de Montpellier 2.
9. Kostic, N. M.; Fenske, R. F. *Organometallics* **1982**, *1*, 974.
10. Silvestre, J.; Hoffmann, R. *Helv. Chim. Acta* **1985**, *68*, 1461.
11. Wakatsuki, Y.; Koga, N.; Yamazaki, H.; Morokuma, K. *J. Am. Chem. Soc.* **1994**, *116*, 8105.
12. García-Yebra, C.; López-Mardomingo, C.; Fajardo, M.; Antiñolo, A.; Otero, A.; Rodríguez, A.; Vallat, A.; Lucas, D.; Mugnier, Y.; Carbó, J. J.; Lledós, A.; Bo, C. *Organometallics* **2000**, *19*, 1749.
13. Stegmann, R.; Frenking, G. *Organometallics* **1998**, *17*, 2089.
14. Wakatsuki, Y.; Koga, N.; Werner, H.; Morokuma, K. *J. Am. Chem. Soc.* **1997**, *119*, 360.
15. Rappert, T.; Nürnberg, O.; Mahr, N.; Wolf, J.; Werner, H. *Organometallics* **1992**, *11*, 4156.
16. Maseras, F.; Morokuma, K. *J. Comp. Chem.* **1995**, *16*, 1170.
17. Peréz-Carreño, E.; Paoli, P.; Ienco, A.; Mealli, C. *Eur. J. Inorg. Chem.* **1999**, 1315.
18. Oliván, M.; Eisenstein, O.; Caulton, K. G. *Organometallics* **1997**, *16*, 2227.
19. Oliván, M.; Clot, E.; Eisenstein, O.; Caulton, K. G. *Organometallics* **1998**, *17*, 3091.
20. Coalter III, J. N.; Spivak, G. J.; Gérard, H.; Clot, E.; Davidson, E. R.; Eisenstein, O.; Caulton, K. G. *J. Am. Chem. Soc.* **1998**, *120*, 9388.
21. Coalter III, J. N.; Bollinger, J. C.; Huffman, J. C.; Werner-Zwanziger, U.; Caulton, K. G.; Davidson, E. R.; Gérard, H.; Clot, E.; Eisenstein, O. *New J. Chem.* **2000**, *24*, 9.
22. Huang, D.; Gérard, H.; Clot, E.; Young, Jr., V.; Streib, W. E.; Eisenstein, O.; Caulton, K. G. *Organometallics* **1999**, *18*, 5441.
23. Gérard, H.; Clot, E.; Giessner-Prettre, C.; Caulton, K. G.; Davidson, E. R.; Eisenstein, O. *Organometallics* **2000**, *19*, 2291.

24. Huang, D.; Bollinger, J. C.; Steib, W. E.; Folting, K.; Young, Jr.; V.; Eisenstein, O.; Caulton, K. G. *Organometallics* **2000**, *19*, 2281.

25. Spivak, G. J.; Coalter III, J. N.; Oliván, M.; Eisenstein, O.; Caulton, K. G. *Organometallics* **1998**, *17*, 999.

26. Coalter III, J. N.; Streib, W. E.; Caulton, K. G. *Inorg. Chem.* **2000**, *39*, 3749.

27. Coalter III, J. N.; Huffman, J. C.; Caulton, K. G. *Organometallics.* **2000**, *19*, 3569.

28. Ferrando, G.; Gérard, H.; Spivak, G. J.; Coalter III, J. N.; Huffman, J. C.; Eisenstein, O.; Caulton, K. G. *Inorg. Chem.* **2001**, *40*, 6610

Chapter 7

Rhodium Diphosphine Hydroformylation

Jorge J. Carbó,[1,2] Feliu Maseras,[2] and Carles Bo[1,*]

[1]Departament de Química Física i Inorgànica, Universitat Rovira i Virgili Pl.Imperial Tarraco, 1, 43005 Tarragona, Spain; [2]Unitat de Química Física, Edifi C.n, Universitat Autònoma de Barcelona, 08193 Bellaterra, Barcelona, Spain.

Abstract: A review of theoretical progress in the modeling of rhodium diphosphine hydroformylation is presented. Early studies using pure quantum mechanics (QM) methods with phosphines modeled as PH_3 were able to provide a general picture of the complicated reaction cycle, but could not tackle more delicate issues as regioselectivity and enantioselectivity. The modeling of real-world phosphines with hybrid quantum mechanics / molecular mechanics (QM/MM) methods, together with the availability of new experimental data, has allowed a more detailed comprehension of the mechanism, as well as an understanding of the role played by steric and electronic effects of the substituents.

Key words: rhodium, hydroformylation, DFT, QM/MM.

1. INTRODUCTION

Hydroformylation is the most successful application of a homogeneous catalytic reaction to industrial processes [1]. Aldehydes are conveniently produced by hydroformylation, in which CO and hydrogen are simultaneously added to an alkene (Figure 1).

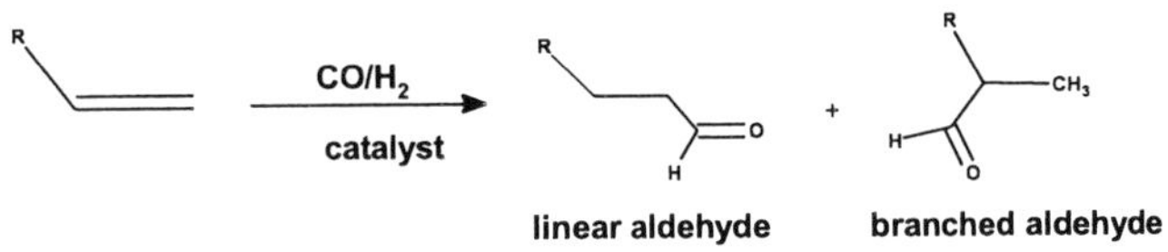

Figure 1. Terminal alkene hydroformylation reaction scheme

F. Maseras and A. Lledós (eds.), Computational Modeling of Homogeneous Catalysis, 161–187.

The original high-pressure cobalt-catalyzed heterogeneous reaction, which was discovered in 1938, was first replaced by the homogeneous process using $HCo(CO)_4$, which provided higher selectivity under milder conditions. Replacing cobalt with the rhodium complex $HRh(CO)_4$, later modified by introducing phosphine ligands, was one of the major breakthroughs in homogeneous catalysis, and introduced the currently known "OXO" process. Several companies manufacture and commercialize rhodium catalysts based on monophosphine ligands such as PPh_3, diphosphine or mixed phosphine-phosphite ligands.

Figure 2 shows the generally accepted dissociative mechanism for rhodium hydroformylation as proposed by Wilkinson [2], a modification of Heck and Breslow's reaction mechanism for the cobalt-catalyzed reaction [3]. With this mechanism, the selectivity for the linear or branched product is determined in the alkene-insertion step, provided that this is irreversible. Therefore, the alkene complex can lead either to linear or to branched Rh-alkyl complexes, which, in the subsequent catalytic steps, generate linear and branched aldehydes, respectively.

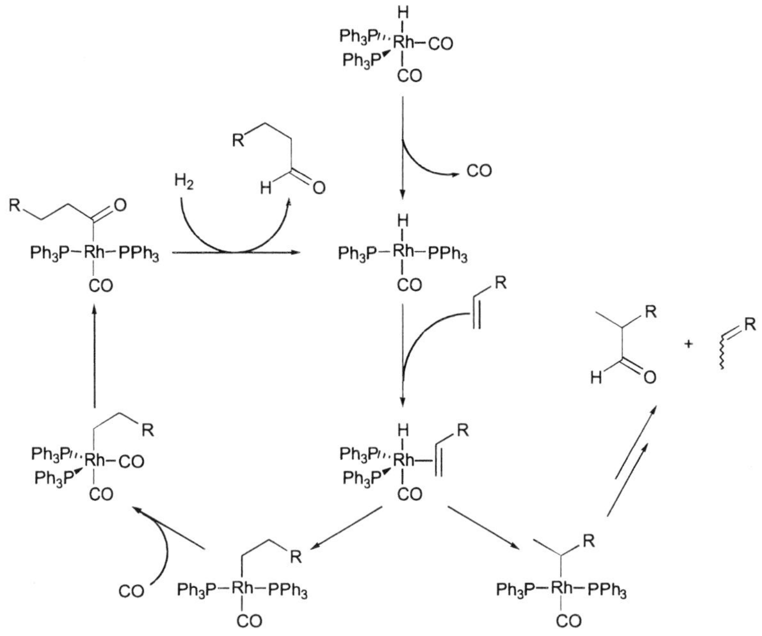

Figure 2. Hydroformylation catalytic cycle

The key issue in hydroformylation is how the ligand is designed. Extensive research has been carried out to develop new ligands with tailored stereoelectronic properties and obtain better activities and better regio- and stereoselectivities [4]. Of the several types of ligand-modified catalysts, those containing phosphines and phosphites have the highest activity and selectivity in the hydroformylation of alkenes. However, the are few systematic studies of how the ligand affects the performance of the catalyst and despite the development of a wide range of ligands, consistent structure-activity relationships are lacking. This is because *a complex web of electronic and steric effects governs selectivity in hydroformylation* [5]. The term *electronic effects* deals with the electronic properties of ligands (basicity, π acceptor/donor capabilities) while *steric effects* usually refers to how bulky the ligand is. Diphosphine steric effects have been rationalized with the help of the natural bite angle concept introduced by Casey and Whiteker [6] as a characteristic of a chelate ligand. The bite angle is the angle between the two phosphorous atoms and the metal, and is partly determined by the ligand backbone. How natural bite angles affect activity and regioselectivity has been studied for several catalytic reactions [7, 8, 9].

Diphosphine ligands generate two kinds of steric effects: those originated by ligand-ligand or ligand-substrate non-bonding interactions (*non-bonding effects),* and those directly related to the bite angle, which we call *orbital effects*. The bite angle determines metal hybridization and this in turn determines metal orbital energies. Therefore, these steric effects are actually electronic effects. The situation gets even more complicated because the bite angle affects the bulkiness of the ligand. Ligand design would certainly benefit from detailed knowledge of the reaction mechanism, the involved intermediates and the factors that influence their stability and reactivity. This is difficult even today when modern high-pressure NMR and IR spectroscopies provide information about species present during catalysis. High reaction rates, rapid equilibria and the coupling between electronic and steric effects make the investigation difficult.

The quantum chemical methods developed in the last ten years, especially those based on density functional theory (DFT), have been widely accepted by the chemical community because they are accurate and efficient [10]. Ziegler and collaborators have extensively studied cobalt hydroformylation [11]. This subject and other organometallic reactions have recently been reviewed [12]. Rhodium-carbonyl $HRh(CO)_4$ catalyzed hydroformylation has also been the object of theoretical research and is still being investigated [13]. As this book comprehensively shows, computational modeling of homogeneous catalysis has evolved in the last five years from the study of extremely simplified ligand models to the study of real catalyst complexes, and from the determination of a limited number of intermediates to simulations of molecular dynamics. This chapter is an overview of recent theoretical contributions to rhodium diphosphine hydroformylation. It emphasizes the significance of theoretical data in contrast to experimental knowledge. We have divided the review in topics, according to the model systems considered.

2.   EARLIEST PHOSPHINE MODELS: PH_3

It is obviously much too simplistic to use PH_3 as model for phosphine ligands. Neither its basicity [14] nor its spatial extent resembles that of a real-world phosphine such as PPh_3. At various levels of theory, Schmid et al. [15] studied ligand dissociation processes of several five-coordinated $HRh(CO)_n(PH_3)_m$ (n,m=1-3, n+m=4) model complexes that may be involved in the hydroformylation catalytic cycle. Their work reached two significant conclusions. First, accurate geometries and even reliable bond energies can be obtained by DFT methods at a fraction of the computational effort required by CCSD(T) calculations, which are known to reproduce experimental dissociation energies. And second, at least $P(CH_3)_3$, as far as dissociation energies are concerned, must be used as a model for PPh_3. Almost simultaneously, Branchadell and col. [16] analyzed the electronic properties of different phosphines in detail and clearly showed that PH_3 is not a good model for PPh_3. Despite this drawback, the practice in computational studies involving transition metal complexes in the last years has been to model phosphines by PH_3, simply because this reduces the costs of computation; the calculations would be impractical otherwise. However, the ever improving performance of computers, the efficiency of DFT codes, and the development of hybrid QM/MM methods are producing a qualitative leap in the modeling of large chemical systems. Studies of model systems pioneered quantum chemical modeling of transition metal complexes. They still serve as tests for theoretical methods and are the essential starting points for more elaborate models.

2.1   Catalytic cycle for ethylene hydroformylation

The potential energy surface for the hydroformylation of ethylene has been mapped out for several catalytic model systems at various levels of theory. In 1997, Morokuma and co-workers [17], considering $HRh(CO)_2(PH_3)$ as the unsaturated catalytic species that coordinates alkene, reported free energies for the full catalytic cycle at the ab initio MP2//RHF level. Recently, in 2001, Decker and Cundari [18] published CCSD(T)//B3LYP results for the $HRh(CO)(PH_3)_2$ catalytic complex, which would persist under high phosphine concentrations. Potential energy surfaces for both Rh-catalyzed model systems were qualitatively very similar. The catalytic cycle has no large barriers or deep thermodynamic wells to trap the hydroformylation process. The preferred pathway for the catalytic cycle originated from the trans isomer of the active catalyst, *trans*-HRh(CO)(PH$_3$)$_2$, which led the equatorial-equatorial alkene complex (Figure 3). Interconversion between trigonal bipyramidal intermediates HRh(CO)$_2$(PH$_3$)$_2$ can take place through Berry pseudorotation processes. Brown and Kent [19] showed that the PPh$_3$ precursor complex HRh(CO)$_2$(PPh$_3$)$_2$ is a combination of two rapidly equilibrating trigonal bipyramidal isomers, diequatorial (**ee**) and equatorial-axial (**ea**) at a ratio of 85:15. Matsubara et al. [17] showed that this process is quite facile, and predicted a modest barrier of about 40 kJ.mol^{-1} for such interconversion in the alkene complex. Unlike the HRh(CO)$_2$L$_2$ precursor complex, the HRh(CO)(alkene)L$_2$ intermediate has not yet been observed directly.

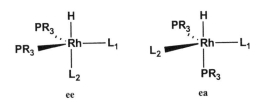

Figure 3. **ee** and **ea** isomers of HRh (PR$_3$)$_2$L$_1$L$_2$ complex

Contrary to experimental evidence, the CO insertion step is predicted as the rate-determining step of the catalytic cycle at all reported levels of theory. The difference between of the computed results and the experiment has been attributed [17] to effects of solvation. Oxidative addition is the only step that involves an unsaturated reactant. The solvent is supposed to stabilize all transition states (TS) in the same extent, but further stabilize the unsaturated complex, which would increase the activation barrier. When a single ethene molecule was used to model the solvent, the activation barrier of H$_2$ oxidative addition increased [17], to almost the same size as the CO insertion barrier. At this point, it seems that theory has not yet managed to distinguish which is the faster step.

It is known that H$_2$ oxidative addition is rate-limiting for the Wilkinson catalyst HRh(CO)$_2$(PPh$_3$)$_2$ and that the rate depends on H$_2$ pressure [2, 20]. Other authors disagree about the rate-determining step for monophosphine ligands, and believe that the increased hydroformylation rate caused by higher H$_2$ pressure is a result of inactive dirhodium species formed under "non-standard", low-pressure conditions. In an early study, Cavalieri d'Oro et al. found that under "standard" conditions, i.e. industrial operating conditions, hydroformylation rate is zero order in H$_2$ concentration [21]. They reported an overall activation energy of 84 kJ.mol^{-1}. In the literature on monodentate ligands, there is much controversy about the rate-determining step, i.e. whether this is insertion (or coodination) of the alkene (type-I kinetics) or the oxidative addition of dihydrogen (type-II kinetics). It has been argued that under "standard" conditions type-I kinetics prevails and that type-II kinetics *is the exception rather than a rule* [22].

The computed olefin insertion barrier in the bisphosphine system $HRh(PH_3)_2(CO)$ [18] is relatively larger than that in the $HRh(PH_3)(CO)_2$ [17] monophosphine catalyst, and is much higher than that in the cobalt system $HCo(CO)_4$. [11]. In the $HRh(PH_3)_2(CO)$ system, the energy barrier for alkene insertion was only 8 kJ.mol^{-1} lower than the barrier of the rate-limiting CO insertion, while in the $HRh(PH_3)(CO)_2$ system the difference was 25 kJ.mol^{-1}. In the cobalt system the barrier is predicted to be very modest (4 kJ.mol^{-1}), while in bisphosphine system it is predicted to be much higher (70 kJ.mol^{-1}). Decker and Cundari [18] attributed the enhanced ethylene insertion barriers to the higher degree of steric crowding in the TS. Theoretical studies by Morokuma [23] also indicated that the coordination of CO or ethene has a low barrier. Moreover, an early TS was found for ethene insertion, which involves a reorganization of the complex. This means that steric hindrance plays a crucial role and that the rotation of the alkene from in-plane coordination to perpendicular coordination determines the barrier.

2.2   Regioselectivity

The theoretical prediction of regioselectivity in catalysis relies on determining relative energies of reactants and competing TS's. A terminal alkene, coordinated in equatorial mode to **ee** HRh(CO)(diphosphine), can adopt four different conformations depending on the face of the double bond that coordinates to the metal and on the orientation of the chain, and assuming H-Rh-CO in only one relative conformation. Figure 4 shows the structure of HRh(CO)(alkene)(diphosphine) schematically. The alkene is in front of the metal center and the diphosphine backbone is at the back. The metal and the carbonyl ligand are omitted for clarity and the labels left/right and up/down, indicating the alkene chain orientation, identify the four conformations. The relative stability of these four conformations is not important for selectivity, since each of them can lead to a pro-linear or pro-branched TS. In fact, if a *left_up* species rotates clockwise (CW), the internal carbon inserts into the metal hydride bond, i.e. the terminal carbon is involved in the formation of the new C-H bond and the product is a branched aldehyde labeled *B_in* in Figure 4. However, if the same *left_up* species rotates counter-clockwise (CCW), the linear aldehyde *L_out* is obtained. Since each conformation generates two products, eight pathways are possible and, therefore, eight different transition states. Figure 4 shows the labels for each TS; *L* for linear and *B* for branched, and *in* and *out* to indicate in which direction the chain points: *in* if it points towards the hydride-Rh-carbonyl axis and *out* if it points towards the phosphine substituents. If the diphosphine moiety can adopt N different conformations, we would obtain 8N TS's.

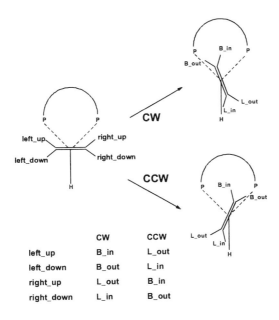

	CW	CCW
left_up	B_in	L_out
left_down	B_out	L_in
right_up	L_out	B_in
right_down	L_in	B_out

Figure 4. Reaction paths for ee HRh(CO)(alkene)(diphosphine)

For an **ea** HRh(CO)(alkene)(diphosphine), in which the hydride is assumed, as in Figure 3, to be in axial position, alkene have two coordination sites available, four conformations for each site, two rotation sides, N ligand conformations, and therefore 16xN TS's. Computation of the full catalytic cycle, all intermediates and TS's, from the entry of the substrate to the departure and regeneration of the catalyst, complemented with IRC calculations to confirm the connection between TS's and intermediates is out of reach for current computational resources. However, suitable modeling strategies can reduce of the problem, and still provide useful insight.

Rocha and de Almeida [24] considered propene to investigate regioselectivity, i.e. the insertion of propene into the Rh-H bond step. Using HRh(CO)(propene)(PH$_3$)$_2$ as the model for the catalytic species, they determined reactants, TS's and products, and the corresponding IRC's from DFT (BP86) calculations, and evaluated energy barriers and reaction energies at the ab initio (MP4//BP86) level. These IRC calculations confirmed for the first time that the TS structures do correspond to the highest energy point on the minimum energy path that connects the reactant $HRh(CO)(propene)(PH_3)_2$ and the product $Rh(CO)(alkyl)(PH_3)_2$. The TS structure is reached by rotating the alkene around the metal-alkene bond by about 70°. This study focused on the **ee** alkene coordination isomer depicted in Figure 5 and considered two possibilities, i.e. two paths that lead the reaction either to linear aldehyde or to branched aldehyde.

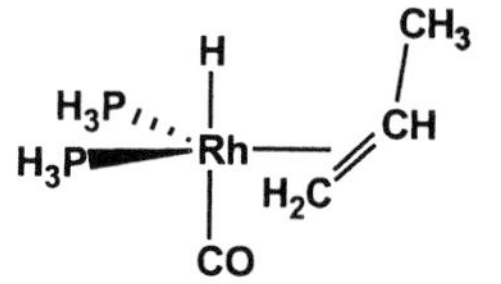

Figure 5. **ee** penta-coordinated propene complex

There was a thermodynamic preference for the reaction to take place at the terminal alkene carbon, which favors the yield of linear aldehyde, but the TS to linear aldehyde path was higher than the TS for the branched aldehyde path. Regioselectivity was evaluated from the products relative stability, i.e. considering that the reaction is under thermodynamic rather than under kinetic control. The linear to branched ratio (l:b) of 94:6 was in excellent agreement with the ratio 95:5 reported for PPh_3 [25]. However, this nice coincidence must be viewed cautiously because the model is simple, reaction paths were partially considered, so a subtle cancellation of errors may have been made.

2.3   Dependence of ee: ea equilibrium on phosphine basicity

Little attention has been paid to the electronic properties of phosphine ligands and how they affect catalytic performance. In one of the few detailed studies, Moser and co-workers investigated the effect of phosphine basicity on hydroformylation using a series of p-substituted triphenylphosphine ligands [26]. Ferrocene based diphosphines of different basicity were studied by Unruh and Christenson [27]. Both studies showed that fewer basic phosphines afford higher reaction rates and higher l:b ratios. In a recent study, Casey and co-workers investigated the electronic effect of equatorial and apical phosphines [28]. Introducing electron-withdrawing groups to the aryl rings of the diequatorial (**ee**) chelate BISBI-(3,5-CF_3) increased the l:b ratio, but decreased the l:b ratio in the equatorial-apical (**ea**) chelate DIPHOS-(3,5-CF_3) compared to the unsubstituted ligands (Figure 6). Based on these results, Casey and co-workers concluded that electron-withdrawing substituents in phosphines in the equatorial position produce high l:b ratios, while in phosphines in the apical position they produce low l:b ratios. The results of this study cannot be described only in terms of electronic effects, since there are major steric differences between the **ee-** and **ea**-coordinating diphosphines.

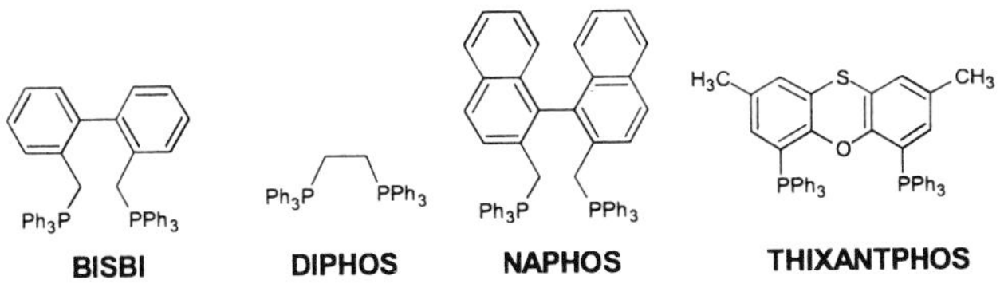

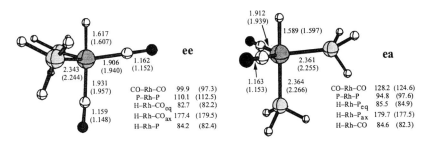

Figure 6. Diphosphine ligands

To study the exact nature of the electronic effect in the rhodium-diphosphine catalyzed hydroformylation, a series of thixantphos ligands (Figure 6) with several electron withdrawing or electron donating substituents on the aryl rings were synthesized. In this series of ligands steric differences are minimal, so purely electronic effects could be investigated. It was found [29] that the equilibrium between **ee** and **ea** coordination in the HRh(CO)$_2$(diphospine) complexes depends on phoshine basicity, and that decreasing phosphine basicity shifts the equilibrium to **ee** coordination.

The enhanced preference of the basic phosphines for apical coordination and the preference of the non-basic phosphines for equatorial coordination was quantitatively demonstrated by our DFT BP86 calculations [29] on the **ee** and **ea** isomers of model complexes HRh(CO)$_2$(PH$_3$)$_2$ and HRh(CO)$_2$(PF$_3$)$_2$.

Figure 7. Selected bond distances (Å) and angles (degrees) for the ee and ea isomers of HRh(CO)$_2$(PH$_3$)$_2$ and HRh(CO)$_2$(PF$_3$)$_2$ (values in parenthesis) at BP86:TZP level.

The difference in energy between the **ee** and **ea** when PH$_3$ was used as the model ligand for a basic phosphine (Table 1) was only 0.5 kJ.mol^{-1} in favour of the **ea** isomer. This small difference corresponds to an **ee:ea** equilibrium composition of approximately 46:54. A preference of the **ea** isomer over the **ee** isomer for the PH_3 ligand was also reported by Schmid et al. [15] and Matsubara et al. [17] When PF_3 was used as the model ligand for a non-basic phosphine, the **ee** isomer was favored. This leads to an **ee:ea** equilibrium ratio of 98:2. The shift in equilibrium that occurred when the basic PH_3 ligand was changed for the less basic PF_3 ligand is consistent with the observed shift in isomer composition in the series of substituted thixantphos ligands. Moreover, the stability of mixed phosphine model systems (Table 1) agrees well with data of Casey and col. [28]

Table 1. Relative energies in kJ.mol^{-1} for some diphosphine model complexes computed at different levels of theory.

Model ligand		ee	Ea
PH_3 , PH_3	BP96//BP96 ref. 29	0.1	0
	BP96//LDA ref. 15	1.7	0
	MP2//MP2 ref. 17	10.5	0
PF_3 , PF_3	BP96//BP96 ref. 29	0	10
PH_3 , PF_3	BP96//BP96	0	1.7 (eq. PF_3)
			12 (eq. PH_3)

2.4   HP-IR spectra of HRh(CO)₂(thixantphos)

High-pressure IR-spectroscopy clearly showed a dynamic equilibrium between **ee** and **ea** isomers. All spectra recorded for the series of HRh(CO)$_2$(thixantphos) complexes, [29] however, showed four absorption bands in the carbonyl region instead of the expected six bands (Figure 8). In order to assign the bands to Rh-H and CO vibrations, the DRh(CO)$_2$(thixantphos) complex was measured for comparison. The spectrum of the deuterated complex also showed four absorption bands in the carbonyl region, two of which shifted to lower frequencies. H/D exchange affects the **ee** isomer because the hydride and a carbonyl ligand are *trans*, which leads to coupling of the vibrations. The disappearance of resonance interaction upon H/D exchange produces frequency shifts of the carbonyl bands of the **ee** isomer. As only two bands (v_1 and v_3) shift upon H/D exchange, these two bands were assigned to the carbonyls of the **ee** complex. The two remaining bands that do not shift (v_2 and v_4) therefore belong to the **ea** complex. The disappearance of a low frequency shoulder upon H/D exchange could mean that one of the rhodium hydride vibrations was partly hidden under v_4.

The empirical peak assignments were fully confirmed by DFT calculations of the IR spectra of $XRh(CO)_2(PH_3)_2$ (X=H and D) complexes. [29] Isotopic effect is clearly seen in Figure 8, which shows the measured spectra and a schematic representation of the frequencies and intensities computed by these model systems.

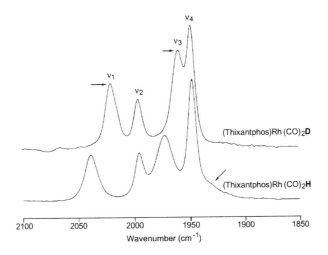

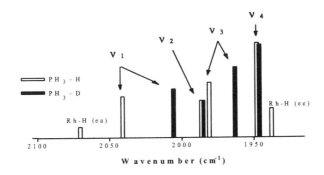

Figure 8. Mesured and computed IR spectra for $XRh(CO)_2(PH_3)_2$ (X=H and D).

v_2 is the symmetric carbonyl vibrational mode of the **ea** isomer (1985 cm^{-1}) and v_4 is the corresponding antisymmetric mode (1948 cm^{-1}). These two bands do not shift upon H/D exchange, and v_4 is by far the most intense. The two other bands (v_1 and v_3) derive from the **ee** isomer, computed at 2041 and 1983 cm^{-1}, respectively, and are combinations of both CO and Rh-H vibrations. They both shift to lower frequencies when hydrogen is substituted by deuterium and increase their intensity, which agrees perfectly with the experimental spectra. The agreement between the absolute values of the computed frequencies and those for the thixantphos complexes is also remarkable [30]. Significantly, the computed frequency corresponding to the rhodium hydride vibration in the **ee** complex can explain the low frequency shoulder of v_4. Moreover, the shift to higher frequencies in thixantphos complexes when phosphine basicity increased [29] was also reproduced when PF_3 was used as model.

3.   MODELING REAL-WORLD PHOSPHINES

The significance of theoretical studies that use model systems to understand catalytic processes is therefore not as limited as one might expect. However, more elaborate computational models that mimic the structure of the catalyst as far as possible are desirable to thoroughly understand how ligand affects reactivity. Nowadays, accurate theoretical treatments have only one limitation: there is not enough computational power to treat moderately sized molecular systems, such rhodium-ligand complexes, in all their isomeric forms, and with all the intermediates and TS's involved in a catalytic cycle. However, some strategies that make good use of approximations (especially those based on hybrid methods) are beginning to fil the gap between real-world catalysts and computer models. Pure MM methods have been used [31] to investigate ligand-substrate interactions in intermediates in the hydroformylation process and how they influence regioselectivity, but definitive conclusions could not be drawn. Gleich et al. [32] made a first contribution to the theoretical description of systems tested experimentally using a combined QM/MM approach. Their method consisted of a full QM treatment of a model system, i.e. $HRh(CO)(alkene)(PH_3)_2$, subsequently substituting the H atoms of PH_3, and adding the rest of the ligand (which are described by MM), while keeping the atoms already described in the QM region frozen. This approach has some limitations because the reactive centers (in the QM region) do not relax under the influence of the ligands. So, although the ligands relax upon complexation, the steric strain induced by the ligands on reactive centers cannot be assessed, and properties such as the relative energies of transition states, which depend on diphosphine structure, cannot be properly evaluated. Decker and Cundari [33] and our group [34] recently reported studies based on hybrid QM/MM approaches that enable QM and MM regions to be simultaneously optimized. This kind of calculations ensures that the steric effects of the bulky diphosphine ligands are transmitted to the reactive center that is being described in the QM region.

3.1   Monophosphines PR$_3$

As we mentioned in section 2, monophosphines have often been modeled by PH$_3$. Decker and Cundari [33] advanced the treatment of real-world phosphines by means of ONIOM two-layer B3LYP:UFF calculations. They studied several PR$_3$ phosphines (R=Me, ^{t}Bu, Ph, m-PhSO$_3^-$ and p-PhSO$_3^-$), including the phosphine substituents in the MM layer and HRh(CO)(alkene)(PH$_3$)$_2$ in the QM layer. They studied the effects of ligands in the insertion of ethylene into the Rh-H bond of HRh(CO)(C$_2$H$_4$)(PR$_3$)$_2$, [33a] and considered three reaction paths, one from the **ee** isomer and other two from the **ea** form (Figure 3), and characterized reactants, TS's and products. The calculations predicted that two reaction paths were operative for PMe$_3$, and that the path from **ee** isomer was thermodynamically favored over the path from one of the **ea**'s. However, with bulkier phosphine substituents there was a clear energetic preference for the insertion to take place, both kinetically and thermodynamically, from the least stable **ee** intermediate rather than from the more stable **ea** isomer.

Decker and Cundari recently studied regioselectivity by considering propene as a model for an alkene and PPh$_3$ as phosphine [33b]. From the complex HRh(CO)(C$_3$H$_6$)(PPh$_3$)$_2$, they studied two **ee** and four **ea** isomers. We recall the reader that for each alkene isomer, two reaction channels are possible to reach the TS through alkene rotation, CW and CCW in Figure 4. TS's for linear aldehydes were located for all paths, whereas TS's for those paths leading to branched aldehydes from **ea** isomers appeared highly destabilised or were even not found. This could mean that **ea** isomers would produce almost exclusively linear aldehydes. **ea** isomers form more stable alkene complexes than those **ee**, thus are present in more concentration ca. 75:25, but the latter present lower energy barriers for the insertion. Decker and Cundari [33b] evaluated that **ee** are fivefold faster than **ea** paths.

3.2   Diphosphines

Diphosphine-based ligands form the basis of current research in hydroformylation. As Figure 9 shows, free energy profiles have been recently proposed [22] to discuss the kinetics and the reaction mechanism. We first see the rapid equilibrium between **ee** and **ea** HRh(CO)$_2$(diphosphine) isomers, followed by CO dissociation. For this uphill process, a very low barrier is proposed. As we know, Morokuma and coworkers, using model systems, computed a barrier for **ee:ea** conversion of 40 kJ.mol^{-1} [17], and a very low barrier for CO dissociation [23]. Alkene association is followed by hydride migration (alkene insertion) and CO coordination to obtain either a linear or a branched (alkyl)Rh(CO)₂(diphosphine) product. Phosphine properties and experimental conditions have a huge impact on the rate-determining step, i.e., the highest point in the energy profile, since it can either be CO dissociation, alkene coordination or alkene insertion. Electron-withdrawing phosphines produce high reaction rates [35, 36] and enhance CO dissociation [37]. In some cases, they also lead to high l:b ratios [38]. Alkene insertion is irreversible for most phosphines, [37] but this conclusion may not be so general. [38]

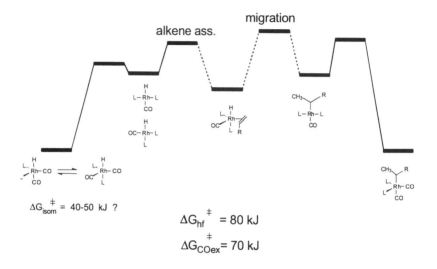

Figure 9. Proposed free energy profile for rhodium diphosphine (ref 22)

The great diversity of the new ligands makes it difficult to identify which effect plays the main role and in which step. It is still not clear whether the rate-determining step and the selectivity-determining step coincide, or whether the selectivity is determined by the HRh(CO)(alkene)(diphosphine) intermediate, species never observed experimentally. High-level quantum mechanical calculations on the whole molecular system are needed to be able to properly describe metal-phosphorous bond properties and its effect on the energy barriers, but this is not possible with the computational resources currently available.

Wide bite angle diphosphines have had an enormous impact in the last few years. These kind of ligands promote the stability and reactivity of a variety of transition metal complexes [39]. Especially important is a series of xantphos-type ligands with natural bite angles ranging from 102 to 121 degrees [40]. Xantphos ligands in Figure 10 have been especially designed to ensure that mutual variation in electronic properties and steric size within the series is minimal and that the bite angle is the only characteristic with a significant variation within the series. In the hydroformylation of 1-octene, selectivity for linear aldehyde formation and activity increased as the natural bite angle increased. For styrene, selectivity for the linear aldehyde followed the same trend, although for that substrate selectivity was found to be also dependent on temperature and CO pressure. These findings suggest that the bite angle affects selectivity in the alkene-coordination and hydride-migration (or alkene-insertion) steps. A plausible explanation for the bite angle effect is that when the natural bite angle increases, the congestion around the metal atom also increases, which in turn favors the less steric demanding transition state (TS), i.e. the TS that drives the reaction towards the linear product.

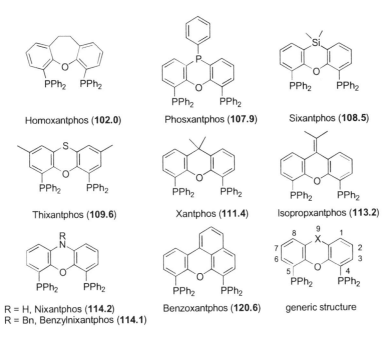

Figure 10. Xantphos diphosphine family. Natural bite angles indicated.

3.2.1 Bite angle effect on regioselectivty

We applied the QM/MM IMOMM method [41] to Rh-diphosphine catalyzed hydroformylation, to provide a quantitative theoretical characterization of the origin of regioselectivity in Rh-diphosphine systems. We focused on the experimentally characterized xantphos ligands, for which variation in electronic properties is minimal. Using the IMOMM method, which only accounts for the steric properties of ligands, was fully justified.

We therefore evaluated how the bite angle affected regioselectivity, and studied the counterbalance of *non-bonding* and *orbital effects*. We choose two diphosphine ligands (benzoxantphos and homoxantphos) which among the series of xantphos ligands represent the extreme cases of natural bite angle, and used propene as a model for terminal aliphatic alkenes and styrene.

Several assumptions were made to reduce the otherwise untreatable number of isomers possibly involved. The alkene complex is obtained from $HRh(CO)_2$(diphosphine) by dissociating one carbonyl molecule. X-ray structures [40] of some $HRh(CO)$(xantphos)(PPh_3) complexes showed an **ee** xantphos coordination, containing PPh_3 in the equatorial position, and H and CO in the apical sites. Carbonyl dissociation rates for (formyl)$Rh(CO)_2$(phosphite) complexes clearly indicate that the equatorial CO dissociates some orders of magnitude faster than the apical CO [37]. These results suggest that the **ee** isomer may be more active than the **ea** one and that alkene may directly coordinate to the tetracoordinated **ee**-resulting complex. The most stable cis should be obtained by twisting the H-Rh-CO moiety, which would involve an energy barrier. The flexibility of the xantphos-type ligands means that they can adopt bite angles up to 164° [42], which are necessary for a quasi-trans square planar structure. Therefore, **ee** equatorial CO dissociation may be quickly followed by alkene coordination/insertion.

All of these arguments indicate that the key intermediate may be the **ee** $HRh(CO)$(alkene)(xantphos) complex, for which two isomers can, be distinguished depending on the relative position of the hydride and carbonyl ligands along the trigonal axis. X-ray structures [40] of some $HRh(CO)$(xantphos)(PPh_3) show the carbonyl ligand to be at the same side as the xantphos ligand backbone. We carried out IMOMM calculations on two isomers (**1a** and **1b**) of the pentacoordinate ethene complex $HRh(CO)(C_2H_4)$(benzoxantphos) (Figure 11) to check whether our computational method was able to reproduce the experimentally reported relative disposition of the hydride and carbonyl axial ligands.

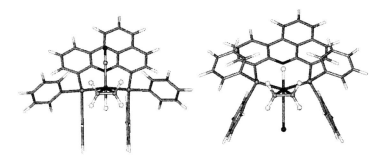

Figure 11. **1a** (left)and **1b** (right) isomers of ee HRh(CO)(ethylene)(benzoxantphos)

In isomer **1a** the carbonyl axial ligand is on the same side of the equatorial plane as the benzoxantphos backbone, while the trans hydride ligand lies between two phenyl moieties on the opposite side. In **1b**, the relative positions of hydride and carbonyl ligands are exchanged. Calculations show that isomer **1a** is more stable than **1b** by 7.5 kJ.mol^{-1}. The accuracy of the computational method to reproduce such small energy differences is obviously questionable, but in any case it is reassuring to be able to reproduce the experimental result. Also, the IMOMM method enables us to analyze the origin of the difference between the two isomers. Decomposition of the energy in QM and MM contributions showed that the differences between **1a** and **1b** were exclusively in the MM part, which clearly indicated a steric origin for the relative stability of the two isomers. The geometry of the isomer that was predicted to be the most stable (**1a**) is completely consistent with the arrangement of axial ligands in X-ray data for the related complex HRh(CO)(PPh$_3$) (benzoxantphos) [40]. Figure 11 shows how the bulkier CO ligand of isomer **1b** lies between two phenyl moieties, which breaks the face-to-face stacking interaction between them and destabilizes the complex. This result illustrates how the IMOMM method can transfer non-bonding interactions between QM and MM parts, since in this case, there is an interaction between the CO (QM part) and the two phenyls (MM part). Stacking interaction in other chemical systems is well characterized, so its is not surprising to find it here [43]. Also, its importance has been evaluated and characterized by the IMOMM method in previous studies [44, 45].

Therefore, as we discussed in 2.2, we took into account that regioselectivity is not determined by the structure of the alkene intermediate but by the relative stabilities of all TS's. Although we admit that it is not the optimal solution, we believe that using the TS energies to calculate selectivities is a substantial improvement over using the energy of intermediates. When using the energy of a particular TS to calculate selectivity, we assume that it indeed corresponds to the step in which this selectivity is decided. Also, the reaction centers of our TS's closely resembles those from previous calculations of propene insertion, for which an IRC calculation was recently performed [24]. If we assume a Boltzmann distribution, we can easily obtain the percentage of linear and branched products from the relative energies of all possible TS's.

Before presenting the results for propene and styrene, we shall first discuss the results for the simplest alkene, ethylene. This simplified system was used to characterize the two transition states for CW and CCW rotations and was the starting point for studying substituted alkenes. For benzoxantphos, the energy barrier reached 58 kJ.mol^{-1}, and the CW and CCW rotations were degenerate. In both TS structures and unlike alkene complexes, the edge-to-face stacking interaction was energetically favored, with a pseudo symmetry plane relating the two TS's. For the homoxantphos conformation considered in this paper, CW and CCW rotations were not degenerate, with CCW as the most stable one by 7 kJ.mol^{-1}. The homoxantphos backbone is based on a seven-membered ring instead of a six-membered one. It can adopt two degenerate conformations that are related by a pseudo-C_2-symmetry operation. Note that diphosphine local symmetry is reflected in the relative stability of the TS. A symmetry plane makes CW and CCW rotations degenerate, whereas both rotation sides are energetically different in the case of a C_2 axis. These two cases correspond to benzoxantphos and homoxantphos ligands, respectively.

Ethylene insertion energy barriers for benzoxantphos and homoxantphos ligands are very similar, and there were no dramatic changes in the geometrical parameters of reaction centers. However, the bite angles were wider in the benzoxantphos than in homoxantphos system for both the alkene complex and the TS structure, which agrees with the previously calculated natural bite angles in Figure 10. The calculated energy barriers for the xantphos systems were of the same order of magnitude as the one previously reported for the insertion process in the related HRh(CO)$_2$(PH$_3$) system [17] where the computed barrier for the most favorable path was about 90 kJ.mol^{-1}. Also, the geometry of the xantphos TS structures showed that in no case was the dihedral angle H-Rh-C-C zero, which as we discussed above supports the hypothesis that for a substituted alkene four TS's are possible for each CW and CCW rotation. For benzoxantphos, since CW and CCW paths are degenerate, only one path must be studied, i.e. four TS's must be characterized. For homoxantphos, eight TS's play a role but as CW is energetically less favored, the four TS's arising from the CCW path are expected to be the lowest. However, we characterized the eight TS's to evaluate how they contribute to regioselectivity.

Table 2. Propene insertion for the benzoxantphos system, energies and geometrical parameters of transition states. Energies in kJ.mol^{-1}, distances in Å and angles in degrees. (a) Dihedral angle $H_{hydride}$-Rh-C_{alkene}-C_{alkene} for defining alkene rotation.

	CW			
	L_out	*L_in*	*B_out*	*B_in*
Total Energy	0.0	1.5	3.7	8.9
MM Energy	0.0	-0.2	2.4	0.5
QM Energy	0.0	1.8	1.3	8.4
P-Rh-P	111.0	112.0	111.7	112.3
H-Rh-C-C^a	18.7	13.4	18.3	12.8
Rh-H$_{hydride}$	1.648	1.639	1.649	1.639
C=C	1.415	1.408	1.413	1.410
Rh-C$_{alkene}$	2.218	2.241	2.264	2.268
C$_{alkene}$-H$_{hydride}$	1.662	1.688	1.619	1.667

Tables 2 and 3 show the relative total energy for all characterized TS's for the two ligands and some geometric parameters. The energies are relative to the energy of the most stable TS for each ligand. The tables also show the decomposition of the total energy into quantum mechanics (QM) and molecular mechanics (MM).

Table 3. Propene insertion for the homoxantphos system, energies and geometrical parameters of transition states. Energies in kJ.mol^{-1}, distances in Å and angles in degrees. (a) The *CW* B_*in* isomer could not be localized (b) Dihedral angle $H_{hydride}$-Rh-C_{alkene}-C_{alkene} for defining alkene rotation.

	CCW				CW			
	L_out	*L_in*	*B_out*	*B_in*	*L_out*	*L_in*	*B_out*	*B_in*^{a}
Total Energy	0.0	6.4	2.4	7.8	5.4	6.7	14.7	
MM Energy	0.0	2.1	1.1	-1.4	5.1	5.5	11.8	
QM Energy	0.0	4.2	1.3	9.2	0.3	1.2	2.9	
P-Rh-P	99.5	99.8	100.6	99.5	100.8	104.3	103.4	
H-Rh-C-C^b	-21.6	-22.3	-21.5	-25.4	18.3	12.1	17.5	
Rh-H$_{hydride}$	1.643	1.638	1.643	1.633	1.645	1.638	1.648	
C=C	1.420	1.418	1.417	1.425	1.413	1.407	1.412	
Rh-C$_{alkene}$	2.198	2.207	2.240	2.213	2.220	2.250	2.267	
C$_{alkene}$-H$_{hydride}$	1.673	1.673	1.633	1.646	1.681	1.672	1.611	

The pattern in the relative energy distribution of transition states was the same for both ligands. The lowest energy transition state was the pro-linear *out* (*L_out*) isomer, and the highest energy TS was the pro-branched *in* (*B_in*). The relative energies of *B_in* isomer were 8.9 and 7.8 kJ.mol^{-1} for the benzoxantphos and homoxantphos systems, respectively. Note that the energies of transition states for the CW rotation of the homoxantphos system also followed the same trend, but their energies shifted to higher values due to the increase in MM energy contribution. This increase is associated with the non-bonding repulsive interactions between the alkene substituent and the phenylphosphino substituents, which were well reproduced by the MM method. The differences in energy between the *L_out* and *B_in* isomers were mostly in the QM part. These differences were 8.4 and 9.2 kJ.mol^{-1} for benzo- and homoxantphos, respectively. The differences in the MM part were much smaller i.e. 0.5 and −1.4 kJ.mol^{-1} for benzoxatphos and homoxantphos, respectively. Therefore, in the context of the IMOMM energy partition scheme and taking into account which atoms are included in each part (QM or MM), a plausible explanation for the energetic differences may come from the reacting centers that are treated quantum mechanically. When we analyzed the TS structures, we saw that in the highest energy isomer (*B_in*), the methyl alkene substituent was parallel to the metal-carbonyl bond. In other words, the methyl and carbonyl ligands were eclipsed, a situation that may destabilize the transition state (see Figure 12).

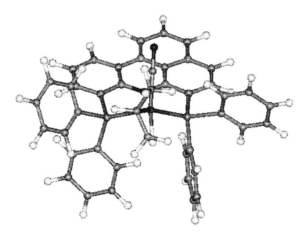

Figure 12. Benzoxantphos CW B_in transition state for propene insertion

There was no clear trend for the geometries of the reaction centers in the series of characterized TS. We also tried to find some correlation between the relative energies and atomic charges from different partition schemes

(Mulliken, NPA), but found no feature of significant magnitude. Since the extreme energetic cases (*L_out* and *B_in*) are similar in both ligand systems, the difference in regioselectivity is governed by the intermediate cases (L_in and B_out), where we found a mixture of MM (steric) and QM (electronic) effects. Table 3 clearly shows that the total energies of CW transition states of homoxantphos ligand are mainly due to an increase in MM energy contributions. This indicates that the energetic differences between the two rotation sides are of steric origin, which could be explained from the structural features of the homoxantphos ligand. The backbone induces a different conformation of the two diphenylphosphine moieties, bringing one of them closer to the metal center and increasing the destabilizing non-bonding interactions with the substrate. In this case, the backbone brings the metal center closer to the diphenylphosphine moiety on the right hand side (front view). This corresponds to CW rotation, which is the energetically disfavored one. Also worth mentioning are the differences in the bite angles for the transition states of the two rotation sides of the homoxantphos system. For CCW rotation, the bite angles ranged from 99.5° to 100.6°, while for the more energetic CW rotation, they ranged from 100.8° to 104.3°. These results are in line with the previous reasoning, that an increase in steric congestion around the metal center is directly related to larger bite angles. Tables 2 and 3 also show that *B_out* isomers in both ligand systems have higher MM energies than their corresponding *B_in* isomers. If we remember the labels used to classify transition states, the *out* notation designated isomers whose alkene substituent pointed towards the phenyl moieties and away from the CO-Rh-H axis. Therefore, in *out* transition states there is a destabilizing interaction between the aliphatic substituent and the phenyl moieties, that is reflected in an increase in MM energy.

Table 4. Calculated vs. experimental regioselectivities

	Exp. (1-octene)	Calculated (propene)	
		Diphenylphosphine model	PH_2 Model
Ligand	% l over b (l:b ratio)	% l over b (l:b ratio)	% l over b (l:b ratio)
Benzox.	98.1 (50.2)	83 (4.9)	74 (2.7)
Homox.	89.5 (8.5)	73 (2.8)	63 (1.7)

For propene, the calculated percentages of linear product over branched product were 83% for benzoxantphos and 73% for homoxantphos system. This agrees with the experimental reported percentages for 1-octene, which were 98.1% and 89.5% for the benzoxantphos and homoxantphos systems, respectively. Despite the small differences in regioselectivities of the two xantphos systems, we succeeded in reproducing the trend. Also, the alkene used in our calculations was propene rather than 1-octene, which may explain why the predicted selectivity for the linear product was lower. The geometries of the reaction centers in the transition states of both systems are similar, but different ranges in bite angle were observed for the two systems (from 111.0° to 112.3° for benzoxantphos and from 99.5° to 104.3° for homoxantphos). Therefore, the bite angle of benzoxantphos, the more selective diphosphine, was larger than that of homoxantphos, the less selective diphosphine. We also proved that the correlation between the bite angle and regioselectivity takes place at the point where the regioselectivity is decided, i.e. the transition state for alkene insertion, and we can exclude the alkene coordination step as the key step for determining regioselectivity. These results show that the IMOMM method is remarkably successful and that it can be used in homogenous catalysis to design ligands.

We used the same method to investigate styrene hydroformylation as we did to investigate propene hydroformylation. We included benzyl alkene substituent in the MM part and all TS states were redetermined, four for benzoxantphos and eight for homoxantphos. The relative energies of *B_in* isomers were substantially lower than those for propene, and were almost the most stable isomers. For styrene, the benzyl substituent eclipsed with the carbonyl group did not seem to destabilize the transition states. For styrene, it was not possible to find any *B_out* isomer and their estimated energies were high. Their structures showed that the phenyl substituent of the alkene pointed towards the diphosphine phenyl substituents of the alkene side (front side), which destabilized the TS. The percentages for styrene were in good agreement with experiment percentages. The predicted percentage of linear product over branched product for homoxantphos was 53%, while the reported experimental percentage is around 40% at 120° C. For benzoxantphos, the predicted percentage of linear product over branched product was 68%, while the experimental percentage was around 65%.

3.2.2   Separation of steric effects

To analyze the specific role of the bite angle and the phenyl substituents of the diphosphine in detail, we determined the regioselectivity that could be afforded by a model system if we replaced each phenyl substituent by hydrogen and maintained the backbone of diphosphine ligands (PH$_2$ model). By removing the phenyl substituents, we can put aside the *non-bonding effects* of the phenyls on regioselectivity. So, by comparing this result with the regioselectivity calculated for the diphenylphosphine catalyst, we evaluated the importance of the interactions between the phenyl substituents and the substrate. Notice that within the IMOMM partition scheme, the phenyls groups are included in the MM part. In this case, we can separate, identify and evaluate the contributions of the xantphos ligands to regioselectivity. Maintaining benzoxantphos and homoxantphos backbones ensures that the transition states of the PH$_2$ model system have different bite angle values for both types of ligands. So, once the phenyls have been removed and by comparing both types of ligands, we can evaluate the effect on regioselectivity that is directly associated to the bite angle (*orbital effect*). Some preference for linear or branched products is expected due to the intrinsic regioselectivity of the substrate.

As in the diphenylphosphine system, there was a different range of bite angle values for the two types of ligands in the model system. The bite angles for the transition states with benzoxantphos ranged from 118.8 to 120.2 degrees and with homoxantphos from 101.2 to 104.9 degrees. Remarkably, bite angles were larger in the PH$_2$ model system than in the diphenylphosphine system for both ligands. This was mainly due to the loss of non-bonding interactions, i.e. the stacking between the two phenyls at the back of the xantphos backbone, and the interactions between the ligand and the substrate.

Table 4 compares the experimentally observed and the predicted regioselectivities for the PH$_2$ and diphenylphosphine model systems. Predicted regioselectivities in the PH$_2$ model system for the linear product over branched product were 74% for benzoxantphos and 63% for homoxantphos. For both ligands, regioselectivities were substantially lower in the PH$_2$ system than in the diphenylphosphine system (83% compared to 73% and 74% compared to 63%, respectively). For the PH$_2$ model system, there was still a difference between benzoxantphos and homoxantphos. However, this difference became proportionally less important. If we translated percentages into l:b ratio, we can see that the difference in the l:b ratio for the diphenylphosphine system is 2.2 (4.9 and 2.7 for benzoxantphos and homoxantphos, respectively) while for the PH$_2$ model system, the difference is only 1.1 (2.8 and 1.7 for benzoxantphos and homoxantphos, respectively).

Most of the differences in energy between the TS's were in the QM part, while in the MM part it was almost the same. This clearly proved that when phenyls are removed, the regioselectivity is not governed by the non-bonding interactions between the diphosphine and the substrate. It therefore seems that the leading role in determining the regioselectivity is played by the phenyl diphosphine substituents. The bite angle seems to have little effect on regioselectivity (the *orbital effect* is low), since the difference in selectivities for both ligands in the PH$_2$ model system is proportionally smaller than in the diphenylphosphine system. This conclusion cannot be extrapolated to other reactions, since for palladium-diphosphine RCN reductive elimination there was a clear orbital effect on the reaction rate [46]. The correlation found experimentally between the bite angle and the regioselectivities had suggested that the bite angle plays an important role in determining regioselectivity. However, this effect has been transferred from the bite angle induced by the ligand backbone to the non-bonding interactions between the phenyls and substrate. Wider bite angles increase the steric interaction of diphosphine substituents with branched species, which become more destabilized. We therefore expect that bulkier groups than phenyl groups would lead to higher l:b ratios.

4. CONCLUDING REMARKS AND FUTURE PERSPECTIVES

The above applications show that computational chemistry has provided the answers to a number of questions. Much work still needs to be done, however. Despite the severe approximations involved in using model systems, a first step has now been taken. From the structure of intermediates and TS's determined for model systems, we have described the main features of the catalytic cycle and laid the ground for the development of more elaborate models. Topics such as **ee:ea** equilibrium and the infrared spectra of $HRh(CO)_2$(diphosphine) have been satisfactorily interpreted.

In 2001, hybrid QM/MM strategies such as those in Section 3, provided a step forward in the modeling of real-world catalysts. However, the scope of these methods is also limited, because they only enable us to properly treat steric effects. Despite this drawback, some important aspects can be studied using these methods. Regioselectivity in diphosphine systems is explained by non-bonding interactions and bite angle effects. However, the role of **ea** complexes has not yet been determined.

Kinetic studies and novel electron-withdrawing phosphines challenge theoretical methods. Difference of one order of magnitude in a rate constant implies only a few kcal.mol^{-1} difference in the activation energy, which is as accurate as current theoretical methods. On the other hand, proper treatment of phosphine basicity requires that we take into account large molecular systems and, therefore, powerful computational resources. Given the rapid progress of the last two years and the challenge the new experimental data presents, we expect that new contributions will soon provide greater insight into the puzzling subtleties of rhodium hydroformylation.

ACKNOWLEDGEMENTS

We wish to thank our supervisors, Prof. Josep M. Poblet (URV) and Prof. Agustí Lledós (UAB) for their continuous support over the years. We are indebted to Prof. Piet van Leeuwen for introducing us to this field, for his very helpful discussions and for providing new results prior to publication. We expect such fertile collaboration between experimental and theoretical groups will promote new questions that will no doubt help to better understand the hydroformylation process. We are also very grateful to Prof. Cundari for communicating the results of their research prior to publication. We gratefully acknowledge financial support from the Spanish DGICYT under projects PB98-0916-C02-01 and PB98-0916-C02-02, and from the CIRIT of the Generalitat de Catalunya under Projects SGR-1999-00089 and SGR-1999-00182. FM thanks DURSI. Finally, we also thank José C. Ortiz (URV) for his technical expertise.

REFERENCES

1   See for instance: van Leeuwen, P. W. N. M.; van Koten, G. *In Catalysis: an integrated approach to homogeneous, heterogeneous and industrial catalysis*; Moulijn, J. A.; van Leeuwen, P. W. N. M.; van Santen, R. A., Ed.; Elsevier: Amsterdam, **1993**, pp 201-202. Beller, M.; Cornils, B.; Frohning, C. D.; Kohlpaintner, C. W. *J. Mol. Cat. A: Chem.* **1995**, *104*,17-85. Frohning, C. D.; Kohlpaintner, C. W. *In Applied Homogeneous Catalysis with Organometallic Compounds: a comprehensive handbook in two volumes*; Cornils, B.; Herrman, W. A., Ed.; VCH: Weinheim, **1996**; Vol. 1, pp 27-104.

2   Evans, D.; Osborn, J. A.; Wilkinson, G. *J. Chem. Soc. (A)* **1968**, 3133.

3   Heck, R. F.; Breslow, D. S. *J. Am. Chem. Soc.* **1961**, *83*, 4023-4027.

4   For a comprehensive review of recent rellevant advances in hydroformylation see *Rhodium Catalyzed Hydroformylation*; van Leeuwen, P.W.N.M; Claver, C., Eds.; Kluwer Academic Publishers: Dordrecht, The Netherlands, **2000** and refereces therein.

5   Casey, C. P.; Paulsen, E. L.; Beuttenmueller, E. W.; Proft, B. R.; Petrovich, L. M.; Matter, B. A.; Powell, D. R. *J. Am. Chem. Soc.* **1997**, *119*, 11817.

6   Casey, C. P.; Whiteker, G. T. *Isr. J. Chem.* **1990**, *30*, 299.

7   Kranenburg, M.; van der Burgt, Y. E. M.; Kamer, P. C. J.; van Leeuwen, P. W. N. M. *Organometallics* **1995**, *14*, 3081.

8   (a) Casey, C. P.; Whiteker, G. T.; Melville, M. G.; Petrovich, L. M.; Gavney, J. A., Jr.; Powell, D. R. *J. Am. Chem. Soc.* **1992**, *114*, 5535. (b) Kranenburg, M.; Kamer, P. C. J.; van Leeuwen, P. W. N. M. *Eur. J. Inorg. Chem.* **1998**, 25.

9   van der Veen, L. A.; Keeven, P. H.; Schoemaker, G. C.; Reek, J. N. H.; Kamer, P. C. J.; van Leeuwen, P. W. N. M.; Lutz, M.; Spek, A. L. *Organometallics*, **2000**, *19*, 872.

10 See for instance issue 100 of Chem. Rev. 2000.

11 (a) Versluis, L.; Ziegler, T.; Baerends, E. J.; Ravenek, W. *J. Am. Chem. Soc.* **1989**, *111*, 2018. (b) Versluis, L.; Ziegler, T. *Organometallics* **1990**, *9*, 2985. (c) Versluis, L.; Ziegler, T.; Fan, L. *Inorg. Chem.* **1990**, *29*, 4530. (d) Ziegler, T. *Pure Appl. Chem.* **1991**,

63, 873. (e) Ziegler, T. Cavallo, L.; Berces, A. *Organometallics* **1993**, *12*, 3586. (f) Sola, M.; Ziegler, T. *Organometallics* **1996**, *15*, 2611.

12   (a) Koga, N.; Morokuma, K. *Chem. Rev.* **1991**, *91*, 823. (b) Musaev, D. G.; Morokuma, K. *Adv. Chem. Phys.* **1996**, *XCV*, 61. (c) Niu, S.; Hall, M. B. *Chem. Rev.* **2000**, *100*, 353. (d) Torrent, M.; Sola, M.; Frenking, G. *Chem. Rev.* **2000**, *100*, 439.

13   Alagona, G.; Ghio, C.; Lazzaroni, R.; Settambolo, R. *Organometallics*, **2001**, *20*, 5394-5404.

14   For a recent discussion about phosphine basicity see Woska, D.; Prock, A.; Giering, W.P. *Organometallics* **2000**, *19*, 4629 and references therein.

15   Schmid, R.; Herrmann, W. A.; Frenking, G. *Organometallics* **1997**, *16*, 701.

16   González-Blanco, O.; Branchadell, V. . *Organometallics* **1997**, *16*, 5556.

17   Matsubara, T.; Koga, N.; Ding, Y.; Musaev, D. G.; Morokuma, K. *Organometallics* **1997**, *16*, 1065.

18   Decker, S. A.; Cundari, T. R. *Organometallics*, **2001**, *20*, 2827.

19   Brown, J. M.; Kent, A. G. *J. Chem. Soc. Perkin Trans. II*, **1987**, 1597.

20   Evans, D.; Yagupsky, G.; Wilkinson, G. *J. Chem. Soc. A* **1968**, 2660.

21   Cavalieri d'Oro, P.; Raimondo, L.; Paggani, G.; Montrasi, G.; Gregorio, G.; Andreetta, A. *Chim. Ind. (Milan)* **1980**, *62*, 389.

22   van Leeuwen, P. W. N. M.; Casey, C. P.; Whiteker, G. T. *In Rhodium Catalyzed Hydroformylation (Chapter 4);* van Leeuwen, P. W. N. M.; Claver, C., Ed.; Kluwer Academic Publishers, **2000**.

23   (a) Musaev, D. G.; Morokuma, K. *Adv. Chem. Phys.* **1996**, *95*, 61. (b) Matsubara, T.; Mebel, A. M.; Koga, N.; Morokuma, K. *Pure Appl. Chem.* **1995**, *67*, 257.

24   Rocha, W. R.; de Almeida, W. B. *Int. J. Quantum Chem.* **2000**, *78*, 42.

25   Herrmann, W.A.; Cornils, B. *Angew. Chem. Int. Ed. Engl.* **1997**, *36*, 1047.

26   Moser, W. R.; Papile, C. J.; Brannon, D. A.; Duwell, R. A.; Weininger, S. J. *J. Mol. Catal.* **1987**, *41*, 271.

27   Unruh, J. D.; Christenson, J. R. *J. Mol. Cat.* **1982**, *14*, 19-34.

28   Casey, C. P.; Paulsen, E. L.; Beuttenmueller, E. W.; Proft, B. R.; Petrovich, L. M.; Matter, B. A.; Powell, D. R. *J. Am. Chem. Soc.* **1997**, *119*, 11817-11825; *J. Am. Chem. Soc.* **1999**, *121*, 63.

29   van der Veen, L. A.; Boele, M. D. K.; Bregman, F. R.; Kamer, P. C. J.; van Leeuwen, P. W. N. M.; Goubitz, K.; Fraanje, J.; Schenk, H.; Bo, C., *J. Am. Chem. Soc.*, **1998**, *120*, 11616.

30   v1: calc. 2041 obs. (2027-2046), v2: calc. 1985 obs. (1983-2004), v3: calc. 1983 obs. (1960-1982) and v4: calc. 1948 obs. (1935-1957).

31   (a) Castonguay, L. A.; Rappé, A. K.; Casewit, C. J. *J. Am. Chem. Soc.* **1991**, *113*, 7177. (b) Paciello, R.; Siggel, L.; Kneuper, H. J.; Walker, N.; Röper, M. *J. Mol. Catal. A* **1999**, *143*, 85. (c) Casey, C. P.; Petrovich, L. M. *J. Am. Chem. Soc.* **1995**, *117*, 6007.

32   (a) Gleich, D.; Schmid, R.; Herrmann, W.A. *Organometallics* **1998**, *17*, 2141. (b) Gleich, D.; Schmid, R.; Herrmann, W.A. *Organometallics* **1998**, *17*, 4828. (c) Gleich, D.; Herrmann, W.A. *Organometallics* **1999**, *18*, 4354.

33   (a) Decker, S. A.; Cundari, T. R. *New. J. Chem.* **2002**, *26*, 129-135. (b) Decker, S. A.; Cundari, T. R., *J.Organomet.Chem.* **2001**, *635*, 132.

34   Carbó, J.J.; Maseras, F.; Bo, C.; van Leeuwen, P.W.N.M. *J. Am. Chem. Soc.* **2001**, *123*, 7630.

35   van der Veen, L. A.; Kamer, P. C. J.; van Leeuwen, P. W. N. M. *Organometallics*, **1999**, *18*, 4765.

36   Magee, M.P.; Luo, W.; Hersh, W.H. *Organometallics*, **2002**, *21*, 362.

37   van der Slot, S.- C.; Kamer, P. C. J.; van Leeuwen, P. W. N. M.; Iggo, J. A.; Heaton, B.
     T.; *Organometallics*, **2001**, *20*, 430.
38   van der Slot, S. C.; Duran, J; Luten, J.; Kamer, P. C. J.; van Leeuwen, P.W. N. M.
     Organometallics, in press.
39   Kamer, P.C.J.; van Leeuwen, P. W. N. M.; Reek, J.N.H. *Acc. Chem. Res.* **2001**, *34*, 895
40   van der Veen, L. A.; Keeven, P. H.; Schoemaker, G. C.; Reek, J. N. H.; Kamer, P. C. J.;
     van Leeuwen, P. W. N. M.; Lutz, M.; Spek, A. L. *Organometallics*, **2000**, *19*, 872.
41   (a) Maseras, F.; Morokuma, K. *J Comput Chem.* **1995**, *16*, 1170. (b) Maseras, F.;
     Chem.Commun. **2000**, 1821.
42   (a) Sandee, A. J.; van der Veen, L. A.; Reek, J. N. H.; Kamer, P. C. J.; Lutz, M.; Spek,
     A. L.; van Leeuwen, P. W. N. M. *Angew. Chem. Int. Ed. Engl.* **1999**, *38*, 3231. (b) van
     Haaren, R. J.; Goubitz, K.; Fraanje, J.; van Strijdonck, G. P. F.; Oevering, H.; Coussens,
     B.; Reek, J. N. H.; Kamer, P. C. J.; van Leeuwen, P. W. N. M. *Inorg. Chem.* **2001**, *40*,
     3363.
43   (a) Jorgensen, W. L.; Severance, D. L. *J. Am. Chem. Soc.* **1990**, *112*, 4768. (b) Graf, D.
     D.; Campbell, J. P.; Miller, L. L.; Mann, K. R. *J. Am. Chem. Soc.* **1996**, *118*, 5480. (c)
     Graf, D. D.; Duan, R. G.; Campbell, J. P.; Miller, L. L.; Mann, K. R. *J. Am. Chem. Soc.*
     1997, *119*, 5888.
44   Ujaque, G.; Maseras, F.; Lledós, A. *J. Am. Chem. Soc.* **1999**, *121*, 1317
45   Vázquez, J.; Pericás, M.A.; Maseras, F.; Lledós, A. *J. Org. Chem.* **2000**, *65*, 7303.
46   Marcone, J. E.; Moloy, K. G. *J. Am. Chem. Soc.* **1998**, *120*, 8527.

Chapter 8

Transition Metal Catalyzed Borations

Xin Huang and Zhengyang Lin*
Deparment of Chemistry, The Hong Kong University of Science and Technology, Clear Water Bay, Kowloon, Hong Kong, People's Republic of China

Abstract: Transition metal catalyzed alkene and alkyne boration reactions are attractive methods for the generation of alkyl- or alkenylboron derivatives with defined regio- and stereochemistry, which increase the potential applications of boron derivatives in synthetic organic chemistry. This chapter reports and discusses theoretical studies on the reaction mechanisms of transition metal catalyzed borations.

Key words: Hydroboration, Diboration, Thioboration

1.   INTRODUCTION

Synthesis of organoboranes is indispensable in organic chemical industries as organoboranes can be converted into a variety of functional groups, including alcohols, amines and halides [1-3]. Hydroboration of unsaturated hydrocarbons is one of the most important reactions in obtaining a variety of organoboranes [4-6]. With respect to uncatalyzed transformations, the transition metal catalyzed hydroborations display a large potential for altering chemo-, regio-, and diastereoselectivities, as well as for achieving enantioselectivities in the presence of chiral transition metal catalysts [7-11].

F. Maseras and A. Lledós (eds.), Computational Modeling of Homogeneous Catalysis, 189–212.

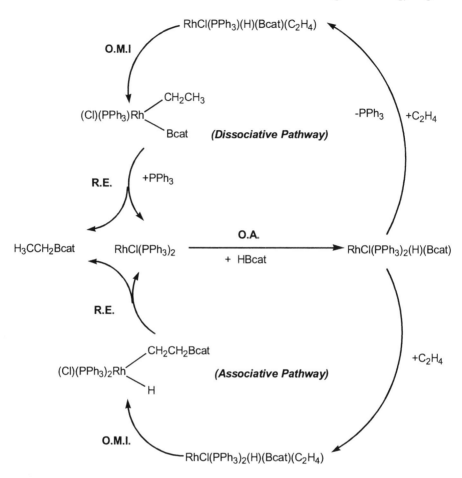

Figure 1.. The two proposed reaction pathways based on experimental results for hydroboration reactions of olefins catalyzed by the Wilkinson catalyst. (O.A.: Oxidative Addition; O.M.I.: Olefin Migratory Insertion; R.E.: Reductive Elimination)

In 1985, the Wilkinson catalyst $Rh(PPh_3)_3Cl$ was first reported by Männig and Nöth to promote the addition of catecholborane to alkenes [9]. Since then, many other related transition metal catalyzed systems have been investigated. Mechanistic studies of the Rh-based systems and other late-transition metal systems have given rise to the proposal of two different reaction pathways (Figure 1). In the dissociative pathway, the reaction starts with the oxidative addition of catecholborane to rhodium, continues with a step invoving phosphine dissociation and olefin addition, and finishes with the olefin migratory insertion and the reductive elimination of the hydroborated product. Based on their deuterium labeling experiments of selected catalytic hydroboration reactions, Evans, Fu and Anderson proposed that the migratory insertion of olefin into the Rh-H bond occurs prior to the reductive C-B bond coupling [12]. In the associative reaction pathway, the dissociation of the phosphine ligand is not considered as an essential step in the addition of the olefin ligand (Figure 1). Another difference is that the migratory insertion of olefin is believed to involve the Rh-B bond instead of the Rh-H bond because the dehydrogenative borylation of olefin was found as a competitive process leading to the formation of side products, i.e., vinylboranes. The formation of these side products is believed to be a result of β-H elimination of the (Cl)(PPh$_3$)$_2$Rh(H)(CH$_2$Ch$_2$Bcat) intermediate [13].

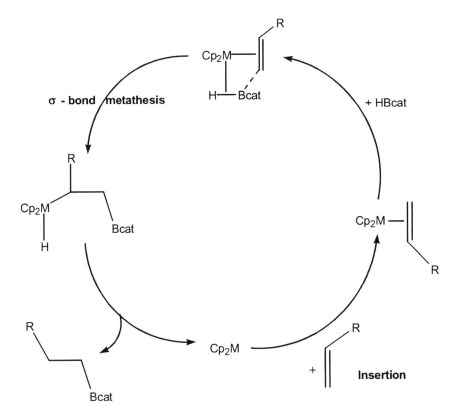

Figure 2. Hydroboration reactions of olefin catalyzed by early transition metal complexes. The proposed reaction mechanism involves a σ-bond metathesis step. (M = Lanthanide or other early transition metals.)

The catalysis of olefin hydroboration by early transition metal complexes, *e.g.*, titanium- and lanthanide-complexes, has also attracted considerable interest in recent years [14-17]. These catalytic systems show different reaction mechanisms to those of late transition metals. In early transition metals the olefin insertion is followed by a σ-bond metathesis step (Figure 2). Because of this reaction mechanism, hydroboration reactions of olefins catalyzed by early transition metal complexes always give anti-Markonikov alkylboronate ester products. In contrast, the Rh-catalyzed reactions give both Markonikov and anti-Markonikov products although some of them have also shown high regioselectivities [5].

In addition to hydroboration reactions, diboration (using R_2B-BR_2) and thioboration (using $RS-BR_2$) reactions have also been reported to produce organoborane compounds useful for further chemical transformations [18, 19]. The corresponding transition metal catalyzed diboration and thioboration reactions have begun to be explored in recent years [20-24]. The systems studied so far mainly involve Pt and Pd complexes, which also show high regio- and stereoselectivities.

In view of the abundant experimental studies on the boration reactions catalyzed by transition metal complexes summarized above, theoretical studies to the detailed reaction mechanisms have also been carried out [25-28]. The main focus of these quantum chemical calculations has been to provide detailed structural and energetic information on the proposed reaction mechanisms.

In this chapter, we will summarize the relevant theoretical studies on the reaction mechanisms of transition metal catalyzed borations. It is our hope that an overall picture can be given in a manner which can be easily understood without detailing all the theoretical aspects. It should be noted here that other comprehensive reviews, both experimental and theoretical, on the topic of catalytic boration reactions can also be found elsewhere in the literature [3-6].

2. HYDROBORATION OF OLEFINS CATALYZED BY THE WILKINSON CATALYST

Hydroboration of olefins catalyzed by the Wilkinson catalyst $Rh(PPh_3)_3Cl$ has been the most studied reaction by quantum chemical calculations [25-27]. In the following, three representative studies are described.

2.1 Dorigo and Schleyer's study

In 1995, Dorigo and Schleyer reported their calculations based on the dissociative mechanism mentioned in the introduction [26]. On the basis of MP2 calculations on the model system $BH_3+C_2H_4+Rh(PH_3)_2Cl$, they confirmed that the dissociative reaction pathway is indeed feasible. In their calculations, the insertion processes of C_2H_4 into both the Rh-H and Rh-B bonds are compared. In agreement with the experimental proposal [9, 12], they found that the insertion into the Rh-H bond is more favorable.

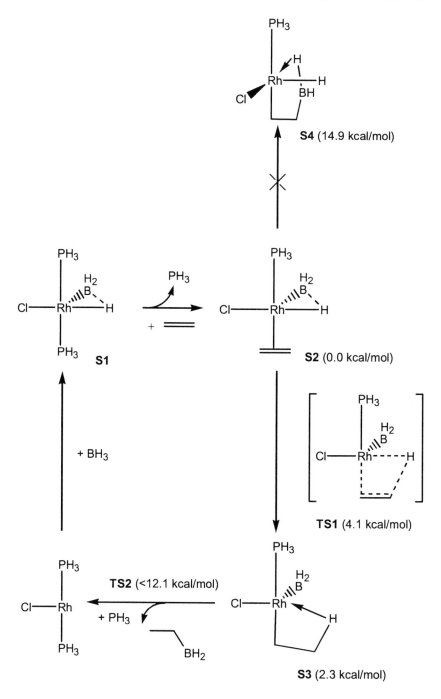

Figure 3. The dissociative reaction pathway calculated by Dorigo and Schleyer.

The computed cycle is shown in Figure 3. Oxidative addition of BH_3 to $[RhCl(PH_3)_2]$ gives complex **S1** with a distorted structure which can be considered as an intermediate structure between a trigonal bipyramid and a square pyramid. In complex **S1**, a significant hydride-boryl interaction is observed. The oxidative addition is then followed by the replacement of PH_3 by the olefin, giving complex **S2**. Starting from complex **S2**, the insertions of the olefin into the Rh-H bond and the Rh-B bond, leading to the formation of the different intermediates **S3** and **S4**, respectively, are possible. The MP2 calculations done by Dorigo and Schleyer show that the insertion into the Rh-H bond is both kinetically and thermodynamically favorable (Figure 3). The instability of **S4** has been explained by the less effective Rh-H-B agostic interaction in comparison to the Rh-H-C interaction. The reason for the stability of **S3** could also be related to the very strong trans influence of the boryl ligand. The **S3** structure can be considered as a square pyramid in which the boryl ligand occupies the apical site. Although the hydride ligand is considered as a strong trans influencing ligand, the boryl ligand is apparently even stronger, making **S3** more stable than **S4**. The reductive elimination of the organoborane molecule from **S3** can occur via two different pathways, depending on how the phosphine addition occurs. If the reductive elimination is prior to the addition, the barrier is found to be 9.8 kcal/mol with respect to **S3**. If the reductive elimination is after the addition, the barrier is much lower.

2.2   Morokuma and co-workers' study

In a related work, Morokuma and co-workers reported their own MP2 calculations on the associative reaction pathway, in which the phosphine dissociation is not considered [25]. In the model system, the $HB(OH)_2$ and $HBO_2(CH_2)_3$ boranes were used, while the catalyst was $RhCl(PH_3)_2Cl$ and the olefin C_2H_4.

The oxidative addition of the B-H bond to the catalyst was again considered as the first step, giving complex **M1**. This step was followed by olefin coordination to **M1** without the dissociation of the PH_3 group. The olefin coordination leads to three possibilities (shown in Figure 4), **M2a**, **M2b**, and **M2c**, where the ethylene ligand takes a position trans to either Cl, $B(OH)_2$ or H, respectively. The MP2 calculations by Morokuma and co-workers show that the **M2a** isomer is the least stable among the three isomers. The instability of **M2a** can be conveniently related to the trans arrangement of the two very strong trans influence ligands hydride and boryl.

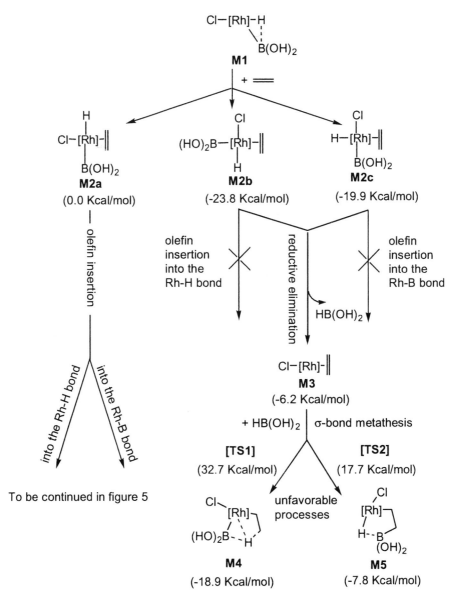

Figure 4. Three possible complexes derived from the coordination of an olefin to the HB(OH)$_2$ oxidative addition product, as calculated by Morokuma and co-workers. Further reactions based on **M2b** and **M2c** are found to be unfavourable. ([Rh] = Rh(PH$_3$)$_2$ The two PH$_3$ ligands are perpendicular to the molecular plane shown in this figure.)

The structural arrangements of **M2b** and **M2c**, suggest that **M2b** can proceed the reaction by olefin insertion into the Rh-H bond while **M2c** can proceed the reaction by olefin insertion into the Rh-B bond. Unfortunately, the relevant transition states for the olefin insertion reactions do not exist according to the MP2 calculations. This unexpected result may be explained as follows. For **M2b**, the likely transition structure for the olefin insertion into the Rh-H bond is the one in which the boryl ligand is trans to an alkyl-like ligand. Such a structure, which was confirmed not to correspond to a stationary point, is expected to be very unstable because the two ligands (boryl and alkyl-like) have both very strong trans-influence. Similarly, the likely transition structure for **M2c** corresponds to a structure in which the hydride and alkyl-like ligands are trans to each other. Apparently, octahedral structures having two strong σ ligands trans to each other do not correspond to stationary points for the Rh(III) complexes studied here.

Instead of having the olefin insertion reactions, the calculations indicate that **M2b** and **M2c** can only proceed uphill with the reductive elimination of $HB(OH)_2$, leading to the formation of **M3**, an olefin complex which could be in principle obtained directly from the addition of olefin to the catalyst Rh $(PH_3)_2Cl$. The olefin complex **M3** then could undergo σ-bond metathesis processes with $HB(OH)_2$, giving two isomeric products **M4** and **M5** depending on the orientation of the $HB(OH)_2$ borane. The σ-bond metathesis processes are however found to be unfavorable because of the very high reaction barriers (Figure 4).

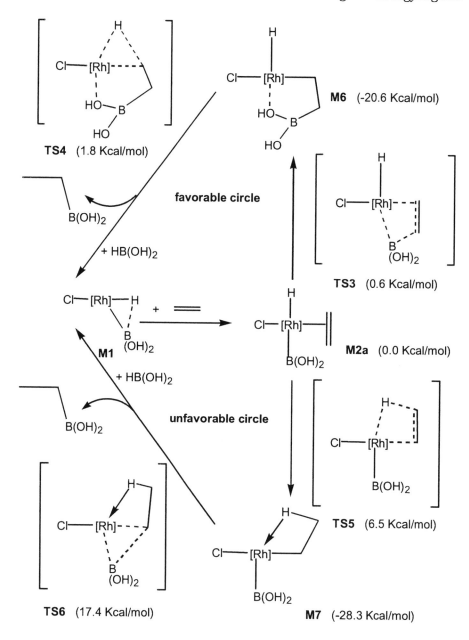

Figure 5. The associative reaction pathways from complex **M2a** calculated by Morokuma and co-workers. ([Rh] = Rh(PH₃)₂. The two PH₃ ligands are perpendicular to the molecular plane shown in this figure.)

Therefore, the favorable reaction pathway will have to start from the complex **M2a** despite of its relatively high energy. Figure 5 illustrates the reaction cycles involving this species. Starting from **M2a**, the olefin insertion reactions give **M6** and **M7**. Both insertion reactions have low reaction barriers. The reductive elimination of the organoborane from **M6** has been calculated to be the most favorable process with the cycle through **M7** having a high barrier (**TS6** is 45.7 kcal/mol above **M7**). It must be mentioned here that this structure **TS6** was obtained by imposing C_s symmetry to the system, which was necessary to keep computer requirements affordable. In conclusion, according to these calculations, the favorable reaction cycle involves the olefin insertion into the Rh-B bond of the **M2a** species. This is consistent with the experimental proposal that by-products (vinylboranes) can be produced by the β-hydrogen elimination of **M6** [13].

2.3   Ziegler and co-workers' study

The two ab initio studies discussed above considered one or the other reaction pathway on different model systems. In 2000, Ziegler and co-workers reported a detailed comparative study of the dissociative and associative mechanisms of the model reaction $HBR_2+C_2H_4 \rightarrow H_3CCH_2BR'_2$ (R= OH, R'= OCH=CHO) catalyzed by $[RhCl(PH_3)_2]$ [27]. This work was carried out with density functional theory (DFT) calculations. The objective of their study was to calculate the energy profiles for the dissociative (Figure 6) and associative (Figure 7) pathways on an equal footing.

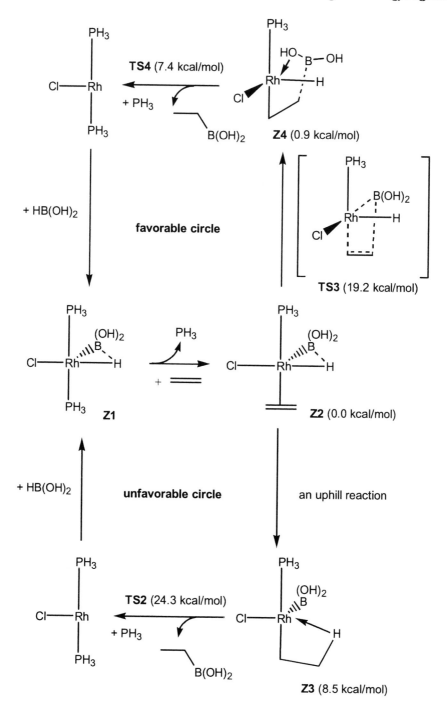

Figure 6. The dissociative pathways calculated by Ziegler and co-workers.

In the dissociative reaction pathway (Figure 6), the overall picture is similar to that found by Dorigo and Schleyer (Figure 3), but there are some significant changes in relative energies. The path going through olefin insertion into the Rh-B bond, which was considered too high in energy in Figure 3, is here found to be the likely event, in contrast to the conclusion made by Evans, Fu and Anderson [12], and by Dorigo and Schleyer [26].

The olefin insertion into the Rh-B bond is calculated to have a higher barrier than the insertion into the Rh-H bond, a result consistent with the calculations by Dorigo and Schleyer. However, the reductive elimination of the organoborane ($CH_3CH_2B(OH)_2$) is found to be much easier from the species (**Z4**) derived from the olefin insertion into the Rh-B bond in comparison to the elimination from **Z3**. The discrepancy between the two sets of calculations can be due to the use of different models for borane (BH_3 *vs.* $BH(OH)_2$) or to the consideration of additional isomers in the more recent calculations. Based on the favorable cycle involving the **Z4** species, Ziegler *et al.* further conclude that the risk for side product formation is lower for the dissociative reaction pathway because **Z4** is difficult to isomerize to a structure containing β-agostic interaction, which can give vinylborane side-products through the β-H elimination reaction.

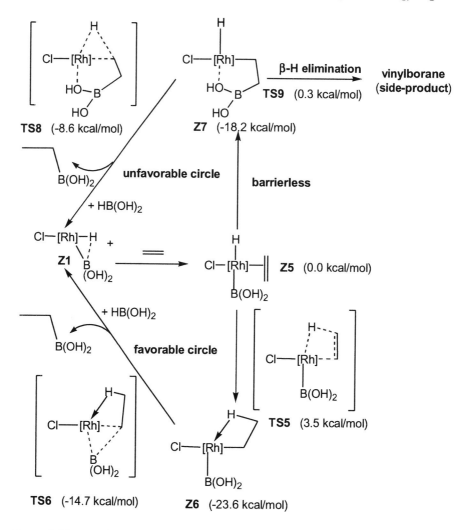

Figure 7. The associative pathways calculated by Ziegler and co-workers. ([Rh] = Rh(PH$_3$)$_2$. The two PH$_3$ ligands are perpendicular to the molecular plane shown in this figure.)

In the associative reaction pathway, the calculations by Ziegler *et al.* suggest that the olefin coordinated species **Z5** (Figure 7), equivalent to **M2a** in Figure 4, is the complex needed to be considered in the catalytic reaction circles. This result is in agreement with the one obtained by Morokuma and co-workers.. The catalytic reaction cycles involving **Z5** calculated by Ziegler *et al.* are given in Figure 7. The overall scheme is similar to the one computed by Morokuma and co-workers (Figure 5). There is however a significant difference in the energy of transition state **TS6**, which affects the nature of the favorable cycle. This transition state had a higher energy in the first set of calculations, which is probably due to the imposition of C_s sysmmetry. In Ziegler's calculations, with no symmetry restrictions, the olefin insertion into the Rh-H bond from **Z5** to **Z6** followed by the reduction elimination of the organoborane ($CH_3CH_2B(OH)_2$) is favored. The olefin insertion either into the Rh-H bond or Rh-B bond leading to **Z6** or **Z7** has a quite low barrier. The reductive elimination from **Z6** is calculated to be easier.

In addition to the favorable reaction cycle, the β-hydrogen elimination from **Z7** leading to the formation of vinylborane side-products is also found to be competitive (Figure 7). In other words, side products are difficult to avoid in the associative reaction pathway.

In view of these results, Ziegler and co-workers suggest that for producing more pure product one should drive the reaction into a dissociative channel in order to suppress the formation of undesired side products. This may be done by employing sterically demanding electron-withdrawing ligands. Steric bulk should lead to a preference of the dissociative reaction mechanism. With the dissociative reaction pathway (Figure 6), one can also drive the reaction into the unfavorable cycle to obtain pure hydroborated product if one can reduce the activation barrier for the reductive elimination of the organoborane from **Z3** since the unfavorable cycle is not able to produce vinylborane side products. The reduction of this activation barrier can be achieved by employing phosphines with electron-withdrawing properties because the oxidation state in the rhodium center during the reductive elimination changes from +3 to +1. Ligands with electron-withdrawing properties can also raise the barrier from **Z2** to **Z4** because the rhodium center oxidation state changes from +1 to +3 in this step. This increase in barrier would reduce the probability of having the reaction to proceed with the olefin insertion into the Rh-B bond, which may have the β-hydrogen elimination giving the undesired side-products.

2.4　Comment

The three theoretical studies presented above give somehow different conclusions regarding the detailed reaction mechanisms. The difference is apparently related to the different models used and different theoretical approaches employed in the calculations. In any case, the emerging overall picture is that the reaction mechanism of olefin hydroborations must be complicated. We would like to see more experimental work done in the future so that the theoretical results can be tested. In particular, it would be nice to evaluate experimentally the utility of the suggestion made by Ziegler and co-workers in the choice of the phosphine ligands in order to produce more pure product.

3.   HYDROBORATION OF OLEFINS CATALYZED BY EARLY TRANSITION METAL COMPLEXES

As mentioned in the introduction, early transition metal complexes are also able to catalyze hydroboration reactions. Reported examples include mainly metallocene complexes of lanthanide, titanium and niobium metals [8, 15, 29]. Unlike the Wilkinson catalysts, these early transition metal catalysts have been reported to give exclusively anti-Markonikov products. The unique feature in giving exclusively anti-Markonikov products has been attributed to the different reaction mechanism associated with these catalysts. The hydroboration reactions catalyzed by these early transition metal complexes are believed to proceed with a σ-bond metathesis mechanism (Figure 2). In contrast to the associative and dissociative mechanisms discussed for the Wilkinson catalysts in which HBR_2 is oxidatively added to the metal center, the reaction mechanism associated with the early transition metal complexes involves a σ-bond metathesis step between the coordinated olefin ligand and the incoming borane (Figure 2). The preference for a σ-bond metathesis instead of an oxidative addition can be traced to the difficulty of further oxidation at the metal center because early transition metals have fewer d electrons.

Sm(III)-catalyzed olefin hydroboration reaction by catecholborane has recently been theoretically investigated [28]. The stationary structures on the model reaction path considering ethylene as the olefin, Cp_2SmH as the active catalyst, and $HB(OH)_2$ as the model borane were obtained at the RHF and MP2 levels. MP4SDQ energy calculations were also carried out at the MP2 structures.

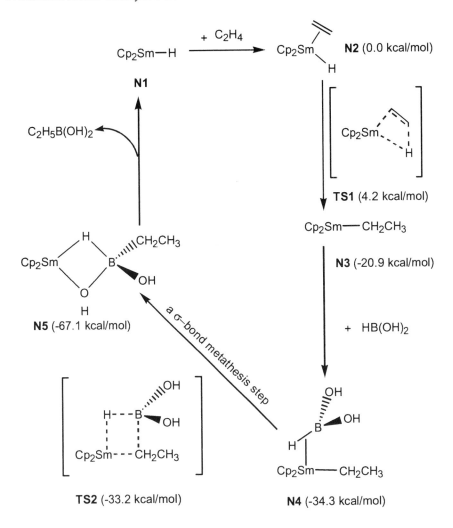

Figure 8. The Sm(III)-catalyzed olefin hydroboration reaction mechanism calculated by Koga and Kulkarni.

The reaction path involves the following elementary steps (Figure 8): olefin coordination to the active catalyst Cp_2SmH to form a π-complex **N2**; insertion of the olefin into the Sm-H bond passing through a barrier of 4.2 kcal/mol leading to $Cp_2SmC_2H_5$; addition of $HB(OH)_2$ to $Cp_2SmC_2H_5$ forming a borane complex **N4**; formation of the product complex **N5** after passing through a barrier of only 1.1 kcal/mol; and finally a highly endothermic dissociation of the product from the product complex. The dissociation step is believed to be the rate-determining step. The small electronegativity and oxophilicity of Sm determine the features of the potential energy profile. The highly ionic Sm-C bond results in the easy formation of the B-C bond in the σ-bond metathesis step. In addition, the oxophilicity of Sm gives the large exothermicity of the step. Apparently, the oxophilicity makes the product dissociation very endothermic.

4.   DIBORATION

Diboration provides another means of obtaining organoboranes. Studies [30-36] have been focused on the diboration reactions of alkynes and olefins with pinacol ester derivatives catalyzed by Pd(0) and Pt(0) metal complexes. Interestingly, it has been shown that Pt (0) complexes catalyze cis addition of the B-B bond in pinacol ester derivatives to alkynes but not to olefins. On the other hand, Pd (0) complexes do not catalyze diboration reactions, neither for alkynes nor for olefins.

The stereoselectivity of the Pt(0)-catalyzed reactions, the inertness of olefins under the Pt(0)/Pd(0) conditions, and the inability of Pd(0) to catalyze diboration reactions have recently been studied theoretically by Morokuma and co-workers [24]. The Pt(0)/Pd(0) catalyzed olefin and alkyne diboration reactions with $(OH)_2B\text{-}B(OH)_2$ as a model for the diboron species were investigated using B3LYP density functional calculations. The computed mechanism, shown in Figure 9, involves several steps. The first step is the coordination of $(OH)_2B\text{-}B(OH)_2$ to $Pt(PH_3)_2$ to give complex **D2**, where the B-B and P-Pt-P axes are perpendicular to each other. The coordination is then followed by the oxidative addition of $(OH)_2B\text{-}B(OH)_2$ to $Pt(PH_3)_2$, passing through **TS1** with a barrier of 14.0 kcal/mol. The oxidative addition product **D3** adopts the square planar structure which is expected for a d^8 Pt(II) metal center. After the oxidative addition, an endothermic dissociation of one of the phosphine ligands gives complex **D4**.

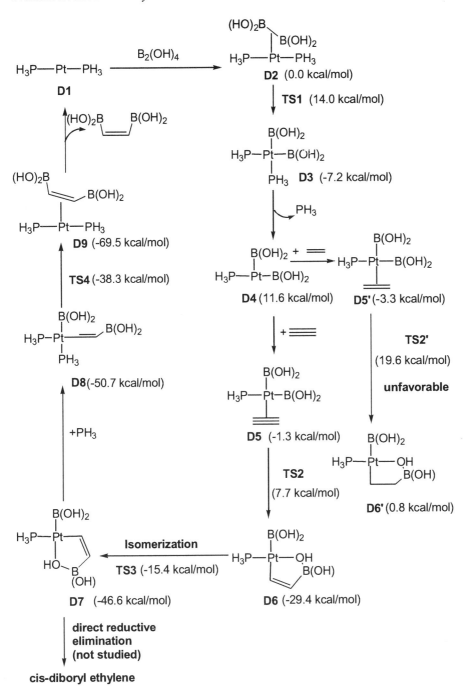

Figure 9. The reaction mechanism of Pt(0)-catalyzed alkyne and alkene diboration reactions calculated by Morokuma and co-workers.

Complex **D4** is considered as the active species for both alkyne and olefin coordinations. Starting from the olefin coordinated complex (**D5'**), the olefin insertion into the Pt-B bond is unfavorable because of a high activation barrier (22.9 kcal/mol). On the contrary, the acetylene insertion from the acetylene coordinated complex (**D5**) occurs easily with a small reaction barrier (9.0 kcal/mol). This significant difference in the reaction barriers has been used to explain the inertness of olefins for diboration reactions. The smaller barrier from **D5** to **D6** coincides with the highly stable insertion product **D6**. In contrast, the olefin insertion product **D6'** is relatively unstable with respect to the olefin coordinated species **D5'**.

Due to the strong trans influence, isomerization of **D6** to **D7**, in which the two strong trans influencing ligands move cis to each other, is a favorable process. Starting from **D7**, a direct reductive elimination can generate a cis-diboryl ethylene, giving the observed stereoselectivity. However, this process was not studied. Instead, the dissociation of the coordinated OH group and the coordination of a PH₃ ligand were considered, giving the **D8** complex. The reductive elimination of diboryl ethylene from **D8** also gives the desired stereoselective product.

The inability of Pd(0) to catalyze diboration reactions of alkynes and olefins was explained in terms of the difficulty in having the oxidative addition of $(OCH_2)_2B-B(OCH_2)_2$ to $Pd(PH_3)_2$. The DFT calculations indicate that the addition process is uphill. The addition product is not stable because the reverse process is very favorable. The different behavior of the Pd(0) catalysts in comparison to Pt(0) species is consistent with the observation that Pt(II) is more common than Pd(II).

5.   THIOBORATION

In comparison with the hydroboration and diboration reactions, thioboration reactions are relatively limited. In 1993, Suzuki and co-workers reported the Pd(0)-catalyzed addition of 9-(alkylthio)-9-BBN (BBN = borabicyclo [3.3.1] nonane) derivatives to terminal alkynes to produce (alkylthio)boranes, which are known as versatile reagents to introduce alkylthio groups into organic molecules [21]. Experimental results indicate that the thioboration reactions, specific to terminal alkynes, are preferentially catalyzed by Pd(0) complexes, e.g. Pd(PPh₃)₄, producing (thioboryl)alkene products, in which the Z-isomers are dominant. A mechanism proposed by Suzuki and co-workers for the reactions involves an oxidative addition of the B-S bond to the Pd(0) complex, the insertion of an alkyne into the Pd-B or Pd-S bond, and the reductive elimination of the (thioboryl)alkene product.

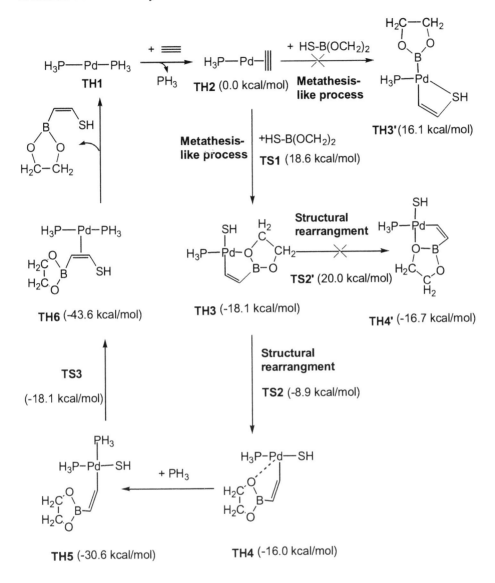

Figure 10. The reaction mechanism of acetylene thioboration reactions catalyzed by Pd(0) complexes calculated by Morokuma and co-workers.

In 1998, Morokuma and co-workers carried out density functional calculations on the following model reaction, $Pd(PH_3)_2 + C_2H_2 + (OCH_2)_2B\text{-}SH \rightarrow Pd(PH_3)_2 + (OCH_2)_2B\text{-}CH=CH\text{-}SH$, to study the detailed reaction mechanism [24]. The theoretical studies suggest that the reaction mechanism involves a metathesis-like process, instead of an oxidative addition, in breaking the B-S bond of the substrate. The reason for not having an oxidative addition of the B-S bond to the Pd(0) center could be attributed to the very weak Pd-SR bonding and relatively less stable Pd(II) center when compared to Pt(II). Figure 10 illustrates the proposed mechanism based on the theoretical calculations. The ligand exchange reaction takes place as the first step, giving complex **TH2**. The metathesis-like process is then followed to give **TH3**. **TH3'** is relatively unstable due to that the two strong bonds (Pd-C and Pd-B) are trans to each other. From **TH3**, a structural rearrangement is necessary in order to achieve the reductive elimination. The calculations suggest that a rotation of 90° along the Pd-C bond is much favorable, giving a structure (**TH4**) in which an oxygen-to-palladium dative bond is approximatively perpendicular to the plane containing the Pd-C, Pd-S and Pd-P bonds. An addition of a PH_3 ligand further facilitates the reductive elimination process. Overall, the metathesis-like process passing through the **TS1** transition state is the rate-determining step.

Can Pt(0) complexes serve as active catalysts for the alkyne thioboration reactions? Morokuma and co-workers also carried out calculations on the mechanism of the Pt(0)-catalyzed acetylene thioboration reaction with a smaller model $HS-B(OH)_2$. They found that the reductive elimination in the last step (from **TH5** to **TH6**) needs to overcome a very high barrier of 27.9 kcal/mol because the Pt(II) analog of **TH5** is calculated to be very stable. It is predicted that the Pt(0) complex is not a good catalyst for thioboration reactions.

6.   SUMMARY

In this chapter, theoretical studies on various transition metal catalyzed boration reactions have been summarized. The hydroboration of olefins catalyzed by the Wilkinson catalyst was studied most. The oxidative addition of borane to the Rh metal center is commonly believed to be the first step followed by the coordination of olefin. The extensive calculations on the experimentally proposed associative and dissociative reaction pathways do not yield a definitive conclusion on which pathway is preferred. Clearly, the reaction mechanism is a complicated one. It is believed that the properties of the substrate and the nature of ligands in the catalyst together with temperature and solvent affect the reaction pathways significantly. Early transition metal catalyzed hydroboration is believed to involve a σ-bond metathesis process because of the difficulty in having an oxidative addition reaction due to less available metal d electrons.

For the diboration reactions of alkynes catalyzed by Pt(0) complexes, the reaction mechanism involves the oxidative addition of diborane to the Pt(0) center, followed by the insertion of alkyne into the Pt-B bond and reductive elimination of a vinylborane. The inability of Pd(0) in catalyzing the diboration reactions was explained in terms of difficulty in having an oxidative addition of diboration to the Pd(0) center.

For the thioboration reactions of alkynes preferentially catalyzed by Pd(0) instead of Pt(0), the reaction mechanism involves a metathesis-like process. The reason for not having an oxidative addition step can be related to the electron richness of the alkylthio group, which prevents the oxidative addition of thioborane to the metal center. Because of the preference for having a metathesis-like process, Pd becomes a better candidate due to its relative less electron richness in comparison to Pt.

In view of the reported theoretical studies, the regio- and diastereoselectivies should be the focus of the future theoretical investigation. Insights into the factors influencing the selectivities are badly required for designing better and useful catalysts.

ACKNOWLEDGEMENTS

The research Grant Council of Hong Kong and the Hong Kong university of Science and Technology are thanked for their financial support.

REFERENCES

1   Wadepohl, H. *Angew. Chem. Int. Ed. Engl.* **1997**, 36, 2441.
2   Braunschweig, H. *Angew. Chem. Int. Ed. Engl.* **1998**, *37*, 1786.
3   Irvine, G. J.; Lesley; M. J. G.; Marder, T.B.; Norman, N.C.; Rice, C.R.; Robins, E.G.;
     Roper, W.R.; Whittell, G.R.; Wright L. J.*Chem.Rev.* **1998**, *98*, 2685.
4   Burgess, K.; Ohlmeyer,M. J. *Chem.Rev.* **1991**, *91*, 1179.
5   Beletskaya, I.; Pelter, A. *Tetrahedron* **1997**, *53*, 4957.
6   Torrent, M.; Solà, M.; Frenking, G. *Chem. Rev.* **2000**, *100*, 439.
7   Mikhailov, B. M.; Bubnov, Y. *'Organoboron Compounds in organic Synthesis'*,
     Harwood Academic Science Publishers, Amsterdam, **1984**
8   Harrison, K. N.; Marks, T. J. *J. Am. Chem. Soc.* **1992**, *114*, 9221
9   Männig, D.; Nöth, H. *Angew. Chem., Int. Ed. Engl.* **1985**, *24*, 878.
10 Westcott, S. A.; Taylor, N. J.; Marder, T. B.; Baker, R. T. *J. Am. Chem. Soc.* **1992**, *114*,
     8863.
11 Doucet, H.; Fernandez, E.; Layzell, T. P.; Brown, J. M. *Chem. Eur. J.* **1999**, *5*, 1320
12 Evans, D. A; Fu, G. C.; Anderson, B. A. *J. Am. Chem. Soc.* **1992**, *114*, 6679
13 Burgess, K.; van der Donk, W. A.; Westcott, S. A.; Marder, T. B.; Baker, R. T.;
     Calabrese, J. C. *J. Am. Chem. Soc.* **1992**, *114*, 9350.
14 Motry, D. H.; Brazil, A. G.; Smith III, M. R. *J. Am. Chem. Soc.* **1997**, *119*, 2743
15 He, X.; Hartwig, J. F. *J. Am. Chem. Soc.* **1996**, *118*, 1696
16 Motry, D. H.; Smith III, M. R. *J. Am. Chem. Soc.* **1995**, *117*, 6615
17 Hartwig, J. F.; Muhoro, C. N. *Organometallics* **2000**, *19*, 30

18   Howarth, J.; Helmchen, G.; Kiefer, M. *Tetrahedron Lett.* **1993**, *34*, 4095.
19   Massey, A. G. *Adv. Inorg. Chem. Radiochem.* **1983**, *26*, 1.
20   (a) Ishiyama, T.; Matsuda, N.; Miyaura, N.; Suzuki, A. *J. Am. Chem. Soc.* **1993**, *115*, 11018. (b). Iverson, C. N.; Smith III, M. R. *J. Am. Chem. Soc.* **1995**, *117*, 4403. (c) Suzuki, A. *Pure Appl.Chem.* **1994**, *66*, 213. (d) Gridnev, I. D.; Miyaura, N.; Suzuki, A. *Organometallics* **1993**, *12*, 589. (e) Lesley, G.; Nguyen, P.; Taylor, N. J.; Marder, T. B.; Scott, A. J.; Clegg, W.; Norman, N. C. *Organometallics* **1996**, *15*, 5137 and references therein. (f) Iverson, C. N.; Smith III, M. R. *Organometallics* **1996**, *15*, 5155.
21   Ishiyama, T.; Nishijima, K.; Miyaura, N.; Suzuki, A. *J. Am.Chem. Soc.* **1993**, *115*, 7219.
22   Baker, R. T.; Nguyen, P.; Marder, T. B.; Wescott, S. A. *Angew. Chem., Int. Ed. Engl.* **1995**, *34*, 1336.
23   Sakaki, S.; Kikuno, T. *Inorg. Chem.* **1997**, *36*, 226.
24   Cui, Q.; Musaev, D. G.; Morokuma, K. *Organometallics* **1997**, *16*, 1355. (b) Cui, Q.; Musaev, D. G.; Morokuma, K. *Organometallics* **1998**, *17*, 742. (c) Cui, Q.; Musaev, D. G.; Morokuma, K. *Organometallics* **1998**, *17*, 1383.
25   Musaev, D. G.; Mebel, A. M.; Morokuma, K. *J. Am. Chem. Soc.* **1994**, *116*, 10693.
26   Dorigo, A. E.;. Schleyer, P.v.R *Angew. Chem., Int. Ed. Engl.* **1995**, *34*, 115.
27   Widauer, C.; Grützmacher, H.; Ziegler, T. *Organometallics* **2000**, *19*, 2097.
28   Kulkarni, S. A.; Koga, N. *J. Mol. Struct. Theochem* **1999**, *461-462*, 297.
29   Lantero, D. R.; Ward, D. L.; Smith III, M. R. *J. Am. Chem. Soc.* **1997**, *119*, 9699.
30   Zhang, J.; Lou, B.; Guo, G.; Dai, L. *J. Org. Chem.* **1991**, *56*, 1670.
31   Baker, R. T.; Ovenall, D. W.; Calabrese, J. C.; Westcott, S. A.; Taylor, N. J.; Williams, I. D.; Marder, T. B. *J. Am. Chem. Soc.* **1990**, *112*, 9399.
32   Baker, R. T.; Ovenall, D. W.; Harlow, R. L.; Westcott, S. A.; Taylor, N. J.; Marder, T. B. *Organometallics* **1990**, *9*, 3028.
33   Westcott, S. A.; Blom, H. P.; Marder, T. B.; Baker, R. T.; Calabrese, J. C. *Inorg. Chem.* **1993**, *32*, 2175.
34   Knorr, J. R.; Merola, J. S. *Organometallics* **1990**, *9*, 3008.
35   Baker, R. T.; Calabrese, J. C.; Westcott, S. A.; Nguyen, P.; Marder, T. B. *J. Am. Chem. Soc.* **1993**, *115*, 4367.
36   Westcott, S. A.; Marder, T. B.; Baker, R. T. *Organometallics* **1993**, *12*, 975.

Chapter 9

Enantioselective Hydrosilylation by Chiral Pd Based Homogeneous Catalysts with First-Principles and Combined QM/MM Molecular Dynamics Simulations

Alessandra Magistrato,[1] Antonio Togni,[1] Ursula Röthlisberger,[1] and Tom K. Woo[2],*

[1]*Laboratory of Inorganic Chemistry, Swiss Federal Institute of Technology, ETH Zentrum, CH-8092 Zürich, Switzerland.;* [2]*Department of Chemistry, The University of Western Ontario, London, Ontario, Canada, N6A 5B7.*

Abstract: The enantioselective hydrosilylation of styrene catalyzed by Pd^0 species generated in situ from dichloro{1-{(R)-1-[(S)2(diphenylphosphino - κP)ferrocenyl]ethyl}-3-trimethylphenyl-5-1H-pyrazole-κN}palladium, **1**, has been investigated in detail by *ab initio* molecular dynamics and combined quantum mechanics and molecular mechanics (QM/MM) simulations. The nature of the unique structural features observed in the pre-catalyst, **1**, and its bis(trichlorosilyl) derivatives have been explored. Using the combined QM/MM method we have been able to pinpoint the steric and electronic influence of specific ligands on these geometric distortions. The whole catalytic cycle of the enantioselective styrene hydrosilylation has been examined in detail with mixed QM/MM Car-Parrinello molecular dynamics simulations. The simulations show that the reaction proceeds through the classical Chalk-Harrod mechanism of hydrosilylation. The rate-determining step was found to be the migration of the silyl ligand to the α-carbon of the substrate. The nature of the regiospecificity and enantioselectivity of the catalysis has been established. In both cases the formation of a η^3-benzyllic intermediate plays a crucial role. The mechanistic detail afforded by the computational study provides a framework for rational ligand design that would improve the enantioselectivity of the catalysis.

Key words: hydrosilylation, enantioselective catalysis, Car-Parrinello, *ab initio* molecular dynamics, combined QM/MM, hybrid methods

F. Maseras and A. Lledós (eds.), Computational Modeling of Homogeneous Catalysis, 213–252.

1. INTRODUCTION

The hydrosilylation reaction involves the addition of a hydrosilane across a carbon-carbon multiple bond as depicted in Figure 1. Hydrosilylation provides a convenient and efficient method of synthesizing organosilicon compounds and organosilyl derivatives. As a result, the reaction is widely used in laboratories and industry for this purpose [1]. Homogeneous transition metal-based catalysts have played an important role in the practical application of the hydrosilylation reaction. One of the best known transition metal hydrosilylation catalysts is Speier's catalyst, H_2PtCl_6 [2, 3]. Incorporation of chiral ligands into the catalyst framework, has lead to the development of enantioselective hydrosilylation catalysis. At present, high enantiometeric excesses (ee) up to 95-99% can be achieved with a broad range of olefinic substrates. Due to this tremendous progress, the hydrosilylation reaction has become an important synthetic route for the preparation of optically active alcohols, amines and alkanes [4, 5] all of which are important targets in the pharmaceutical and agrochemical industries. Transition metal catalyzed enantioselective hydrosilylation is also of some commercial importance for the production of silicon rubbers.

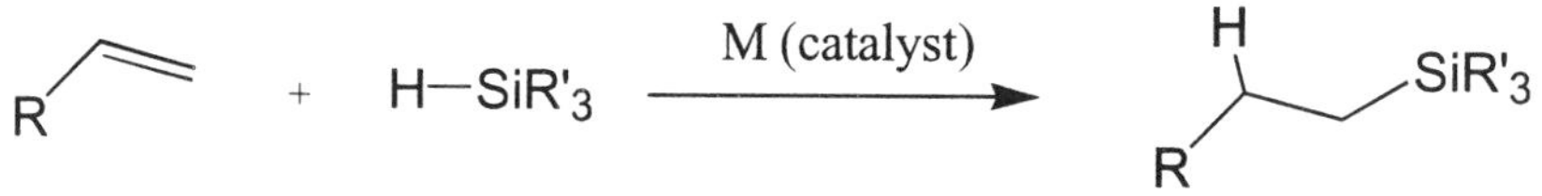

Figure 1. Generalized hydrosilylation reaction catalyzed by a transition metal complex

Catalytic hydrosilylation with trichlorosilane offers a powerful tool for a one-pot conversion of olefins into chiral alcohols [6]. The alkyltrichlorosilane formed as product of the catalytic reaction is subsequently oxidized following a method developed by Tamao [7-9]. Because of the complete retention of configuration during the oxidation step, this synthetic route allows the stereoselective generation of alcohols. Recently, Togni and co-workers have achieved asymmetric hydrosilylation of olefins [6] using chiral ferrocenyl ligands [10]. The catalyst precursor dichloro {1-{(R)-1-[(S)-2(diphenylphosphino -κP)ferrocenyl] ethyl}-3-trimethylphenyl- 5-1H pyrazole -κN}palladium **1** (see Figure 2) has been utilized for the enantioselective conversion of norbornene [6], producing after oxidative workup, exo-norborneol with an high enantiomeric excess of 99.5%. This is indeed impressive, but in the hydrosilylation of styrene derivatives the observed ee's do not exceed 67% [6, 11]. More interestingly, an inversion in selectivity from the R form to the S form for a series of *para*-substituted styrenes as detailed in Table 1 is observed. Empirically it was found that the electron withdrawing or releasing nature of the para-substituent plays a crucial role in determining both the level and the sense of chiral induction. From Table 1 it is apparent, the highest selectivity (64% ee) for the *R* form is achieved for 4-(dimethylamino)styrene, whereas the corresponding chloro derivative leads predominantly to the *S*-product (67.2 % ee). This constitutes a rare example in which a reversal of the enantioselectivity is induced by the electronic properties of the substrate, as modified by a peripheral substituent. Although the stereoselectivity of the hydrosilylation exhibits a strong variation, the regioselectivity in all cases highly favors the branched product (Figure 2). Indeed the unbranched product is not observed providing a regioselectivity greater than 99%.

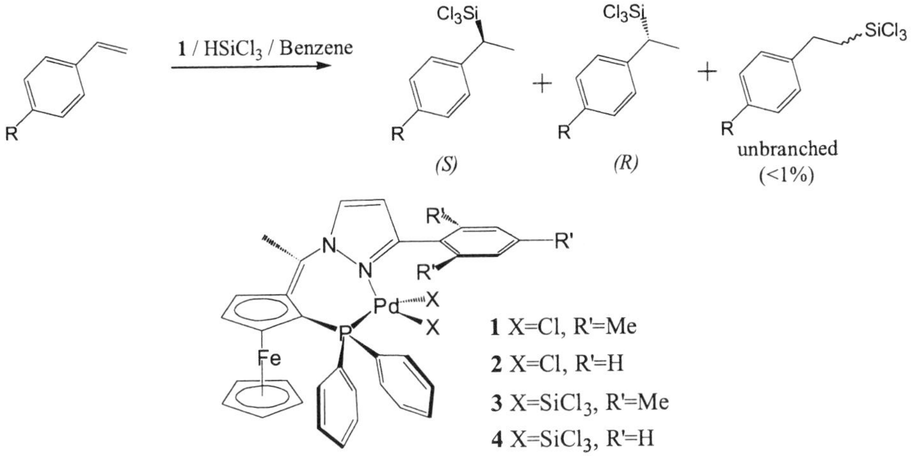

Figure 2. The Pd-catalyzed hydrosilylation of *para*-substituted styrenes with the structure of the precatalysts and their derivatives.

A number of interesting and fundamental questions regarding the nature of the pre-catalyst and the mechanism of hydrosilylation have arisen during the development of catalyst system [11]. In fact, very little is known about the mechanistic details of the catalysis. As it is usually not possible to isolate unstable reaction intermediates, almost no experimental data is available about the species involved in the catalytic cycle. Therefore, a theoretical approach seems particularly well suited to obtain direct mechanistic information at the atomic scale that might help rationalize the experimental observations. However, a theoretical characterization of transition metal catalyzed reactions is far from trivial since the catalysts involved are rather large organometallic complexes. Moreover, a theoretical investigation of enantioselective reactions is an even more difficult task since the difference in free energies between stereomorphous pathways are generally small [12].

Table 1. Results of the hydrosilylation of styrene and its derivatives with trichlorosilane with **1** as a catalyst precursor.

Para-substituent[a] of styrene substrate	%ee Product and absolute configuration
Cl	67% S
H	64% S
CF$_3$	59% S
Me	47% S
OMe$_3$	6% R
Nme$_2$	64% R

[a]substituent R in Figure 2.

In this chapter we describe how we have used quantum chemical simulation techniques to help answer some of the puzzling and fundamental questions regarding the enantioselective hydrosilylation of styrenes by the palladium catalyst complex, **1** (Figure 2) [6]. This chapter has been arranged according to the specific questions we have tried to answer with computer simulation. Each of the sections, 2 through 5, is intended to be independent of one another. Consequently, some of the sections include summaries of results first presented in an earlier section. In Section 2, we use molecular modeling to help rationalize some of the unique structural features observed in the precatalysts and their derivatives, **1-4**. For example, why does the catalyst derivative **3** possess the world's longest known Pd-P bond distance, whereas its close analogue **2** has a rather modest Pd-P bond distance.   In Section 3, we have examined the mechanism of hydrosilylation in detail at the atomic level. We have tried to answer some basic mechanistic questions such as whether the hydrosilylation more likely proceeds through the classic Chalk-Harrod mechanism or through what is known as the modified or anti-Chalk-Harrod mechanism. Section 4 describes how molecular modeling has been used to rationalize why the hydrosilylation of styrenes by these catalysts are highly regiospecific. Finally, a detailed mechanistic study of the enantioselective hydrosilylation of styrene is given in Section 5.

We have used a variety of quantum chemical techniques - some standard and some rather novel. The general features of these techniques are described in a non-technical fashion in the sections as we use them. We provide technical details of the calculations in the Appendix for the interested reader.

2.   RATIONALIZATION OF THE STRUCTURE OF THE CATALYST AND ITS DERIVATIVES

The Pd dichloro complexes **1** and **2** (Figure 2) are precatalysts for the enantioselective hydrosilylation reaction developed by Togni and co-workers

[6]. The precatalysts are found to smoothly react with an excess of HSiCl₃ to give the corresponding *cis*-bis(trichlorosilyl) derivatives, **3** and **4**, respectively. Since isolated and characterized Pd trichlorosilyl complexes of this kind are extremely rare, the solid state structure of **3** and **4** were determined by X-ray crystallography. In comparison to analogous dichloro complexes and other similar compounds, the bis(trichlorosilyl) derivatives posses some unexpected structural features. For example, the bis(trichlorosilyl) complex **3**, possesses the longest known Pd-P bond distance of 2.50 Å [13], whereas similar dichloro complexes, such as **2**, have rather modest Pd-P bond distances of 2.26 Å. (Suitable crystals of **1** could not be isolated.) An obvious explanation of this bond elongation is the strong trans-influence of the silyl groups [14-17]. On the other hand, the Pd-Si distances fall in the lower range of metal-silicon bond lengths [18-20], which suggests that other factors might be involved.

Another prominent feature of the two bis(trichlorosilyl) complexes is the severe distortion of the coordination about the Pd center from the ideal square planar geometry that is expected for Pd⁰ complexes of this type. The deviation from the ideal square planar geometry can be quantified by the angle θ between the P-Pd-N and Si-Pd-Si planes [21]. The distortion angle, θ, was determined to be 34° and 18° for **3** and **4**, respectively, whereas it is only 9° in the analogous dichloro complex **2**. It is natural to attribute the tetrahedral distortion in **3** and **4** to the greater steric demands of the trichlorosilyl moieties as compared to the chloro moieties. However, it is also plausible that there might be an electronic contribution of the trichlorosilyl ligands to this distortion.

In view of the magnitude of the geometric distortions, particularly in complex **3**, and with the aim of ascertaining their exact origin, we have performed a computational investigation of these derivatives. Here we demonstrate a straightforward computational method for separating steric from electronic ligand effects in the catalyst structure. We believe this is of interest in catalysis research, because in principle the fine tuning of catalytic processes can be achieved by the systematic manipulation of the steric and electronic properties of the ancillary ligands.

2.1   The Combined QM/MM Method

In order to examine the nature of the unusual geometric distortions observed in the X-ray structures of complexes **3** and **4**, we have performed a series of density functional theory (DFT) and combined quantum mechanics and molecular mechanics (QM/MM) calculations [22-27]. Although the computational resources to wholly treat both **3** and **4** at the DFT level are available, we have employed the combined QM/MM method to unravel the steric factors from the electronic effects that influence the distortions observed. With the combined QM/MM method, part of the molecular system is treated with an electronic structure calculation, in this case DFT [28, 29], while the remainder of the complex is treated with a molecular mechanics approach. The different models **A-F** that have been utilized in this study are illustrated in Figure 3. For models **B**, **D** and **E** the QM/MM partitioning is also indicated in Figure 3, as a dotted polygon.

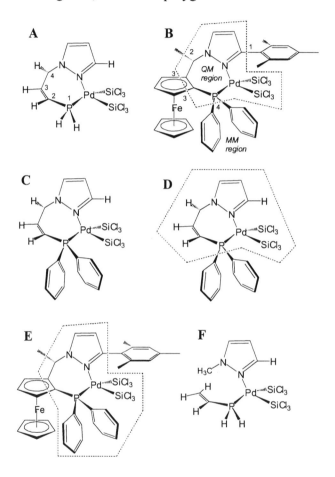

Figure 3. Structural representation of the computational models A-F. Models **B**, **D**, and **E** are combined QM/MM models where the regions enclosed in the dotted polygons represent the QM region. The regions outside the polygons are treated by a molecular mechanics force field. For the electronic structure calculation of the QM region, the covalent bonds that traverse the QM/MM boundary (the dotted polygon), have been capped with hydrogen atoms. In model **A** the atoms labelled 1 through 4 are the atoms that have been fixed in the calculations of models **A** through **E**.

There exist many varieties of combined or hybrid QM/MM methods. In the IMOMM [26, 30] approach that we have utilized, electronic effects across covalent bonds that traverse the QM/MM boundary are for the most part neglected. The primary reason for this is that the electronic structure calculation is performed on a truncated "QM model system" whereby the boundary bonds are terminated with 'capping' atoms, usually hydrogens. The capping atoms satisfy the valence requirements across the boundary bonds such that a standard electronic structure calculation can be performed on the QM model system. The important point is that with the QM/MM approach that has been utilized, only the steric influences of the fragments contained in the MM region are modeled. For example, in the QM/MM model **B**, only the steric influence but not the electronic donor/acceptor properties of the phosphine phenyl groups are accounted for (more precisely, the electronic effects of the phenyl rings are made equivalent to those of the capping atoms). Although this is often a limitation with the QM/MM method, it can be advantageous in view of separating as much as possible the steric and electronic effects of specific molecular fragments [31, 32].

2.2   Nature of the Pd-P elongation

The nature of the extreme lengthening of the Pd-P bonds in the bis(trichlorosilyl) complexes **3** has been examined with a particular emphasis on separating the steric from the electronic influences. Although an obvious rationalization for the anomalous Pd-P distances is the strong *trans*-influence of the silyl group (*vide supra*), another plausible explanation involves the steric pressure that is exerted on the Pd-P bond due to the interaction of the phenyl groups with the bulky trichlorosilyl fragments. To pinpoint the nature of the elongation we have generated a systematic series of model systems, in which the steric and electronic influences of the phenyl ligands have been selectively altered. We start with the QM model **A**, which is depicted in Figure 3.

Table 2. Comparison of selected geometric parameters derived from the experimental X-ray structure of **3** and those of the various calculated models.[c]

parameter[a]	exp. X-ray	full QM[b]	A	B	C	D	E	F
Pd-P	2.50	2.53	2.39	2.46	2.42	2.40	2.50	2.39
Pd-N	2.21	2.25	2.18	2.20	2.19	2.19	2.21	2.19
P-Pd-N / Si-Pd-Si plane angle	34.0	35.7	11.8	30.4	18.6	15.8	34.6	1.4

[a]Distances reported in Å, angles reported in degrees. [b] This refers to the fully optimized structure of **3** that was wholly calculated at the DFT level. [c]See Figure 3 and text for description of model systems **A** through **D**.

Key geometric parameters for the optimized structure of model **A** are detailed in the fourth column of Table 2. The removal of the phenyl substituents of the phosphine and the trimethylphenyl substituent of the pyrazole ring results in a significant contraction of the Pd-P bond from 2.504 Å to 2.39 Å. We note that a distance of 2.39 Å for a Pd-P bond still lies in the upper end of the literature range, with a distance of 2.437 Å being recently reported [33]. From model **A** it can be concluded that both sets of phenyl substituents in **3** are principal components in the extreme lengthening of the Pd-P bond. Thus, the 'world record' lengthening can not be fully explained in terms of the *trans*-influence of the silyl groups.

Next let us examine a QM/MM model system of complex **3** where the steric influence of the phenyl substituents and of the ferrocene is accounted for but the electronic effects have been largely eliminated (model **B**). In other words, the peripheral groups have been delegated to the MM region, while keeping the molecular system used for electronic structure calculation identical to that in model **A**.

Table 2 reveals that compared to model **A**, the Pd-P bond distance in model **B** is elongated from 2.39 Å to 2.46 Å. Thus, it can be concluded that the steric influence of the phenyl substituents plays a substantial role in the elongation of the Pd-P bond in **3**. The trimethylphenyl substituent of the pyrazole ring in **3** is oriented perpendicular to its host. This results in a strong steric interaction between the trimethylphenyl group and one of the phenyls of the phosphine, thus putting pressure on the Pd-P bond to elongate. Although the results of these two model systems suggest that the steric influence of the phenyl substituents is a significant factor in the long Pd-P bond distance observed in **3**, they do not account for it completely. For example, the Pd-P bond distance of 2.46 Å in **4** is also extremely long, even though the phenyl substituent of the pyrazole ring in **3** lies almost coplanar to its host. Perhaps part of the lengthening observed in model **B**, as compared to model **A**, is a result of the interaction between the phosphine phenyl rings and one of the bulky trichlorosilyl groups (see also model **C** below).

With the next model, we attempt to isolate the effect of the phosphine phenyl substituents from that of the trimethylphenyl group on the pyrazole ring. Thus, in the pure QM model **C** the electronic structure of the phenyl rings of the phosphine is treated explicitly, but the trimethylphenyl group is completely removed (replaced by hydrogen). Table 2 shows that compared to model **A**, explicit treatment of the phosphine phenyl groups results in a moderate elongation of the Pd-P distance from 2.39 Å to 2.42 Å. This lengthening of the Pd-P bond in model **C** could be attributed to the electronic and/or steric influence of the PPh_2 group compared to the PH_2 in model **A**.

To separate the steric effects of the phenyl phosphine groups from their electronic effect, model **D** was constructed. This model is similar to model **C** except that the phenyl rings are treated by a molecular mechanics potential. Thus, with model **D**, only the steric influence of the phenyl phosphines is modeled. The Pd-P distance in model **D** is 2.40 Å, revealing that the steric lengthening of the Pd-P bond due to the interaction between the SiCl₃ fragments and the phosphine phenyl groups is minimal. Thus, it can be concluded that compared to PH_2 the PPh_2 group plays a role in elongating the Pd-P bond through electronic effects, as one would qualitatively expect based on the donor ability of these two fragments. An examination of the Mulliken charges on the P atom in model **A** compared to **C** is consistent with this picture. In model **A** the phosphorous has a calculated Mulliken charge of +0.12 e, whereas it is determined more electron deficient in model **C** with a Mulliken charge of +0.25 e.

Comparison of the Pd-P bond distance in model **B** with that of model **D**, suggests that much of the lengthening of the Pd-P bond in complex **3** results from the steric interaction between the phenyl groups of the phosphine and the trimethyl phenyl group of the pyrazole ring. As discussed above, this interaction is strong in complex **3** due to the orientation of the trimethylphenyl group. The primary difference between models **B** and **D** is that in the latter the steric interaction between the two substituent groups is absent. Therefore, it can be concluded that this steric interaction approximately accounts for an elongation of the Pd-P bond from 2.40 Å (model **D**) to 2.46 Å (model **B**). Additionally, there is minimal elongation due to the interaction between the phosphine phenyl groups and one of the bulky trichlorosilyl groups.

Finally, we have constructed a QM/MM model of complex **3** whereby the phenyl phosphine groups are contained in the QM region, while the ferrocenyl and phenyl substituents on the pyrazole is accounted for on a steric basis only (model **E**). The agreement between the X-ray structure of **3** and model **E** is remarkable. For example, both the Pd-P distance and the twist of the coordination plane of the Pd center, θ, are virtually identical to those of the X-ray structure. In fact, the selected parameters displayed in Table 2 are generally better than those of the full QM calculation. The good agreement between the calculated and experimental structures is important for the detailed mechanistic study of the hydrosilylation that is presented in later sections of this chapter.

2.3   Nature of the non-square planar coordination of the Pd center

Both complexes **3** and **4** possess an anomalous non-square planar coordination geometry of the Pd center, as indicated by the angle θ between the planes defined by the P-Pd-N and Pd-Si-Si atoms. Model **A** has a slightly distorted square planar geometry as demonstrated by the θ angle of 11.8° (Table 2). This is somewhat surprising since the pendant phenyl groups on the pyrazole ring and the phosphine are not present in this model. Thus, even without the presence of these groups there is a slight preference away from the ideal square planar geometry. Detailed examination of the geometry reveals that the structure of Pd-ferrocenyl-phosphine-pyrazole chelate ring orients the pyrazole such that there is a notable steric interaction between the substituent in the 3 position (in this case H) of the heterocycle with the trichlorosilyl group *trans* to the phosphine [20]. Figure 4 depicts this interaction for model **A** if an ideal square planar coordination is enforced.

In model **F**, the chelate ring is broken thereby removing the imposed orientation on the pyrazole fragment. Optimization of model **F** reveals that a near ideal square planar structure is favoured with a θ angle of 1.4°. Interestingly, when the optimization of model **F** is initiated from a distorted structure, a non-square planar stationary point with an θ angle of 25° is located 4.5 kcal/mol above that of the planar structure. Thus, it can be concluded that in spite of the electronic preference for the square planar coordination of the Pd center, the steric effects imposed by the structure of the chelate ring result in a slight distortion of the structure.

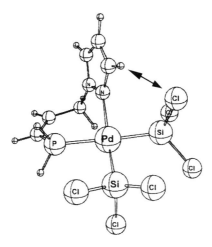

Figure 4. Geometry of model **A** showing the steric interaction between the trichlorosilyl ligand and the hydrogen at the 3 position of the pyrazole ring that results in a slight deviation from the intrinsic square planar coordination geometry at Pd.

In the X-ray structures of both **3** and **4**, the tetrahedral distortion is greater than that found in model **A**. Since there is an electronic preference for a square planar coordination, the pendant phenyl substituents in **3** and **4** likely result in further steric crowding and therefore in more distorted structures compared to model **A**. In the QM/MM model **B** an optimized θ angle of 30° is found, close to the 34° angle of the X-ray structure. Since the phenyl and trimethyl phenyl groups are accounted for on a steric basis only in model **B**, the result supports the notion that the severely distorted coordination of the Pd center in **3** is due to a steric effect.

In models **C** and **D** the phenyl substituent of the pyrazole ring is not present and here the θ angles are 19° and 16°, respectively. Even without the trimethylphenyl moiety, the Pd coordination is more distorted than in model **A**, suggesting that the phenyl rings of the phosphine enhance the distortion. Furthermore, since the magnitude of the distortion observed in model **C** is almost the same as that in model **D**, it can be concluded that the effect of the phenyl phosphine groups on the coordination geometry is steric in nature and not electronic.

2.4 Examining Steric and Electronic Structural Effects of Ligands. A Summary

Using the combined QM/MM method a systematic series of model systems have been constructed in which the steric and electronic influences of the substituent groups were selectively removed or altered. The method has been effectively used as an analytical tool to probe the nature of the unique structures observed in complex **3**. The extreme lengthening of the Pd-P bond in **3** are a combination of electronic and steric effects. In this respect, the phosphine phenyl substituents contribute to the Pd-P bond lengthening on both a steric and electronic level. The strong trans influence of the trichlorosilyl ligand is enhanced by the presence of the phosphine phenyl groups. The "world record" Pd-P bond distance can be primarily attributed to the strong steric interaction between the PPh$_2$ group with the trimethylphenyl fragment on the pyrazole ring.

The deviation from the square planar geometry at the Pd center in complexes **3** can be attributed wholly to steric factors. Calculations have shown that there is a moderate electronic preference favouring the planar coordination amounting to approximately 4.5 kcal/mol. However, steric interactions between the bulky trichlorosilyl ligands with both phenyl substituents of the phosphine and pyrazole ring result in a distortion away from the ideal square planar geometry. We further have found that the framework of the specific chelate backbone positions the pyrazole ring such that these interactions are enhanced.

3.   HYDROSILYLATION MECHANISM: CHALK-HARROD VS. MODIFIED-CHALK-HARROD

The conventional hydrosilylation of alkenes catalyzed by homogeneous transition metal catalysts such as Speier's catalyst, are generally assumed to proceed by the Chalk-Harrod mechanism [34] as shown on the right hand side of Figure 5. Oxidative addition of a hydrosilane, in this case HSiCl$_3$, gives a hydrido-silyl complex which is depicted in the center of Figure 5. Coordination of this complex with the olefinic substrate produces a π-complex, which to date are very seldom isolated. Olefin insertion into the metal-hydride bond (*hydrometallation*) gives a silyl-alkyl species which followed by Si-C reductive elimination yields the hydrosilylation product, which is highlighted with a box in Figure 5. An alternative proposed mechanism, that is usually termed the 'modified-Chalk-Harrod' mechanism [35, 36] is depicted on the left-hand-side of Figure 5. With the modified-Chalk-Harrod mechanism, the olefin inserts into the M-Si bond (*silylmetallation*) instead of the M-H bond as with the Chalk-Harrod mechanism. Following olefin insertion, C-H reductive elimination yields the hydrosilylation product. Although several experimental studies addressing the mechanism of hydrosilylation have been performed for ruthenium [37], rhodium [38-40], iridium [41-42], palladium [43], and zirconium [44] based catalysts, it is still not clear which one of the two possible mechanisms is the most favorable one and also, if depending on the particular catalyst [37, 38] a different reaction pathway is preferred. Only few theoretical studies [45, 46] exist so far aiming at a better mechanistic understanding of this reaction. However, these theoretical studies, performed on small models of a platinum catalyst, confirm the general validity of the classical Chalk-Harrod mechanism.

Using quantum chemical molecular modelling tools we have examined the reaction mechanism of palladium catalyzed hydrosilylation of styrene by the precatalyst system, **1**, developed by Togni and co-workers. One fundamental question that we have focused on is whether the reaction proceeds by the classical Chalk-Harrod mechanism or by an alternative mechanism such as the modified-Chalk-Harrod mechanism. In this section, the general features of the catalytic cycle are examined.

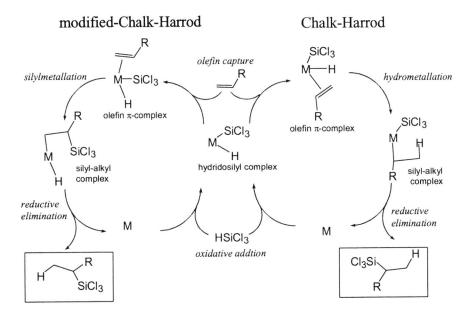

Figure 5. General schemes for the Chalk-Harrod (right) and the modified-Chalk-Harrod (left) mechanisms for the metal catalyzed hydrosilylation of olefins.

3.1   Computational Approach: General Considerations

In this and subsequent sections, we investigate the reaction mechanism of the palladium catalyzed hydrosilylation of styrene via *ab initio* molecular dynamics and combined quantum mechanics and molecular mechanics (QM/MM) techniques. Both methodologies constitute powerful approaches for the study of the catalytic activity and selectivity of transition metal compounds [13, 22, 47-49]. Although complexes of the size of **1** can be treated entirely at the first-principles level (as we have done in the previous section), these calculations are particularly time consuming. With the computing resources available, we felt that it was too time consuming to examine, in detail, the potential energy surfaces of the multiple reaction channels possible at this level. Thus, we have used series of model systems to reduce the total amount of time required to perform the study. We have started with a minimal quantum mechanical model system which we have labelled model **A** as shown in Figure 3. Compared to the real catalyst complex, **1**, the bulky phosphine phenyl groups, the ferrocenyl ligand and the mesityl substituent on the pyrazole ring have been stripped off in model **A**. As a result, both on a steric and electronic effects of these ligands are not accounted for in model **A**. This model is clearly a drastic approximation of the real catalyst. Nevertheless, due to its low computational cost, it allows an efficient exploration of a vast number of possible isomeric structures and reaction pathways. The potential energy scans using model **A** served as a guide for more complex and time consuming potential energy studies of the complete catalyst system. In most cases, the results using model **A** are not presented here, but are described elsewhere [58a].

We have performed calculations on the whole catalyst system using more realistic QM/MM models **B** and **E** as shown in Figure 3 and described in Section 2.1. In models **B** and **E** in Figure 3, the dashed polygons represent the QM/MM boundary with the quantum mechanical region being located inside the polygon. The important difference between models **B** and **E** is that in model **B** the phosphine phenyl groups are treated only on a steric level since they are part of the molecular mechanics region. In model **E**, these phenyl groups are included in the electronic structure calculation and therefore both the steric and electronic effects are accounted for. A comparison of the calculations performed with the minimal QM model **A** and the two larger QM/MM models **B** and **E**, can be a useful tool to rationalize the steric or the electronic influence of the large substituents in a similar manner that was introduced in Section 2 of this chapter.

In some situations we have performed finite temperature molecular dynamics simulations [50, 51] using the aforementioned model systems. On a simplistic level, molecular dynamics can be viewed as the simulation of the finite temperature motion of a system at the atomic level. This contrasts with the conventional 'static' quantum mechanical simulations which map out the potential energy surface at the zero temperature limit. Although 'static' calculations are extremely important in quantifying the potential energy surface of a reaction, its application can be tedious. We have used *ab initio* molecular dynamics simulations at elevated temperatures (between 300 K and 800 K) to more efficiently explore the potential energy surface.

Since the catalyst is large and has a high degree of conformational variability, the *ab initio* molecular dynamics simulations have been used to provide an initial scan of the potential energy surfaces. This approach has been previously applied to study homogeneous catalytic systems [47, 52-54].

3.2 The Coordination of Styrene

Following olefin coordination, the Chalk-Harrod mechanism proceeds by olefin insertion into the M-H bond, whereas with the modified Chalk-Harrod mechanism, olefin coordination is followed by insertion into the M-Si bond. This step distinguishes the two mechanisms. Thus, the coordination of styrene to the hydridosilyl complex to form an olefin π-complex may be the first step of the catalytic cycle that discriminates between the two mechanisms. We have examined this coordination process as well as the relative energies of the many isomers of the π-complex that are possible.

3.2.1 The Hydridosilyl Complex

As a starting point for studying the coordination process, we have examined the hydridosilyl complex shown in the center of Figure 5. Two different structural isomers have been considered which are shown in Figure 6 - a form with the hydride *cis* (**5a**) and *trans* (**5b**) to the phosphine ligand. Within the minimal model **A**, the *cis* isomer is slightly favored with respect to the *trans* counterpart by 2.2 kcal/mol. When considering the whole catalyst structure by using model **B**, the relative energies of the *cis* and *trans* isomers are inverted. In fact, due to steric interactions between the mesityl and the silyl ligand, the *cis* isomer **5a-B** is less stable than the *trans* isomer **5b-B** by 4.9 kcal/mol.

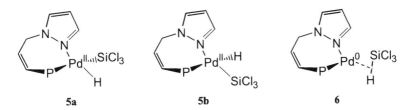

Figure 6. Pd^{II}(hydrido)silyl *cis* and *trans* isomers **5a** and **5b**, respectively. A proposed Pd^0-η^2-trichlorosilane complex, **6**. Selected substituents have been removed for clarity.

Although we have assumed this complex to be a Pd^{II}(hydrido)silyl complex it is plausible that this complex may be characterized as a Pd^0-η^2-

trichlorosilane complex where the H-Si bond is not completely broken. However, using model systems **A** and **B** attempts to find a η^2 complex failed, with optimization always leading to a hydridosilyl complex, **5**.

3.2.2   Ligand Dissociation Before Styrene Coordination?

In their initial studies of this catalyst system Pioda and Togni [6] proposed that olefin coordination would likely require a ligand dissociation. They proposed that the opening of the chelate ring, with the pyrazole nitrogen dissociating as shown in the lower pathway in Figure 7, could fulfill this condition. However, no experimental evidence could be obtained for the ligand dissociation accompanying the styrene coordination.

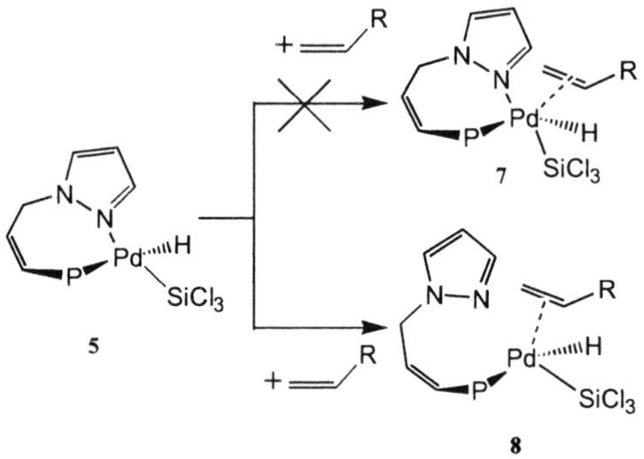

Figure 7. Coordination of Styrene. The lower pathway shows olefin coordination accompanied by detachment of the pyrazole ligand. The upper pathway with retention of the coordinated pyrazole ligand was not observed.

Our computational studies suggest that olefin coordination must be accompanied by ligand dissociation. A penta-coordinated π-complex, **7**, could not be found. Any attempt to coordinate the styrene substrate, without ligand dissociation lead to immediate ejection of the substrate. This is the case for the minimal model **A** and the full QM/MM model **B**. High temperature molecular dynamics simulations on the hydridosilyl complex, **5a**, using both models **A** and **B**, revealed that the pyrazole ligand was labile and if olefin coordination required ligand dissociation that it would likely be the pyrazole ligand.

The coordination of styrene with the pyrazole detached resulted in the formation of a stable π-complex (*vide infra*). Therefore, throughout the catalytic cycle the pyrazole plays the role of a hemilabile ligand [55]. This is an important result because experimental studies [11] suggest that olefin coordination is slow and reversible. Moreover, the tuning of the steric bulk of the pyrazole substituents could be used to enhance the activity of the catalysis if the attachment or detachment of the ligand is involved in the rate determining step.

3.2.3   Styrene Coordinated Complexes

The coordination of styrene is expected to be strongly influenced by substituents that are neglected in the minimal QM model **A**. Thus, for sake of clarity, we do not present the styrene coordination results using model **A**. Depicted in Figure 8 are the three most stable styrene coordinated isomers, **8a-c**. The coordination energies, which are also shown in Figure 8 in kcal/mol, reveal that the initial formation of the π-complex is slow and reversible. In fact, only for isomer **8a** is the styrene coordination exothermic and here it is only exothermic by 0.5 kcal/mol. Isomers **8a-c** all have the olefinic bond of the styrene lying parallel to the plane defined by the P-Pd-Si atoms. No other sterically accessible isomers could be located where this bond lies parallel to this plane. Due to steric reasons, complexes with the olefinic bond perpendicular to this plane were found to be at least 8 kcal/mol less stable.

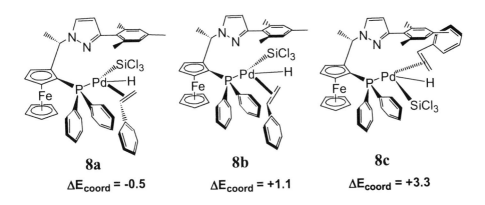

8a	**8b**	**8c**
ΔE_{coord} = -0.5	ΔE_{coord} = +1.1	ΔE_{coord} = +3.3

Figure 8. Most stable styrene coordinated complexes. The styrene coordination energies are reported in kcal/mol. The coordination energies are reported as the energy of the π-complex relative to the energy of the free styrene molecule and the respective hydrido-silyl complexes.

The fact that isomers **8a**, **8b** and **8c** are the lowest energy styrene coordinated complexes have potentially important ramifications that concern the modified-Chalk-Harrod mechanism and the regioselectivity observed in the hydrosilylation. With the modified-Chalk-Harrod mechanism olefin insertion into the M-Si bond follows styrene coordination. However, in all three isomers depicted in Figure 8, the coordinated hydride lies between the olefin and the M-Si bond. Thus, the structure of the π-complexes shown in Figure 8 would appear to better facilitate the olefin insertion into the M-H bond and therefore favour the Chalk-Harrod mechanism. This does not rule out the possibility of the modified-Chalk-Harrod mechanism if there is a high electronic barrier to olefin insertion into the M-H bond as compared to olefin insertion into the M-Si bond. We will discuss the insertion barriers in more detail in the next two sub-sections (3.3 and 3.4).

If we were to assume that the reaction followed the Chalk-Harrod mechanism, then insertion of the olefin into the Pd-hydride bond in all three isomers **8a-c** would lead to the correct regioisomer product. Thus, to some degree the regioselectivity of the hydrosilylation in this catalyst system is determined in the styrene coordination step. We will discuss the origin of the regioselectivity in more detail in Section 4.

3.3   A Computational Exploration of the Chalk-Harrod Mechanism

We first explore possible reaction pathways that follow the Chalk-Harrod mechanism, and shall in Section 3.4 discuss pathways associated with the modified Chalk-Harrod mechanism. According to a classic Chalk-Harrod mechanism, the insertion of the styrene into the palladium-hydride bond constitutes the next step of the catalytic cycle following olefin coordination. We have examined the relative stabilities of the migratory insertion products as well as the insertion barriers. Using the full QM/MM model **B**, we have limited our study to the migratory insertion processes commencing from the three lowest energy π-complexes **8a-c** (Figure 8).

3.3.1   Migratory Insertion of Styrene into the Pd-Hydride Bond

The product of the migratory insertion commencing from the three lowest energy π-complexes, **8a**, **8b** and **8c**, leads directly to three distinct silyl-alkyl complexes **9a**, **9b** and **9c**, respectively, which are represented in Figure 9. Among these isomers, complex **9a** turns out to be the thermodynamically most stable. Complex **9a** and **9b** differ by 5.1 kcal/mol, while the energy difference increases drastically to 23.9 kcal/mol for **9c**. A possible rationale of the high thermodynamic instability of **9c** is the occurrence of close steric contacts of 2.1 Å between a hydrogen of the substrate and a hydrogen of the chelating ligand. Shown in Figure 9 are the calculated insertion energies, ΔE_{ins}, defined simply as the energy difference between the silyl-alkyl complexes, **9**, and their corresponding π-complexes, **8**. The insertion process was found to be endothermic for all isomers. For isomers **8a** and **8b** it was found to be only slightly endothermic, with ΔE_{ins}=0.9 and 3.6

kcal/mol, respectively, while it was significantly larger for **8c** where it was found to be ΔE_{ins}=15.2 kcal/mol.

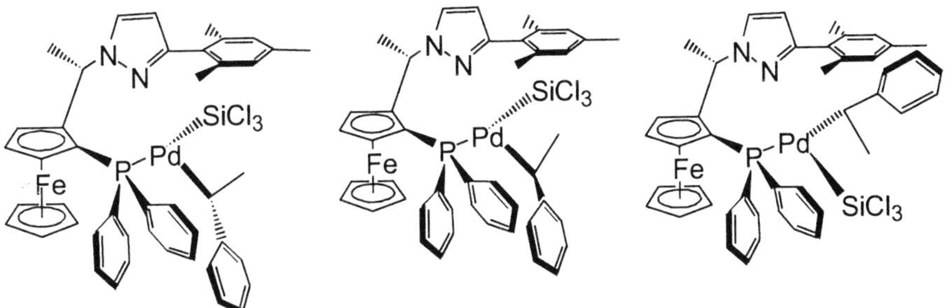

9a

$\Delta E_{ins} = +0.9$

9b

$\Delta E_{ins} = +3.6$

9c

$\Delta E_{ins} = +15.2$

Figure 9. Silyl-alkyl isomers, the migratory insertion products.. The migratory insertion energies, ΔE_{ins}, are reported as the difference in energy (kcal/mol) between the silyl-alkyl complex and the respective π-complexes.

The calculated activation barrier for migratory insertion of the substrate into the palladium-hydride bond was determined to be 4.2 kcal/mol for the pathway **8a** to **9a**. For isomer **8c**, a large thermodynamic insertion barrier of $\Delta E_{ins} = +15.2$ kcal/mol exists, so the activation barrier transforming **8c** to **9c** was not examined further.

3.3.2 Formation of a Silyl(η^3-Benzyl) Complex

It has been proposed that when styrene is the substrate in transition metal catalyzed hydrosilylation, the silyl-alkyl complex can isomerize to form a η^3 allyl complex. As depicted in Figure 10, the rearrangement can occur in two ways, depending on which of the two diastereotopic *ortho*-carbons of the phenyl group coordinates to the Pd center. This leads to the formation of two different isomers in which the Pd-coordinated ortho-carbon of the phenyl group is *syn* or *anti* to the methyl group of the benzylic fragment. (The methyl group of the benzylic fragment is generated upon migratory insertion of the hydride to the β-carbon.) It is important to realize that the *syn-anti* pairs are stereochemically equivalent, as they will lead to the same stereoisomer of the product.

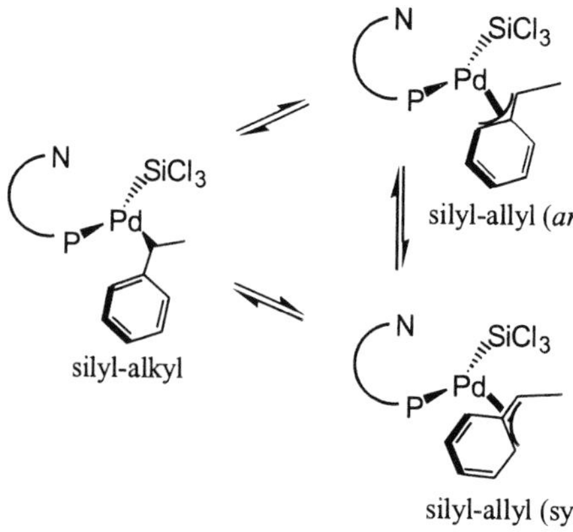

Figure 10. A rearrangement of the silyl-alkyl insertion products leads to the formation of two distinct η^3-silyl-allyl intermediates (*syn* and *anti*).

Our calculations show that the isomerization of the silyl-alkyl complex to form a η^3-allyl complex affords a significant stabilization as summarized in Figure 11. The η^1 to η^3 isomerization of **9a** to the *anti* silyl-allyl complex, **10a-*anti***, results in a 9.5 kcal/mol stabilization and isomerization to the *syn* isomer, **10a-*syn***, results in a 7.2 kcal/mol stabilization. Isomerization of **9b** to the *anti* silyl-allyl complex, **10b-*anti***, results in a 6.1 kcal/mol stabilization and isomerization to the *syn* isomer, **10b-*syn*** results in a 5.4 kcal/mol stabilization. High temperature (500 °C) molecular dynamics simulations initiated at the η^1 complex, **9a**, reveal that the η^1 to η^3 isomerization has a minimal barrier and occurs in the sub-pico time frame. The inter-conversion between the *syn* and *anti* isomers has not been examined since both isomers are stereochemically equivalent, however, we expect the barrier to be small.

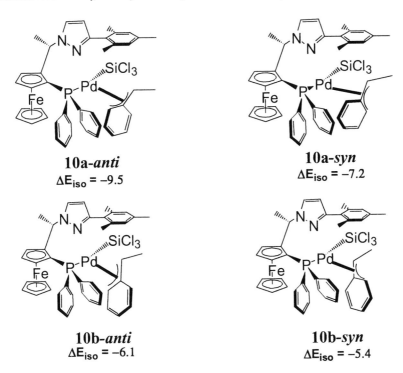

10a-*anti*
$\Delta E_{iso} = -9.5$

10a-*syn*
$\Delta E_{iso} = -7.2$

10b-*anti*
$\Delta E_{iso} = -6.1$

10b-*syn*
$\Delta E_{iso} = -5.4$

Figure 11. Silyl-allyl complexes as a result of the η^1 to η^3 isomerization from **9a** and **9b**. The isomerization energies, ΔE_{iso}, are reported in kcal/mol.

The formation of the highly stabilized η^3-allyl complexes have important implications to the regioselectivity of the hydrosilylation. The reason for this is that the η^3-allyl complex cannot form if the hydride is transferred to the β-carbon of the styrene which leads to the unbranched regioisomer of the product. We will discuss the regioselectivity in more detail in Section 4.

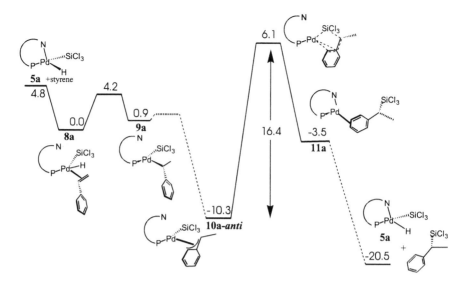

Figure 12. Calculated reaction profile with QM/MM model **B** of the catalytic cycle for the most favoured pathway.

3.3.3   Reductive Elimination

The final step of the Chalk-Harrod mechanism that is distinct from the modified Chalk-Harrod mechanism is the reductive elimination to yield the product. For this step, we have only examined the reductive elimination from the most stable η^3-allyl complex **10a-*anti***.

Just as detachment of the pyrazole ligand was associated with the styrene coordination, one might expect recoordination to be associated with the elimination of the product. To help us examine this potentially complex process we have performed a series of molecular dynamics simulations at 500°C with the Si-C$_\alpha$ bond distance constrained at various lengths. Through these simulations it is possible to observe the detachment of the silyl ligand from the Pd center and its transfer to the α-carbon of the benzylic fragment. Moreover, in these runs we observe that after the migration of the silyl ligand, the (1-phenylethyl)trichlorosilane moiety still remains coordinated to the Pd center through its phenyl group in a η^2-mode. The metal center migrates along the ring, and eventually forms a η^2 complex, **11a**, with a simultaneous reattachment of the pyrazole nitrogen. With further examination of this process with conventional static calculations, we determined the migration process to form **11a** from **10a-*anti*** has a barrier of about 16.1 kcal/mol and is endothermic by 6.8 kcal/mol.

Final removal of the product to form the Pd^0 complex as shown on the lower portion of the catalytic cycle in Figure 5 was found to be endothermic by 12 kcal/mol. This would make the Pd^0 complex the highest lying complex on the potential energy surface of the catalytic cycle. We therefore studied the concerted ejection of the product by addition of another trichlorosilane molecule to the η^2-complex, **11a**. Molecular dynamics simulations show that the addition of the trichlorosilane molecule leads to immediate elimination of the product. We find that the associative elimination of the product from complex **11a** is exothermic by 17 kcal/mol. Thus, we believe addition of trichlorosilane is responsible for the liberation of the product with concomitant regeneration of the catalyst.

Figure 12 shows the reaction profile for the hydrosilylation process involving the most stable η^3-silyl-allyl complex, **10a-*anti***, calculated with model **B**. Examination of the reaction profile suggests that the rate determining step of the catalytic cycle is the reductive elimination. More specifically, the transfer of the silyl moiety to the β-carbon of the styrene. Since recoordination of the pyrazole ligand occurs in this step, it is possible that enhancement of this ligands ability to recombined with the Pd center may lead to improved activities.

3.4 A Computational Exploration of the Modified-Chalk-Harrod Mechanism

Despite efforts to characterize the general features of hydrosilylation mechanism [37-42, 44], it still remains ambiguous if initial coordination of the olefinic substrate is followed by insertion into the metal-hydride bond or the metal-silicon bond. The second of these two possibilities, which is shown on the left-hand-side of Figure 5, is usually termed the modified Chalk-Harrod mechanism [35, 36]. For the specific catalytic system we are studying, our calculations on the styrene coordinated π-complex already suggest that the modified Chalk-Harrod mechanism is not favourable because a π-complex that was amenable to insertion of styrene into the M-Si bond could not be located. Only three sterically accessible styrene coordinated π-complexes could be optimized as shown in Figure 8, and all of these isomers appear to direct the styrene to insert into the M-H bond. (This was first discussed in subsection 3.2.3.) Nonetheless we wanted to determine the intrinsic electronic barrier to the styrene insertion into the Pd-Si bond. To do this we used the minimal QM model **A**. In this model the bulky substituents that disfavour the modified-Chalk-Harrod mechanism from proceeding are not present, thereby allowing us to find a stable π-complex amenable to styrene insertion into the Pd-Si bond.

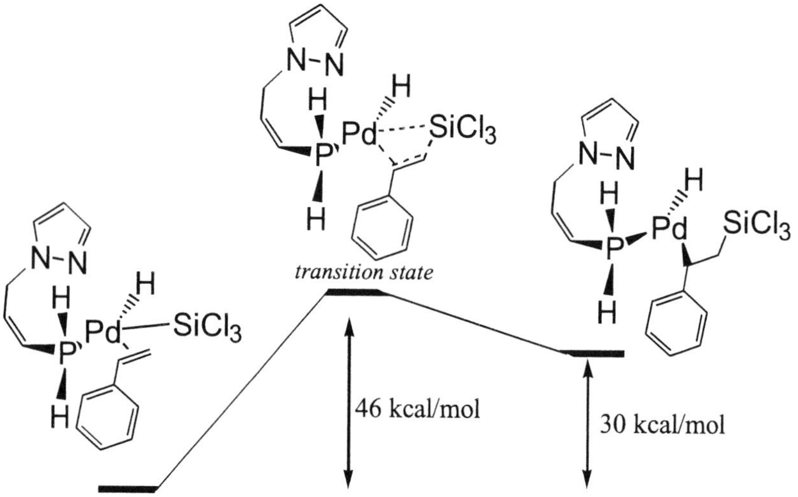

Figure 13. Calculated reaction profile for the silylmetallation process required for the modified-Chalk-Harrod hydrosilylation mechanism.

With the minimal QM model **A**, a stable π-complex, **8d**, with the coordinated olefin lying adjacent to the trichlorosilyl group was found. This π-complex, which is sketched in Figure 13, was found to be 10 kcal/mol less stable than the most stable π-complex with the hydride lying between the styrene and trichlorosilyl moieties. We have examined the silylmetallation process commencing from the π-complex **8d**. The transition state that we have located lies 46 kcal/mol higher in energy than the π-complex **8d**. The resulting alkyl-silyl intermediate is also 30 kcal/mol higher in energy than the π-complex **8d**. The high activation energy provides further explanation as to why the (2-phenylethyl)trichlorosilane regio-product is never observed. Thus, the modelling studies provide strong evidence that a modified Chalk-Harrod mechanism is not operative, a result that is consistent with previous theoretical studies of similar processes [46].

3.5   General Features of the Hydrosilylation Mechanism: A Summary

According to our simulations, hydrosilylation reaction proceeds through the classic Chalk-Harrod mechanism as depicted on the right-hand-side of Figure 5. The modified or anti-Chalk-Harrod mechanism is hindered by a rather large silylmetallation barrier which is calculated to be 46 kcal/mol for the minimal QM model **A**.

Following the classic Chalk-Harrod mechanism, we find that the coordination of styrene is thermoneutral and is likely a slow and reversible process. Interestingly, the coordination of the olefin requires the detachment of the pyrazole ligand. Three stable styrene coordinated π-complexes could be located, **8a**, **8b** and **8c** (Figure 8). Interestingly, each of these isomers has the hydride sandwiched between the styrene and trichlorosilyl groups, and therefore is poised to facilitate olefin insertion into the Pd-hydride bond, rather than insertion into the Pd-Si bond. Insertion of the olefin into the Pd-H bond to form a silyl-alkyl complex, **9** (Figure 9), is slightly endothermic and has a small activation barrier of 4.2 kcal/mol. Isomerization of the η^1-silyl-alkyl complex to a η^3-silyl-allyl complex affords a stabilization of about 5 to 10 kcal/mol depending on the isomer (Figure 11) and occurs with a minimal barrier. Reductive elimination of the silyl ligand onto the β-carbon of the styrene has a moderate 16.4 kcal/mol barrier which points to this step being the rate-determining step in the catalytic cycle. Molecular dynamics simulations suggest that addition of another trichlorosilane molecule associatively liberates the product and regenerates the catalyst with a minimal barrier.

4. NATURE OF THE REGIOSELECTIVITY

The hydrosilylation of *para*-substituted styrenes catalyzed by complex **1**, is regiospecific giving rise to product that is greater than 99% the Markovnikov regioisomer. In other words, the hydrosilylation highly favours the branched product as depicted in Figure 2. The regioselectivity of styrene hydrosilylation has often been attributed to the formation of a η^3-silyl-allyl complex [6] as shown in Figure 10. Our detailed mechanistic study is in general agreement with this and in this section we specifically discuss our computational results with respect to the nature of the regioselectivity.

In our detailed examination of the hydrosilylation process presented in Section 3, we found that the catalysis most likely proceeds via a generic Chalk-Harrod mechanism [34]. This general mechanism is depicted in the right-hand-side of Figure 5 and the computed reaction profile is displayed in Figure 12. Starting from the hydridosilyl complex, the first step of the catalysis is the coordination of the styrene substrate to form a olefinic π-complex. We have examined the styrene coordination using a combined QM/MM model **B** shown in Figure 3. In this model the whole catalyst is treated, however, the two phosphine phenyl groups, the mesityl substituent on the pyrazole ring and the ferrocenyl moiety are only treated on a steric level.

Only three stable styrene coordinated π-complexes, which are labeled **8a-c** and are shown in Figure 8, could be located. Of importance to the regioselectivity, is that the unsubstituted end of the olefinic group is directed towards the hydride in all three isomers. Thus, subsequent insertion of the styrene into the Pd-hydride bond would lead to the observed regioisomer of the product. This suggests that the origin of the regioselectivity begins with the coordination of the styrene, but as we will discuss shortly, the definitive regioselective lock comes after the initial insertion of the styrene into the Pd-hydride bond.

We have found that the insertion process to form the silyl-alkyl complex is endothermic by +0.9, +3.6 and +15.2 kcal/mol for **8a**, **8b** and **8c**, respectively. Thus, one might expect the migratory insertion to be reversible, meaning that the possibility of eventual insertion leading to 'wrong' unbranched regioisomeric product exists. However, we have found that the η^1-silyl-alkyl insertion product can isomerize with minimal barrier to form a highly stabilized η^3-silyl-allylic complex as depicted in Figure 14 (and also in Figure 10). The isomerization affords a stabilization energy in the range of 5 to 10 kcal/mol. The isomerization can occur in two ways, depending on which of the two *ortho*-carbons of the phenyl group coordinates to the Pd center (This is demonstrated in Figure 11). This leads to the formation of two different isomers in which the Pd-coordinated ortho-carbon of the phenyl group is *syn* or *anti* to the methyl group of the benzylic fragment. Figure 14, shows the formation of the *anti* η^3-silyl-allylic complex from the most stable styrene π-complex, **8a**.

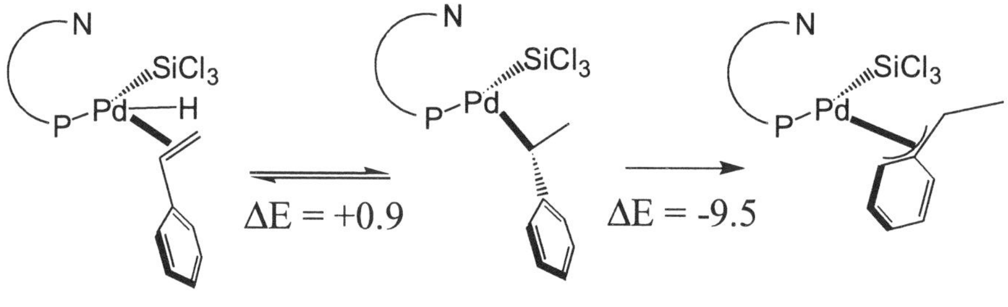

$$\Delta E = +0.9 \qquad \Delta E = -9.5$$

8a	**9a**	**10a-*anti***
π-complex	η^1-silyl-alkyl	η^3-silyl-allyl
	complex	complex

Figure 14. Formation of a stabilized η^3-silyl-alkyl complex which acts as a regiochemical lock in the hydrosilylation process.

The formation of highly stabilized η^3-silyl-allyl complex is only possible if the hydride ligand is transferred to the β-carbon of styrene, explaining why the (2-phenylethyl)trichlorosilane product is never observed experimentally.

In fact, due to the substantial energy stabilization associated with its formation, the η^3 intermediate locks in the correct regioisomer, resulting in the observed regioselectivity. These results provide theoretical evidence for what has been previously speculated. It is important to realize that the *syn-anti* pairs are stereochemically equivalent, as they will lead to the same stereoisomer of the product. We will discuss the nature of the enantioselectivity in the next section.

Using the QM/MM model **B**, we have demonstrated that the formation of the η^3-silyl-allyl intermediate plays a crucial role for the regioselectivity of the reaction. In this model, the electronic effect of the phosphine phenyl groups is neglected. Due to the proximity of this groups to the Pd center we have as a last check, examined the formation of the η^3-silyl-allyl intermediate with the QM/MM model **E**. In this model, the phosphine phenyl groups are included in the quantum mechanical region. We have found that the electronic effect of the phenyl substituents enhances the stabilization due to the η^3-coordination mode of the benzylic fragment by about 5 kcal/mol with the relative stability of the *syn* and *anti* isomers remaining the same. The regioselective lock, as we have termed it, is in fact stronger with the more realistic QM/MM model **E**, thus reinforcing our conclusions.

5.   ENANTIOSELECTIVITY OF STYRENE HYDROSILYLATION

The great practical importance of asymmetric hydrosilylation of olefins, motivates a characterization of the factors determining the inversion of the enantioselectivity that could establish a basis to design a catalyst with improved selectivity properties. In view of the small differences in activation (free) energies that are usually involved in the discrimination of two stereoisomers, the characterization of enantioselective reactions and the determination of the factors that govern the stereoselectivity is a highly challenging task for a computational study [56, 57]. These energy differences are usually close to the limit of accuracy of first-principles calculations. However, because the stereoisomeric complexes that form during the catalytic cycle have usually very similar structures, a fortuitous error cancellation can be highly effective leading to an unusual high accuracy in predicting relative stabilities.

The enantioselective hydrosilylation catalyst system based on the chiral pre-catalyst dichloro {1-{(R)-1- [(S)-2 (diphenylphosphinoκP) ferrocenyl] ethyl}-3- trimethyl- phenyl-5-1H- pyrazole-κN} palladium, **1** (shown in Figure 2) exhibits a interesting inversion of selectivity with a series of *para-*

substituted styrene substrates. As it is apparent from Table 1, the highest selectivity (64% ee) for the *R* form is achieved for 4-(dimethylamino)styrene, whereas the corresponding chloro derivative leads predominantly to the *S*-product (67.2 % ee).

In order to rationalize the factors determining the enantioselectivity of the hydrosilylation of the *para*-substituted styrenes, we have calculated the relative thermodynamic stabilities of all the intermediates of the catalytic cycle that are precursors of the two enantiomeric products as a function of the *para*-substituted substrates. Since, the *S* configuration product was formed in 64% ee from styrene, whereas 4-(dimethylamino)styrene afforded the *R* product with 64% ee [6], we have performed all calculations with these two different substrates. We shall demonstrate, in fact, that the relative thermodynamic stabilities of the η^3-allylic complexes are decisive for both the regio and the stereoselectivity.

The results of our detailed mechanistic study presented in Sections 3 and 4 can be summarized in the pathway depicted in Figure 12. According to our results, the reaction proceeds in general agreement with a classical Chalk-Harrod mechanism, shown in the right-hand-side of Figure 5. The active catalyst for hydrosilylation is a silyl(hydrido) complex and the coordination of the olefin represents the first step of the catalytic cycle. This is followed by the migratory insertion of styrene into the Pd-H bond with the initial formation of a η^1-alkyl compound. A rearrangement of the substrate converts the η^1-alkyl intermediate into a η^3-allylic form. The significant energy stabilization due to the formation of the η^3 intermediate offers a rationalization of the observed formation of the Markovnikov regioisomers (Section 4). The reaction proceeds through reductive elimination of the silane and the oxidative addition of a new molecule of trichlorosilane leads to regeneration of the active catalyst with concomitant ejection of the product.

In this section we focused our attention to a rationalization of the factors determining the stereoselectivity through *ab initio* mixed quantum/classical (QM/MM) Car-Parrinello molecular dynamic simulations. We have used the same basic computational approach used in Section 3 to explore the potential energy surface of the reaction. Since the catalyst system, **1**, is relatively large, we have used the combined QM/MM model system **B** as shown in Figure 3 and described in subsections 2.1 and 3.1.

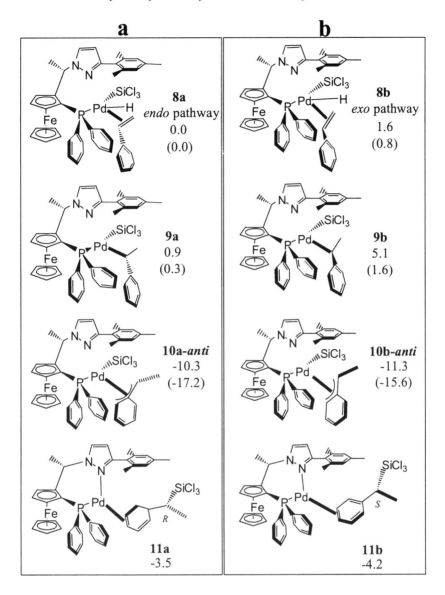

Figure 15. Intermediates of the catalytic cycle for the two sterochemically distinct pathways. a) shows what we have termed the endo pathway, which gives rise to the *R* form of the product, while b) shows the exo pathway which gives rise to the *S* form of the product. Relative energies in kcal/mol of each of the intermediates with respect to the *cis-endo* π-complex (**8a**) are shown for styrene as the substrate and in paranthesis for 4-(dimethylamino)styrene as the substrate.

5.1   Hydrosilylation of Styrene

An inversion of enantioselectivity was observed experimentally for the hydrosilylation of a series of *para*-substituted styrenes as shown in Table 1. We intend to examine the nature of the enatioselectivity by studying the catalytic cycle for styrene which reacts to give predominately the S form of the product (64% *S* ee) and for 4-(dimethylamino)styrene, which gives predominately the stereochemical opposite product (64% *R* ee). Although we have already examined the hydrosilylation of styrene in Section 3 and 4, in this section we focus on enantioselectivity of the catalytic process for comparison to the hydrosilylation with 4-(dimethylamino)styrene.

The first step of the catalytic cycle for the hydrosilylation of olefins involves the coordination of the substrate to the silyl(hydrido) form of the active catalyst (Figure 6). The two most stable π-complexes that form, **8a** and **8b**, are distinct on a stereochemical level and are shown at the top of Figure 15 (They are also shown in Figure 8). To better discuss the stereochemistry of the isomers, we will label **8a** the *endo* isomer and **8b** the *exo* isomer. The isomers where the phenyl of the substrate is oriented in the same or in the opposite direction of the ferrocene (with respect to the plane defined by the Pd-Si-P atoms) will be termed *endo* or *exo*, respectively. These two isomers are extremely important since they are precursors of the two different enantiomeric products, by virtue of the coordination of the two different enantiofaces of the olefin (*re* for *endo* and *si* for *exo*). Depicted in Figure 14 are selected intermediates along the Chalk-Harrod hydrosilylation mechanism with the relative thermodynamic stabilities of intermediates relative to the *endo* π-complex **8a** reported in kcal/mol. The paranthetic values are the relative stabilities for 4-(dimethylamino)styrene as a substrate, which we will discuss in Section 5.1. The left-hand-side of this figure follows the endo pathway that produces the *R* form of the product and the right-hand-side of the figure follows the *exo* pathway that produces the *S* form of the product.

In the calculations performed with model **B** the *exo*-π-complex is destabilized by 1.6 kcal/mol with respect to the corresponding *endo* isomer. If we were to follow the more stable *endo* π-complex isomer through the Chalk-Harrod mechanism, this would lead to the *R* form of the product as shown on the left-hand-side of Figure 15. Since it is actually the *S* form of the product that dominates when styrene is the substrate, the formation of the π-complex cannot be the stereodetermining step of the catalytic cycle.

The same order of the relative stability is observed also for the initial product of the migratory insertion step, with the *endo* alkyl(silyl) complex, **9a**, being more stable than the *exo* one , **9b**, by -4.2 kcal/mol. A rearrangement of the η^1-alkyl ligand leads to a η^3-coordination of the benzylic fragment and the formation of such an allylic intermediate can occur in two different ways, depending on which one of the two diastereotopic *ortho*-carbon coordinates to palladium. The resulting isomers are termed *syn* or *anti* if the coordinated *ortho*-carbon is *trans* or *cis* with respect to the methyl group, respectively. Only the *anti*-isomer is shown in Figure 15 because in all cases the *anti* isomer is more stable than the *syn* isomer and both *syn* and *anti* isomers lead to the same enantiomer of the product.

We have already demonstrated in section 4 that the significant energy stabilization due to the formation of the η^3 intermediate is at the origin of the observed high regioselectivity of > 99%. In parallel, an inversion in the relative thermodynamic stabilities of the *endo* and *exo* isomers occurs upon formation of the allylic complexes. As the *endo* and *exo* isomers are precursors of *R* and *S* enantiomeric products, respectively, this inversion suggests that the allylic form is a key factor in determining the enantioselectivity. We find that the *exo-anti* η^3-complex is slightly more stable ~ 1 kcal/mol than the *endo-anti* η^3-complex. Since the *exo* isomers are precursor of the *S* enantiomeric product, their higher thermodynamic stability is in agreement with the enantiomeric *S*-excess of 64 % observed experimentally (Table 1).

The following step of the catalytic cycle involves the reductive elimination of the silane with concomitant regeneration of the catalyst. This step is demonstrated to be rate determining. The activation energy barriers calculated with respect to the corresponding *anti*-allylic forms, are ~16.4 and ~16.1 kcal/mol for the *endo* and the *exo* isomers, respectively. The similar activation energies suggest that the enantioselectivity is controlled by thermodynamic factors (i.e. by the relative thermodynamic stabilities of the η^3-benzylic intermediates). However, considering the relative energies of the *anti*-allylic isomers our calculations slightly overestimate the ee.

The transfer of the silyl ligand onto the α-carbon of the substrate is followed by the formation of an intermediate in which the (1-phenylethyl)trichlorosilane product still weakly coordinated in a η^2-fashion. The corresponding *endo* and *exo* intermediates (**11a** and **11b**, respectively) are 3.5 and 4.2 kcal/mol, respectively, more stable than the *endo* π-complex, **8a**. Therefore the *exo*-stereoisomer is again thermodynamically more stable than the *endo* form. Finally, the oxidative addition of a molecule of trichlorosilane occurs with concomitant liberation of the products. The formation of both *R* and *S* products is exothermic by ~20 kcal/mol.

5.2   Hydrosilylation of para-Dimethylaminostyrene

As mentioned above, the calculations performed for styrene as a substrate suggests that the enantioselectivity can be directly correlated with the relative thermodynamic stabilities of the η^3-allylic complexes. Indeed, the *exo* stereoisomer, precursor of the enantiomeric product found in excess experimentally, becomes favoured with respect to the *endo* one upon η^3-coordination, and remains thermodynamically more stable until product release. However, the observed energy differences in the relative stabilities of the different allylic forms (1-2 kcal/mol) are certainly at the limit of accuracy of density functional calculations.

As an inversion of enantioselectivity was observed experimentally for 4-(dimethylamino)styrene, (64% *R* ee) as compared to styrene (64% *S* ee), we have recalculated the relative thermodynamic stabilities of *endo* and *exo* isomers for each step of the catalytic cycle using this second substrate. These calculations allow us to verify the quality of our findings by checking if an inversion in the relative stabilities of the *endo* and the *exo*-η^3-silyl-allyl intermediates (with the *endo* being more stable than the *exo*) is observed with 4-(dimethylamino)styrene. Using 4-(dimethylamino)styrene as the substrate, the calculated relative stabilities of the intermediates in the Chalk-Harrod mechanism are shown as paranthetic values in Figure 15.

Considering this second substrate, the first part of the reaction pathway is very similar to the one computed for styrene. In fact, both the *endo*-π and alkyl complexes are thermodynamically more stable than the corresponding *exo* species by -0.8 and -0.7 kcal/mol, respectively. Interestingly, the relative thermodynamic stabilities still favour the *endo* isomers even upon formation of the η^3-silyl-allyl intermediates. This is opposite to that observed when unsubstituted styrene is the substrate where we have calculated an inversion in the relative stabilities of the *exo* and *endo* pathways at this point.

The *endo-anti* is stabilized by -17.2 kcal/mol, while the *exo-anti* form is stabilized by -15.5 kcal/mol, respectively. Thus, the thermodynamically most stable allylic diastereoisomer is the *endo* derivative (precursor of the *R* product) as opposed to what observed for styrene. This result agrees with the *R*-enantiomeric excess observed experimentally with 4-(dimethylamino)styrene. Additionally, the observed inversion in the relative thermodynamic stabilities of the allylic forms as a function of the *para*-substituent of the substrate suggests that the corresponding energy difference is the very factor correlating with the observed ee. In fact, the similar activation energies (16.4 and 16.1) calculated for the transfer of the silyl ligand onto the α-carbon of the *endo-anti* and *exo-anti* intermediates (obtained for styrene) suggests that the stereoselectivity is thermodynamically and not kinetically controlled. We are currently performing further calculations to confirm this result.

5.3   Tuning of the Enantioselectivity

Our calculations suggest that the stereoselectivity of the hydrosilylation is determined by the thermodynamic stability of the η^3-allylic complex that forms after styrene insertion. This opens up the possibility of improving the enantioselectivity by modifying the catalyst framework to alter the stability of the *exo* versus the *endo* η^3-allylic intermediate.

We have inspected the *exo* and *endo-anti* allylic intermediates for possible discriminating steric interactions. For example, in the *endo*-allyl complex **10a** (*syn* or *anti*), the methyl group of the benzylic fragment points toward the pyrazole group whereas in the *exo*-allyl complex **10b**, the same methyl group is directed away from that group. This is shown in Figure 16. We note that this is true no matter what the *para* substituent is on the styrene substrate. We have found that one of the two *ortho* methyl groups of the mesityl substituent displays short intramolecular contacts to hydrogens of the coordinated benzyl ligand. Moreover, these close contacts are significantly shorter for the *exo* intermediate than for the *endo* intermediate. With NMe$_2$ as the *para*-substituent, the *exo-anti* allylic complex, **10b-anti**, exhibits a close interatomic contact of 2.47 Å of one hydrogen of the methyl group (Figure 16b). On the other hand, in the *endo-anti* intermediate, **10a-anti**, the closest contact of 2.63 Å concerns the hydrogen of the benzylic carbon (Figure 16a).

Since the *ortho* groups of the mesityl groups interact sterically with the substrate in the enantio-determining η^3-silyl-allyl intermediate, the enantioselectivity of the hydrosilylation could be tuned by modification of the *ortho* substituents on the pyrazole phenyl group. Increasing the steric bulk of the *ortho*-substituent should destabilize the *exo*-η^3-allyl intermediate more than the *endo* intermediate. Therefore, without considering other ramifications of the ligand modification, this should steer the reaction to favour the R enantiomer of the product. We are currently examining this possibility and other catalyst improvements based on these computational study [58].

a)　　　**10a-*anti***
　　　　　　(endo-pathway)

b)　　　**10b-*anti***
　　　　　　(exo-pathway)

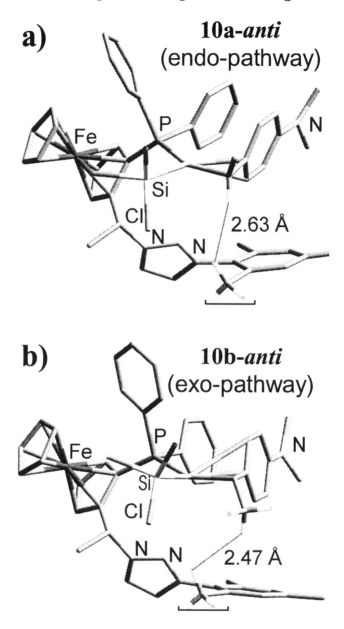

Figure 16. Optimized structures of the *endo* (**10a-*anti***) and *exo* (**10b-*anti***) η^3-silyll allyl complexes with 4-(dimethylamino)-styrene as a substrate. Highlighted in (a) is the steric contact between the hydrogen of the benzylic carbon of the substrate and a hydrogen of the mesityl substituent (2 63 Å) for the *endo-anti* intermediate, **10a-*anti***. Highlighted in (b) is the steric contact between on hydrogen of the methyl group of the substrate and a hydrogen of the mesityl substituent (2 47 Å) for the *exo-anti* silyl-allyl intermediate, **10b-*anti***. Hydrogen atoms are not shown except those whose interactions are highlighted.

6.   CONCLUSIONS

Computer simulations play an important role in many areas of science and engineering.   Chemistry is no different, and molecular modeling involves simulating chemical systems at the atomic and electronic level.   In the area of catalysis it is often difficult to establish the mechanism of a reaction through experiment, even though the reactants, products and catalyst are well characterized.   We used molecular modelling and computer simulation to answer a number of fundamental questions that have arisen during the laboratory development of an enantioselective hydrosilylation catalyst system based on the precatalyst dichloro{1-{(R)-1-[(S)2(diphenylphosphino-κP) ferrocenyl] ethyl}-3-trimethylphenyl-5-1H-pyrazole-κN}Pd, **1**.   In this contribution we highlight how we have applied the *ab initio* molecular dynamics and combined QM/MM *ab initio* molecular dynamics methods to help resolve these fundamental questions.

The solid state structures of the precatalyst, **1**, and particularly its *bis*-trichlorosilyl derivative, **3**, exhibit some unique structural features.   Most notably, **3** possesses the world's longest known Pd-P bond of 2.50 Å.   Instead of using the QM/MM approach to simply handle the large size of the compounds studied, the method has been effectively used as an analytical tool to probe the nature of the unique structures observed in the complexes.   A systematic series of combined QM/MM model systems have been constructed in which the steric and electronic influences of specific ligands were selectively removed or altered. The technique has allowed the exact nature of the geometric distortions to be pinpointed.   The extreme lengthening of the Pd-P bonds is combination of electronic and steric effects.   More specifically, we have found that the phosphine phenyl substituents contribute to the Pd-P bond lengthening on both a steric and electronic level.   The strong trans influence of the trichlorosilyl ligand is enhanced by the presence of the phosphine phenyl groups.   In **3**, the "world record" Pd-P bond distance can be primarily attributed to the strong steric interaction between the PPh_2 group with the trimethylphenyl fragment on the pyrazole ring.   The deviation from the square planar geometry at the Pd center in complexes **3** can be attributed wholly to steric factors.

A detailed study of the mechanism of the enantioselective palladium catalyzed hydrosilylation of styrene with trichlorosilane was carried out with combined QM/MM *ab initio* molecular dynamics simulations.   A number of fundamental mechanistic questions have been addressed, including the main features of the catalytic cycle, as well as the specific nature of the regioselectivity and enatioselectivity.

The conventional hydrosilylation of alkenes catalyzed by late transition metal catalysts such as Speier's catalyst are generally assumed to proceed by the Chalk-Harrod Mechanism (Figure 5). However, some experimental results suggest that alternative mechanisms such as the modified Chalk-Harrod mechanism might be at play [59-61]. The catalytic cycle has been examined in detail using a hybrid quantum mechanics and molecular mechanics model where the whole catalyst was treated explicitly. The calculations show that the styrene coordinates in *cis* position to the phosphine and that the *endo*-π-complex 8a is the most favorable one. The coordination of the substrate involves the detachment of the pyrazole ligand and no barrier has been found for this step. The most likely subsequent step is the migratory insertion of the hydride onto the β-carbon of the styrene. This slightly endothermic process is associated with a small activation energy (4.2 kcal/mol). The modified Chalk-Harrod mechanism is not operative because of the high activation energy (~47 kcal/mol) involved in the migration of $SiCl_3$ to the α-carbon of styrene.

The observed regioselectivity of >99% can be explained by the pronounced stabilization of the η^3-allylic coordination mode of the 1-phenylethyl fragment. In agreement with what has been speculated, this allylic intermediate is a definitive lock for the Markovnikov regioisomer. The high activation energy calculated for the transfer of the silyl ligand onto the β-carbon of styrene points to the reductive elimination as the rate-determining step. Moreover, molecular dynamics runs on a structure obtained by relaxing the transition state show that the product still remains coordinated with its phenyl in an η^2-mode. The addition of trichlorosilane to this complex has been found to display almost no activation barrier, and is thus responsible for the liberation of the product with concomitant regeneration of the catalyst.

To examine the nature of the enatioselectivity of the catalysis we have examined the catalytic cycle with two substrates, styrene which leads to predominantly the R form of the product (64% ee) and 4-(dimethylamino)styrene which gives predominantly the S form of the product (67% ee). Our simulations suggest that the η^3-allylic coordination of the styrene substrate plays an important role in defining the enatioselectivity of the hydrosilylation. As a first step, this theoretical study constitutes a valid contribution in rationalizing the enantioselective determining factors and possibly in designing a new catalyst with improved enantioselective properties. We are currently examining nature of the enantioselectivity in more detail as well as the dependence of the enantioselectivity on the electronic nature of the substrates [58].

ACKNOWLEDGEMENTS

The authors are indebted to Dr. Giorgio Pioda for the experimental work on the catalytic system studied here as well as for the many valuable discussions. Michael Wörle, Arianna Martelletti, and Diego Broggini for the X-ray crystallographic studies. We would like to acknowledge the Swiss National Science Foundation and NSERC of Canada for Financial support.

APPENDIX. COMPUTATIONAL DETAILS

The study of the enantioselective hydrosilylation reaction was performed with a series of combined quantum mechanics/molecular mechanics (QM/MM) calculations [26, 30] within the computational scheme of *ab initio* (AIMD) (Car-Parrinello) [62] molecular dynamics. The AIMD approach has been described in a number of excellent reviews [63-66]. AIMD as well as hybrid QM/MM-AIMD calculations [26, 47] were performed with the *ab initio* molecular dynamics program CPMD [67] based on a pseudopotential framework, a plane wave basis set, and periodic boundary conditions. We have recently developed an interface to the CPMD package in which the coupling with a molecular mechanics force field has been implemented [26, 68].

An analytic local pseudopotential was used for hydrogen and nonlocal, norm-conserving soft pseudopotentials of the Martins-Trouiller type [69] for the other elements. Angular momentum components up to $l_{max} = 1$ have been included for carbon and nitrogen and of $l_{max} = 2$ for silicon, phosphorus and chlorine. For palladium, we have constructed a semicore pseudopotential [19] where all the 18 valence electrons of the $4s^2 4p^6 4d^8 5s^2$ shells have been treated explicitly. This pseudopotential incorporates also scalar relativistic effects. For all light elements, the pseudopotentials have been transformed to a fully nonlocal form using the scheme developed by Kleinman and Bylander [70], whereas for palladium, the nonlocal part of the pseudopotential has been integrated numerically using a Gauss-Hermite quadrature. One electron wave functions have been expanded up to a kinetic energy cutoff of 70 Ry.

The exchange-correlation functional used in the calculations is of the gradient corrected type due to Becke [71] and Perdew [72] for the exchange and correlation parts, respectively. All calculations have been performed in periodically repeated face centered cubic super cells. For the small QM model system (model **A**) a cell of edge a=20 Å has been used, whereas to account also for possible structural modifications of the catalyst along the reaction pathway, we used larger cells of edge of a= 23 and 24 Å for model **B** and **E**, respectively. The geometry optimization runs have been performed with a preconditioned conjugated gradient procedure. For the MD runs, the classical equations of motion were integrated with a velocity Verlet algorithm with a time step of 0.145 fs and a fictitious mass for the electronic degrees of freedom of $\mu = 800$ au. All the transition states have been determined performing constraint molecular dynamics runs with increments of 0.1 Å along appropriately chosen reaction coordinates in combination with local optimization techniques.

The Tripos [73] force field was used to perform the molecular mechanics calculations, augmented with parameters developed by Doman *et al.* [74] for the ferrocenyl ligand. In the QM/MM hybrid AIMD simulations, the electronic structure calculation was performed on a reduced system in which each of the substituents that have been removed from the QM part was replaced by a hydrogen atom, in order to saturate the valence of the QM boundary atoms.

As a consequence, the electronic properties of these ligands are replaced by the electronic properties of hydrogen atoms, while their steric influence is taken into account by the empirical force field approach.

REFERENCES

1   *The Chemistry of Organic Silicon Compounds*, Patai, S.; Rappoport, Z., Eds.; Wiley: New York, 1989.
2   Speier, J. L.; Webster, J. A.; Barnes, G. H. *J. Am. Chem. Soc..* **1957**, *79*, 974.
3   Speier, J. L. *Adv. Organomet. Chem.* **1979**, *17*, 407.
4   Uozumi, Y. *J. Am. Chem. Soc.* **1991**, *113*, 9887.
5   Uozumi, Y.; Kitayama, K.; Hayashi, T.; Yanagi, K.; Fukuyo, E. *Bull. Chem. Soc. Jpn.* **1995**, *68*, 713-722.
6   Pioda, G.; Togni, A. *Tetrahedron: Asymmetry* **1998**, *9*, 3903-3910.
7   Tamao, K.; Kakui, T.; Kumada, M. *J. Am. Chem. Soc.* **1978**, *100*, 2268-2269.
8   Tamao, K.; Yoshida, J.; Takahashi, M.; Yamamoto, H.; Kakui, T.; Matsumoto, H.; Kurita, A.; Kumada, M. *J. Am. Chem. Soc.* **1978**, *100*, 290-292.
9   Tamao, K.; Ishida, N.; Tanaka, T.; Kumada, M. *Organometallics* **1983**, *2*, 1694-1696.
10 Togni, A.; Bieler, N.; Burckhardt, U.; Kollner, C.; Pioda, G.; Schneider, R.; Schnyder, A. *Pure Appl. Chem.* **1999**, *71*, 1531-1537.
11 Pioda, G ETH Ph.D. Thesis No. 13405, "Catalisi Asimmetrica Idrosililazione Enantioselettiva e Ciclizzazione via Trimetilenmetano", Zürich, 1999.
12 Blöchl, P. E.; Tongi, A. *Organometallics* **1996**, *15*, 4125.
13 Woo, T. K.; Pioda, G.; Röthlisberger, U.; Togni, A. *Organometallics* **2000**, *19*, 2144-2152.
14 Brost, R. D.; Bruce, G. C.; Joslin, F. L.; Stobart, S. R. *Organometallics* **1997**, *16*, 5669-5680.
15 Hofmann, P.; Meier, C.; Hiller, W.; Heckel, M.; Riede, J.; Schmidt, M. U. *J. Organomet. Chem.* **1995**, *490*, 51-70.
16 Levy, C. J.; Puddephatt, R. J.; Vittal, J. J. *Organometallics* **1994**, *13*, 1559-1560.
17 Yamashita, H.; Hayashi, T.; Kobayashi, T.; Tanaka, M.; Goto, M. *J. Am. Chem. Soc.* **1988**, *110*, 4417-4418.
18 Koga, N.; Morokuma, K. *J. Am. Chem. Soc.* **1993**, *115*, 6883-6892.
19 Koloski, T. S.; Pestana, D. C.; Carroll, P. J.; Berry, D. H. *Organometallics* **1994**, *13*, 489-499.
20 Lichtenberger, D. L.; Rai-Chaudhuri, A. *Inorg. Chem.* **1990**, *29*, 975-981.
21 Tsuji, Y.; Nishiyama, K.; Hori, S.-i.; Ebihara, M.; Kawamura, T. *Organometallics* **1998**, *17*, 507-512.
22 Maseras, F. *Chem. Commun.* **2000**, 1821-1827.
23 Singh, U. C.; Kollmann, P. A. *J. Comput. Chem.* **1986**, *7*, 718-730.
24 Gao, J. *Rev. Comput. Chem.* **1996**, *7*, 119-185.
25 Field, M. J.; Bash, P. A.; Karplus, M. *J. Comput. Chem.* **1990**, *11*, 700-733.
26 Woo, T. K.; Cavallo, L.; Ziegler, T. *Theor. Chem. Acc.* **1998**, *100*, 307-313.
27 Gao, J.; Thompson, M., Eds. *Methods and Applications of Combined Quantum Mechanical and Molecular Mechanical Methods*; American Chemical Society: Washington, DC, 1998.
28 Ziegler, T. *Chem. Rev.* **1991**, *91*, 651.

29 *A Chemist's Guide to Density Functional Theory*, Koch, W.; Holthausen, M. C.; Wiley-VCH: Weinheim, Germany, 2000.

30 Maseras, F.; Morokuma, K. *J. Comput. Chem.* **1995**, *16*, 1170-1179.

31 Ogasawara, M.; Maseras, F.; Gallego-Planas, N.; Kawamura, K.; Ito, K.; Toyota, K.; Streib, W. E.; Komiya, S.; Eisenstein, O.; Caulton, K. G. *Organometallics* **1997**, *16*, 1979-1993.

32 Woo, T. K.; Ziegler, T. *J. Organomet. Chem.* **1999**, *591*, 204-213.

33 Drago, D.; Pregosin, P. S.; Tschoerner, M.; Albinati, A. *J. Chem. Soc., Dalton Trans.* **1999**, 2279-2280.

34 Chalk, A. J.; Harrod, J. F. *J. Am. Chem. Soc.* **1965**, *87*, 16-21.

35 Millan, A.; Fernandez, M. J.; Bentz, P.; Maitlis, P. M. *J. Mol. Catal.* **1984**, *26*, 89.

36 Ojima, I.; Fuchikami, T.; Yatabe, M. *J. Organomet. Chem.* **1984**, *260*, 335-346.

37 Esteruelas, M. A.; Herrero, J.; Oro, L. A. *Organometallics* **1993**, *12*, 2377-2379.

38 Duckett, S. B.; Perutz, R. N. *Organometallics* **1992**, *11*, 90-98.

39 Hostetler, M. J.; Butts, M. D.; Bergman, R. G. *Organometallics* **1993**, *12*, 65-75.

40 Bosnich, B. *Acc. Chem. Res.* **1998**, *31*, 667-674.

41 Hostetler, M. J.; Bergman, R. G. *J. Am. Chem. Soc.* **1990**, *112*, 8621-8623.

42 Esteruelas, M. A.; Olivan, M.; Oro, L. A. *Organometallics* **1996**, *15*, 814.

43 LaPointe, A. M.; Rix, F. C.; Brookhart, M. *J. Am. Chem. Soc.* **1997**, *119*, 906-917.

44 Takahashi, T.; Hasegawa, M.; Suzuki, N.; Saburi, M.; Rousset, C. J.; Fanwick, P. E.; Negishi, E. *J. Am. Chem. Soc.* **1991**, *113*, 8564-8566.

45 Sakaki, S.; Ogawa, M.; Musashi, Y.; Arai, T. *J. Am. Chem. Soc.* **1994**, *116*, 7258-7265.

46 Sakaki, S.; Mizoe, N.; Sugimoto, M. *Organometallics* **1998**, *17*, 2510-2523.

47 Woo, T. K.; Margl, P. M.; Deng, L.; Ziegler, T. *Catalysis Today* **1999**, *50*, 479-500.

48 Deng, L.; Woo, T. K.; Cavallo, L.; Margl, P. M.; Ziegler, T. *J. Am. Chem. Soc.* **1997**, *119*, 6177.

49 Carbó, J. J.; Maseras, F.; Bo, C.; van Leeuwen, P. W. N. M. *J. Am. Chem. Soc.* **2001**, *123*, 7630-7637.

50 van Gunsteren, W. F.; Berendsen, H. J. C. *Angew. Chem. Int. Ed. Engl.* **1990**, *29*, 992-1023.

51 *Molecular Modelling: Principles and Applications*, Leach, A. R.; 2nd ed.; Prentice Hall: Harlow, England, 2001.

52 Woo, T. K.; Margl, P. M.; Lohrenz, J. C. W.; Blöchl, P. E.; Ziegler, T. *J. Am. Chem. Soc.* **1996**, *118*, 13021-13030.

53 Margl, P. M.; Woo, T. K.; Bloechl, P. E.; Ziegler, T. *J. Am. Chem. Soc.* **1998**, *120*, 2174-2175.

54 Woo, T. K.; Patchkovskii, S.; Ziegler, T. *Comput. Sci. Eng.* **2000**, *2*, 28-37.

55 Braunstein, P.; Naud, F. *Angew. Chem. Int. Ed.* **2001**, *40*, 680.

56 Ujaque, G.; Maseras, F.; Lledós, A. *J. Am. Chem. Soc.* **1999**, *121*, 1317-1323.

57 Vazquez, J.; Pericàs, M. A.; Maseras, F.; Lledós, A. *J. Org. Chem.* **2000**, *65*, 7303-7309.

58 (a) Magistrato, A.; Woo, T. K.; Togni, A.; Röthlisberger, U. submitted . (b) Magistrato, A.; Togni, A.; Röthlisberger, U. submitted.

59 Bergens, S. H.; Noheda, P.; Whelan, J.; Bosnich, B. *J. Am. Chem. Soc.* **1992**, *114*, 2128.

60 Duckkett, S.; Perutz, R. N. *Organometallics* **1992**, *11*, 90.

61 Brookhart, M.; Grant, B. E. *J. Am. Chem. Soc.* **1993**, *115*, 2151.

62 Car, R.; Parrinello, M. *Phys. Rev. Lett.* **1985**, *55*, 2471.

63 Galli, G.; Parrinello, M. In *Computer Simulation in Material Science*; Meyer, M. P., V., Ed.; Kluwer: Dordrecht, The Netherlands, 1991; p 238.

64 Remler, D. K.; Madden, P. A. *Mol. Phys.* **1990**, *70*, 921-966.

65   Marx, D.; Hutter, J. In *Modern Methods and Algorithms of Quantum Chemistry*; Grotendorst, J., Ed.; Forschungzentrum Juelich, NIC Series, 2000; Vol. 1, p 301.

66   Parrinello, M. *Solid State Comm.* **1997**, *102*, 107-120.

67   Hutter, J.; Ballone, P.; Bernasconi, M.; Focher, P.; Fois, E.; Goedecker, S.; Parrinello, M.; Tuckerman, M. In;; CPMD, Max-Planck-Institut für Festkörperforschung and IBM Zurich Research Laboratory, 1995-1999.

68   Woo, T. K.; Röthlisberger, U. unpublished work.

69   Trouiller, M.; Martins, J. L. *Phys. Rev. B* **1991**, *43*, 1993.

70   Kleinman, L.; Bylander, D. M. *Phys. Rev. Lett.* **1982**, *48*, 1425.

71   Becke, A. D. *Phys. Rev. A: Gen. Phys.* **1988**, *38*, 3098-3100.

72   Perdew, J. P. *Phys. Rev. B* **1986**, *33*, 8822.

73   Clark, M.; Cramer, R. D., III; Van Opdenbosch, N. *J. Comput. Chem.* **1989**, *10*, 982-1012.

74   Doman, T. N.; Landis, C. R.; Bosnich, B. *J. Am. Chem. Soc.* **1992**, *114*, 7264-7272

Chapter 10

Olefin Dihydroxylation

Thomas Strassner

Institut für Anorganische Chemie, Technische Universität München, Lichtenbergstrasse 4, D-85747 Garching, Germany

Abstract: The reaction mechanism of the olefin dihydroxylation by transition metal complexes has been the subject of a controversy in recent years, with two major mechanistic proposals, the so called [2+2] and [3+2] pathways, supported by different experimental evidences. Theoretical calculations have been essential in solving this controversy, and their contribution is summarized here, with special emphasis on the results with the two most common oxidants, osmium tetraoxide and permangate. Other related issues, like enantioselectivity and kinetic isotope effects, have also been the subject of computational studies and are briefly discussed.

Key words: Dihydroxylation, osmium tetraoxide, permanganate, olefin

1. INTRODUCTION

Stoichiometric and catalytic transition-metal oxidation reactions are of great interest, because of their fundamental role in industrial and synthetic processes [1]. The introduction of oxygen atoms into unsaturated organic molecules via dihydroxylation reactions leading to 1,2-diols is one of these examples. 1,2-diols can be synthesized by the reaction of alkenes with organic peracids via the corresponding epoxides and subsequent hydrolysis or metal-catalyzed by strong oxidants such as OsO_4, RuO_4, MnO_4^- and chromium(VI) compounds.

F. Maseras and A. Lledós (eds.), Computational Modeling of Homogeneous Catalysis, 253–268.
© 2002 *Kluwer Academic Publishers. Printed in the Netherlands.*

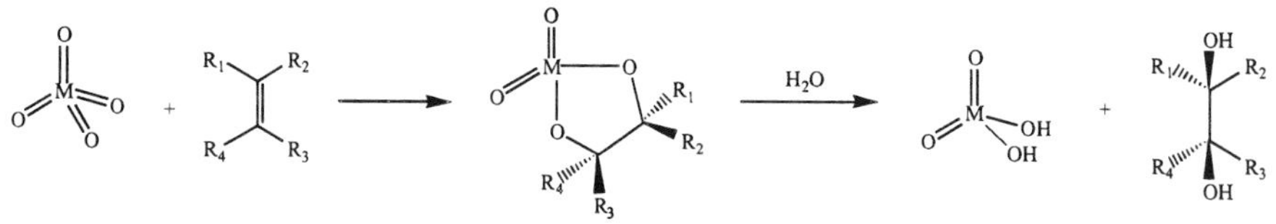

Figure 1. Dihydroxylation reaction (M=Os, Ru, Mn).

Osmium tetraoxide and permanganate are the textbook example reactants for the direct addition of the hydroxyl function to double bonds as shown in Figure 1. Several reagents such as hydrogen peroxide, periodate, hexacyanoferrate(III) or recently also molecular oxygen [2-6] have been used to reoxidize the different metal-oxo compounds.

In the case of prochiral alkenes the dihydroxylation reaction creates new chiral centers in the products and the development of the asymmetric version of the reaction by Sharpless was one of the very important accomplishments of the last years. He received the Nobel Price in Chemistry 2001 for the development of catalytic oxidation reactions to alkenes.

2.   OSMIUM TETRAOXIDE

The reaction mechanisms of these transition metal mediated oxidations have been the subject of several computational studies, especially in the case of osmium tetraoxide [7-10], where the controversy about the mechanism of the oxidation reaction with olefins could not be solved experimentally [11-20]. Based on the early proposal of Sharpless [12], that metallaoxetanes should be involved in alkene oxidation reactions of metal-oxo compounds like CrO_2Cl_2, OsO_4 and MnO_4^- the question arose whether the reaction proceeds via a concerted [3+2] route as originally proposed by Criegee [11] or via a stepwise [2+2] process with a metallaoxetane intermediate [12] (Figure 2).

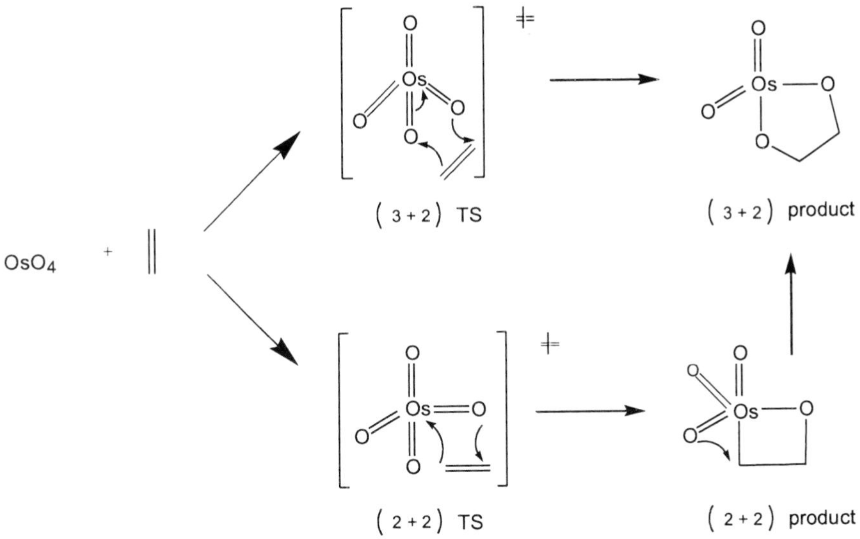

Figure 2. [3+2] vs. stepwise [2+2] reaction.

Already in 1936 Criegee had observed that the reaction rate increases when bases like pyridin are added [11]. Sharpless recognized that the chirality of bases is transferred to the substrates which allowed the development of the asymmetric version of the reaction. The cinchona amines (like $(DHQ)_2PHAL$, Figure 3) have been shown to give high enantiomeric excess and are part of the commercially available AD-mix (0,4% $K_2OsO_2(OH)_4$, 1,0% $(DHQ)_2PHAL$, 300% $[K_3Fe(CN)_6]$, 300% K_2CO_3).

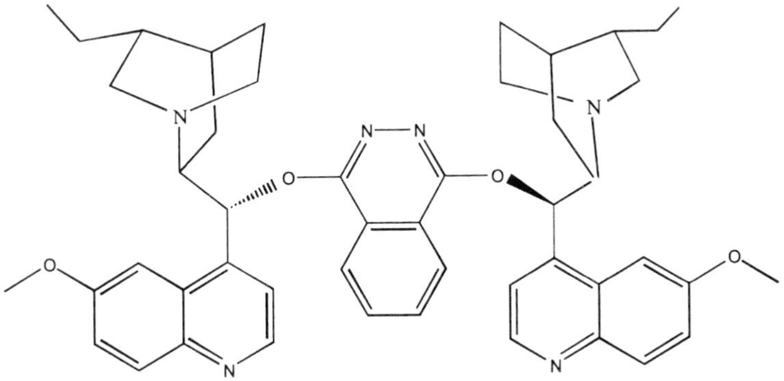

Figure 3. Chinchona-Base $(DHQ)_2PHAL$.

Kinetic data on the influence of the reaction temperature on the enantioselectivity using chiral bases and prochiral alkenes revealed a nonlinearity of the modified Eyring plot [16]. The observed change in the linearity and the existence of an inversion point indicated that two different transition states are involved, inconsistent with a concerted [3+2] mechanism. Sharpless therefore renewed the postulate of a reversibly formed oxetane intermediate followed by irreversible rearrangement to the product.

This mechanistic question is one of the examples of the success of density functional theory methods in organometallic chemistry. Earlier work on the reaction mechanism could not discriminate between the two alternatives. Analysis of the different orbitals based on extended Hückel calculations came to the result that the [3+2] pathway is more likely, but could not exclude the possibility of a [2+2] pathway [13]. Similar conclusions where obtained from the results of Hartree-Fock calculations in combination with QCISD(T) single point calculations [21]. Attempts to use RuO_4 as a model for osmium tetraoxide indicated that the formation of an oxetane is less favorable compared to the [3+2] pathway, but still possible [22, 23].

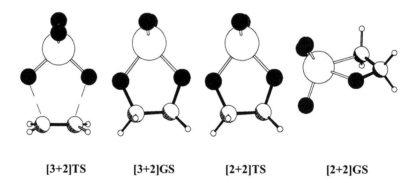

| [3+2]TS | [3+2]GS | [2+2]TS | [2+2]GS |

Figure 4. Calculated transition states (TS) and local minima (GS) for the reaction of OsO_4 with ethylene [24].

Table 1. Optimized values (Å and kcal/mol) for the calculated geometries shown in Figure 4 for the reaction of OsO_4 with ethylene. Relative energies are referred to the separate reactants.

	[3+2]TS	[3+2]GS	[2+2]TS	[2+2]GS
Os-O	1.76	1.90	1.82	1.98
O-C	2.09	1.45	1.92	1.42
C-C	1.38	1.53	1.41	1.52
Os-C	-	-	2.44	2.16
ΔH	+3.9	-34.5	+43.8	+5.1

The picture was clarified by DFT calculations, which allowed accurate geometry optimizations of all significant structures for both pathways in the reaction of osmium tetraoxide with ethylene. Within a short period of time four different groups published DFT results on the reaction using different quantum chemical packages and different levels of theory [7-10]. These studies could show that the barrier for the [2+2] addition of OsO_4 to ethylene and the ring expansion are significantly higher (~35 kcal/mol) than the activation enthalpy needed for the [3+2] pathway. Geometries from one of these sets of calculations [10, 24] are shown in Figures 4 and 5, and selected geometrical parameters are collected in Tables 1 and 2.

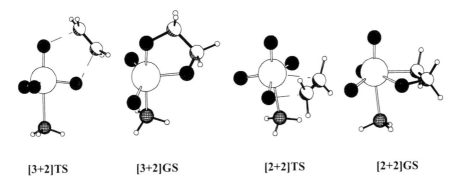

[3+2]TS   [3+2]GS   [2+2]TS   [2+2]GS

Figure 5. Calculated transition states (TS) and intermediates (GS) for the reaction of $OsO_4(NH_3)$ with ethylene [24].

Table 2. Optimized values (Å and kcal/mol) for the calculated geometries shown in Figure 5 for the reaction of $OsO_4(NH_3)$ with ethylene. Relative energies are referred to the separate reactants.

	[3+2]TS	[3+2]GS	[2+2]TS	[2+2]GS
Os-O	1.76	1.93	1.83	2.01
O-C	2.11	1.43	1.96	1.43
C-C	1.38	1.52	1.40	1.52
Os-N	2.44	2.27	2.42	2.42
Os-C	-	-	2.49	2.21
ΔH	+3.2	-42.0	+44.1	+5.4

As pointed out above, the addition of a base to the reaction can increase the reaction rate and induce enantioselectivity. Therefore, this process was also interesting to study. Though the available computational power has increased significantly during the last years, it is still not feasible to evaluate a potential energy surface and to optimize complexes with bases of the size used experimentally, like that shown in Figure 3. The bases have been modeled in the quantum mechanical (QM) studies by NH_3. Results with the same method used above [10, 24] for the reaction with base are shown in Figure 5 and Table 2. Comparing the energy profiles with and without base ligand it becomes obvious that they are very similar as can be seen from Figures 4, 5 and Tables 1, 2. The activation enthalpy differences are smaller than 1 kcal/mol and the [3+2] pathway is always significantly lower in energy

Table 3. Energetics (kcal/mol) computed by different authors for the reaction between $OsO_4(NH_3)$ and ethylene. Relative energies are referred to the separate reactants.

	[3+2]TS	[3+2]GS	[2+2]TS	[2+2]GS
Frenking [7]	+4.4	-39.8	+44.3	+5.1
Houk [10, 24]	+3.2	-42.0	+44.1	+5.4
Morokuma [8]	+1.4	-23.5	+50.4	+13.1
Ziegler [9]	+0.8	-28.4	+39.1	+3.6

Table 3 collects the results obtained by different authors on the reaction of $OsO_4(NH_3)$ with ethylene. Though the methods were not identical, the results are very similar, and in all cases there is a clear preference of the [3+2] over the [2+2] pathway. Additional confirmation was provided by a combined experimental and theoretical study using kinetic isotope effects (KIEs) to compare experiment and theory. Kinetic isotope effects were measured by a new NMR technique [25] and compared to values, which are available from calculated transition states. It showed that indeed only the [3+2] pathway is feasible [10].

Table 4. Calculated and experimental enantioselectivities in the asymmetric dihydroxylation with different alkenes and bases (adapted from Ref. 28).

Alkene	DHQD Ligand	ee_{calc}	ee_{exp}	Ref.
1-Phenyl-cyclohexene	CLB	91%	91%	31
Styrene	CLB	70%	74%	31
β,β-Dimethyl styrene	CLB	72%	74%	31
β-Vinyl naphthalene	CLB	94%	88%	31
trans-Stilbene	CLB	98%	99%	31
t-Butyl ethene	CLB	70%	44%	31
α-Methyl styrene	CLB	65%	62%	28
cis-β-Methyl styrene	CLB	78%	35%	28
Styrene	MEQ	94%	87%	31
Styrene	PHN	98%	78%	31
t-Butyl ethene	PHN	89%	79%	31
β-Vinyl naphthalene	PHAL	100%	98%	32
Styrene	PHAL	97%	97%	33
α-Methyl styrene	PHAL	99%	94%	33
trans-Stilbene	PHAL	100%	100%	33

The disadvantage of using NH_3 as a model for the large cinchona base are the missing steric effects. Therefore QM/MM-studies [26-28] and molecular dynamics calculations [29] were conducted combining the advantages of the QM treatment with the possibility of treating a large number of atoms [30]. The origin of the experimentally observed enantioselectivity in the dihydroxylation of styrene was investigated and it was found that π-interactions between the aromatic rings are responsible for the observed enantioselectivity. Norrby [28] parametrized a Macromodel force field for the reaction and was able to reproduce the experimentally observed enantioselectivities, several examples are shown in Table 4.

3.   PERMANGANATE

It was generally accepted that the oxidation of alkenes by permanganate proceeds via a concerted mechanism with a cyclic ester intermediate, until the suggestion was made of a stepwise mechanism involving a metallaoxetane intermediate which is supposed to rearrange to a cyclic ester before the hydrolysis takes place [12]. Hydrolysis of the intermediate is generally accepted to give rise to the observed diols. Similar to the osmium tetraoxide discussion the two mechanistic proposals could not be distinguished on the basis of experimental data. Several groups tried to identify the elusive metallaoxetane and a wealth of kinetic data provided no indication for the existance of the species. But the possibility that it might be a non rate-determining intermediate could not be excluded. Different mechanisms were proposed to explain the variety of experimental results available, but the mechanistic issues remained unresolved.

DFT-calculations (B3LYP/6-31G*) on the reaction of permanganate with ethylene [34] show that permanganate is a very similar case to osmium tetraoxide. The activation energy for the [3+2] pathway is a little higher in energy (+ 9.2 kcal/mol) compared to osmium tetraoxide, while the barrier for the [2+2] pathway is more than 40 kcal/mol higher in energy (+50.5 kcal/mol) [34]. These results indicate that also in the case of permanganate the dihydroxylation proceeds via a [3+2] transition state to a cyclic ester intermediate which is hydrolized in the course of the reaction.

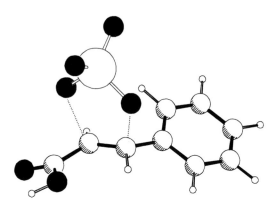

Figure 6. [3+2] transition state for the oxidation of cinnamic acid by permanganate [40].

Experimental studies of the oxidative cleavage of cinnamic acid by acidic permanganate [35] resulted in secondary kinetic isotope effects, k_H/k_D, of 0.77 (α) and 0.75 (β), while another paper from the same group on the same reaction with quaternary ammonium permanganates [36] reported very different isotope effects of 1.0 (α) and 0.91 - 0.94 (β) depending on the counterion. Different mechanisms were discussed in the literature [37, 38] to explain the variety of experimental results available, but the mechanistic issues were unresolved. The reported activation energy for the oxidation of cinnamic acid, 4.2 ± 0.5 kcal/mol [39], is very well reproduced by B3LYP/6-311+G** calculations [40], which predict an activation enthalpy of 5.1 kcal/mol for the transition state shown in Figure 6. Two sets of ^{13}C kinetic isotope effects were experimentally determined by the NMR method of Singleton [25] and compared to theoretically predicted values. They agree very well within the experimental uncertainty [40].

Freeman has experimentally studied the oxidation by permanganate [41, 42] and the influence of substituents on the rate of the oxidation by permanganate in phosphate buffered solutions (pH 6.83 ± 0.03). Nine unsaturated carboxylic acids have been used as substrates (Figure 7) and free energy values derived from experimental kinetic data have been published [43].

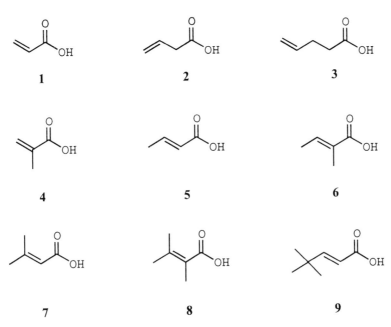

Figure 7. Substrates used in the experimental study of the oxidation by permanganate [43]

In compounds **1** to **3** the chain length and position of the double bond is varied, while compounds **4** to **9** are all α, β unsaturated carboxylic acids with a changing number and position of alkyl substituents at the double bond. The rates of reaction were measured by monitoring spectral changes with a stopped flow spectrometer. Using pseudo-first-order conditions they recorded the disappearance of the permanganate ion at 526, 584, or 660 nm and/or the rate of formation of colloidal manganese dioxide at 418 nm. The authors concluded, that according to the measured reaction rates the rate of oxidation is more sensitive to steric factors than to electronic effects.

Regarding the reaction mechanism they suggested that both the (3 + 2) and the (2 + 2) transition states are in agreement with their results [43].

From the data provided by the systematic experimental study at standardized conditions the free energy of activation ($\Delta G^{\neq}_{exp.}$) was calculated from the experimental rate constant and compared to calculated $\Delta G^{\neq}$ values. Two different basis sets have been employed in the DFT calculations: the split valence double-ξ (DZ) basis set 6-31G(d) with a triple-ξ (TZ) [44, 45] valence basis set for manganese (we will refer to this combination as basis set I (BS1)) and the triple-ξ basis set 6-311+G(d,p), which will be denoted basis set II (BS2). The BS1-results for transition states and intermediates are shown in Table 5, a comparison of the free activation energies is shown in Figure 8 [46].

Table 5. BS1-calculated free energies $\Delta G^{\neq}$ (kcal/mol) for intermediates (GS) and transition states (TS) for the reaction between the substrates shown in Figure 7 and permanganate [46].

Compound	$\Delta G^{\neq}_{exp.}$	$\Delta G^{\neq}$[3+2] TS	$\Delta G^{\neq}$[2+2] TS	ΔG[3+2] GS	ΔG[2+2] GS
1	17.1	10.8	53.6	-35.3	26.8
2	17.8	17.7	58.0	-37.8	25.0
3	17.6	17.2	56.1	-38.1	24.7
4	17.1	11.8	55.1	-34.5	27.7
5	17.4	12.7	57.2	-32.9	29.8
6	17.0	13.5	57.5	-32.6	29.7
7	18.3	14.7	59.4	-30.0	33.6
8	18.6	14.6	58.9	-35.7	32.5
9	18.0	11.9	57.4	-33.9	30.8

The geometries of the transition states differ significantly. While a rather symmetrical transition state is calculated for 2_{TS} and 3_{TS} with bond lengths of 2.05/2.06 Å (2_{TS}) and 2.04/2.09 Å (3_{TS}) for the forming bonds between the permanganate oxygens and the alkene carbons, 1_{TS} shows an unsymmetrical, but still concerted transition state with calculated bond lengths of 2.29 and 1.90 Å. Cartesian coordinates of all geometries are given in the supplementary material of Ref. 46. The symmetrical transition states 2_{TS} and 3_{TS} are very similar to the transition state calculated for the permanganate oxidation of ethylene [34], indicating that the substitutent does not play a major role. On the other hand, the geometries of the transition states of $4_{TS} - 9_{TS}$ are similar regardless of the number of substituents. All transition states are very unsymmetrical, but concerted. The bond lengths of the C-O forming bonds do not change significantly, whether the substituent is a methylgroup in α-position (5_{TS}, 2.30/1.91 Å), β-position (4_{TS}, 2.31/1.92 Å) or a t.-butyl group (9_{TS}, 2.29/1.93 Å). Even three methyl groups as in 8_{TS} do not distort the geometry more than what was observed for 1_{TS}.

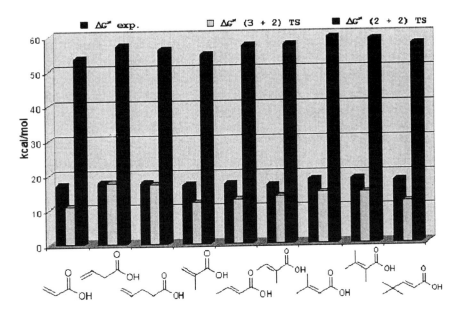

Figure 8. Comparison of computed and experimental free activation energies for substrates **1** -**9**.

The differences between the geometries calculated with the two different sets BS1 and BS2 are very small, deviations in the bond lengths of 0.01 Å as well as changes of torsional angles of up to 4° are observed. Energetics present however some differences between the two basis sets, as seen in Table 6. The agreement between experimental data and computational results improves with the larger basis set.

Table 6. : Influence of the basis set on the calculated free energies (in kcal/mol) related to the reaction of substrates **1-9** with permanganate [46].

Compound	$\Delta G^{\neq}_{exp.}$	$\Delta G^{\neq}[3+2]$ TS BS2	$\Delta G^{\neq}[3+2]$ TS BS1	$\Delta G^{\neq}[2+2]$ TS BS2	$\Delta G^{\neq}[2+2]$ TS BS1
1	17.1	15.7	10.8	61.5	53.6
2	17.8	22.0	17.7	63.4	58.0
3	17.6	21.3	17.2	62.0	56.1
4	17.1	18.3	11.8	63.7	55.1
5	17.4	18.3	12.7	64.8	57.2
6	17.0	19.6	13.5	65.9	57.5
7	18.3	21.8	14.7	66.7	59.4
8	18.6	21.4	14.6	67.2	58.9
9	18.0	20.3	11.9	67.3	57.4

An additional factor that should be taken into account is the effect of solvents. Certainly the interaction is weaker in solution than in gas phase, where the negative charge of the permanganate ion is going to be solvated, especially in aqueous solution. Therefore we conducted solvent calculations using the PCM model of Tomasi and coworkers. As can be seen from Table 7 the [3+2] pathway is still hugely favored and the activation energies for the solvated [3+2] transition states are in reasonable agreement, although they are in general lower than the experimental values with the exception of compound **8**. Compared to the gas phase calculations, the deviations are much smaller in the PCM calculations.

Table 7. Free activation energies (BS1, PCM solvatation model) for the transition states (in kcal/mol) of the reaction of substrates **1-9** with permanganate.

Comp.	1	2	3	4	5	6	7	8	9
$\Delta G^{\neq}_{exp}$	17.1	17.8	17.6	17.1	17.4	17.0	18.3	18.6	18.0
$\Delta G^{\neq}$ [3+2]	15.4	15.0	16.3	16.2	17.3	15.5	17.4	20.6	17.0
$\Delta G^{\neq}$ [2+2]	52.6	61.4	54.4	54.1	53.5	54.3	57.4	61.4	55.5

Substrates **4** - **9** were chosen by Freeman to study the influence of steric bulk on the free activation energy. As discussed before the substituents also show an electronic effect and it is hard to separate both effects, but at least some comparisons can be made, *e.g.*, for trans-crotonic acid **4** and 4,4-dimethyl-trans-2-pentenoic acid **9**. The steric bulk of the t.-butyl group compared to a methyl group should be by far more important than the difference in the electronic effect.

It can be concluded that the [3+2] pathway seems to be the only feasible reaction pathway for the dihydroxylation by permanganate. The study on the free activation energies for the oxidation of α, β unsaturated carboxylic acids by permanganate shows that the [3+2] mechanism is in better agreement with experimental data than the [2+2] pathway. Experimentally determined kinetic isotope effects for cinnamic acid are in good agreement with calculated isotope effects for the [3+2] pathway, therefore it can be concluded that a pathway via an oxetane intermediate is not feasible.

The situation changes if one of the oxygen atoms is replaced by a chlorine. Limberg could show that the product of the addition of MnO_3Cl to ethylene is more stable in the triplet state and that the product distribution can be explained in terms of reaction channels [47].

4.   OTHER TRANSITION METAL OXIDES

The interaction of other transition metal oxides with alkenes has also been studied in detail. Some significant examples are the $LReO_3$ compounds which were experimentally studied by Herrmann [48-50] and theoretically by Rappé (L=Cp, Cp*, CH₃, OH, ...) [51] and Rösch (L=CH₃) [52-58].

Rappé points out that the metal ligand π bonding interaction determines whether dihydroxylation or epoxidation is the preferred pathway. The selectivity depends critically on the ligand L. For the reaction of CpReO₃ and ethylene, the calculated and experimental [59] value agree very well indicating that the reaction leading to the dioxylate structure is the thermodynamically lower pathway [51]. A similar study by Frenking confirmed that the predicted pathway depends strongly on the nature of the ligand L [60]. The first [2+2]-cycloaddition of a transition-metal oxide compound with a C=C double bond was published by Frenking [60, 61] for the reaction of (R₃PN)ReO₃ with ketenes of the type R₂C=C=O. Recently Rösch succeeded in correlating more than 25 different activation energies and concludes that they depend in a quadratic fashion on the reaction energies as predicted by Marcus theory [62].

Chromylchloride, CrO₂Cl₂, the main subject of the publication which led to the original discussion about the mechanism [12], shows a very different reactivity compared to the other transition metal oxides discussed above. Even in the absence of peroxides, it yields epoxides rather than diols in a complex mixture of products, which also contains cis-chlorohydrine and vicinal dichlorides. Many different mechanisms have been proposed to explain the great variety of products observed, but none of the proposed intermediates could be identified. Stairs et al. have proposed a direct interaction of the alkene with one oxygen atom of chromylchloride [63-65], while Sharpless proposed a chromaoxetane [12] formed via a [2+2] pathway.

Early calculations using the general valence bond approach by Rappé came to the conclusion that the chromaoxetane is a likely intermediate [66-68], 70 kcal/mol more stable than the [3+2] intermediate. Their conclusions are nevertheless not confirmed by more recent studies disfavoring the [2+2] intermediate, which because of the more accurate level applied are probably correct. Concise theoretical studies of Ziegler et al. [69, 70] analyzed all possible pathways for the two electron oxidation from Cr^{VI} to Cr^{IV} including the crossover from the singlet to the triplet surface with the transition state on the singlet surface when the formed product is a triplet species. It could be shown that the epoxide precursor is formed via a [3+2] addition of ethylene to two Cr=O bonds followed by rearrangement to the epoxide product. The activation barriers were computed as 13.9 and 21.8 kcal/mol for these two steps. The alternative pathways, direct addition of an oxygen to the ethylene and the [2+2] addition both have a higher activation energy of 29.8 kcal/mol. The formation of the chlorohydrine products could also be explained by a [3+2] addition to one Cr-Cl and one Cr=O bond. Still, all these calculations cannot explain, why no diols are found experimentally.

5. CONCLUSIONS

Theoretical calculations have been fundamental in solving the controversy on the mechanism for the dihydroxylation of double bonds by transition metal oxo complexes. Nowadays, this topic which was the subject of a controversy just a few years ago seems to be solved in favor of the [3+2] pathway, at least in a vast majority of the cases. Despite this spectacular success there are still a number of open issues for this particular reaction which have not been solved, and which continue to be a challenge for computational chemists. Among this, one can mention the correlation between the nature of the substrate and its reactivity with permanganate, and the mechanisms leading to the proportion of products experimentally observed when CrO_2Cl_2 is applied. Hopefully, these issues will be solved in the future with the help of theoretical calculations.

REFERENCES

1 Torrent, M.; Solà, M.; Frenking, G. *Chem. Rev.* **2000**, *100*, 439-493.
2 Doebler, C.; Mehltretter, G. M.; Sundermeier, U.; Beller, M. *J. Organomet. Chem.* **2001**, *621*, 70-76.
3 Mehltretter, G. M.; Doebler, C.; Sundermeier, U.; Beller, M. *Tetrahedron Lett.* **2000**, *41*, 8083-8087.
4 Doebler, C.; Mehltretter, G. M.; Sundermeier, U.; Beller, M. *J. Am. Chem. Soc.* **2000**, *122*, 10289-10297.
5 Doebler, C.; Mehltretter, G.; Beller, M. *Angew. Chem., Int. Ed. Eng.*. **1999**, *38*, 3026-3028.
6 Kolb, H. C.; Sharpless, K. B. *Trans. Met. Org. Synth.* **1998**, *2*, 219-242.
7 Pidun, U.; Boehme, C.; Frenking, G. *Angew. Chem., Int. Ed. Engl.* **1997**, *35*, 2817-2820.
8 Dapprich, S.; Ujaque, G.; Maseras, F.; Lledós, A.; Musaev, D. G.; Morokuma, K. *J. Am. Chem. Soc.* **1996**, *118*, 11660-11661.
9 Torrent, M.; Deng, L.; Duran, M.; Solà, M.; Ziegler, T. *Organometallics* **1997**, *16*, 13-19.
10 DelMonte, A. J.; Haller, J.; Houk, K. N.; Sharpless, K. B.; Singleton, D. A.; Strassner, T.; Thomas, A. A. *J. Am. Chem. Soc.* **1997**, *119*, 9907-9908.
11 Criegee, J. *Justus Liebigs Ann. Chem.* **1936**, *522*, 75.
12 Sharpless, K. B.; Teranishi, A. Y.; Baeckvall, J. E. *J. Am. Chem. Soc.* **1977**, *99*, 3120-3128.
13 Jorgensen, K. A.; Hoffmann, R. *J. Am. Chem. Soc.* **1986**, *108*, 1867.
14 Corey, E. J.; Noe, M. C. *J. Am. Chem. Soc.* **1993**, *115*, 12579-12580.
15 Becker, H.; Ho, P. T.; Kolb, H. C.; Loren, S.; Norrby, P.-O.; Sharpless, K. B. *Tetrahedron Lett.* **1994**, *35*, 7315-7318.
16 Goebel, T.; Sharpless, K. B. *Angew. Chem., Int. Ed. Engl.* **1993**, *32*, 1329.
17 Corey, E. J.; Noe, M. C.; Grogan, M. J. *Tetrahedron Lett.* **1996**, *37*, 4899-4902.

18 Nelson, D. W.; Gypser, A.; Ho, P. T.; Kolb, H. C.; Kondo, T.; Kwong, H.-L.; McGrath, D. V.; Rubin, A. E.; Norrby, P.-O.; Gable, K. P.; Sharpless, K. B. *J. Am. Chem. Soc.* **1997**, *119*, 1840-1858.

19 Corey, E. J.; Sarshar, S.; Azimiora, M. D.; Newbold, R. C.; Noe, M. C. *J. Am. Chem. Soc.* **1996**, *118*, 7851.

20 Vanhessche, K. P. M.; Sharpless, K. B. *J. Org. Chem.* **1996**, *61*, 7978-7979.

21 Veldkamp, A.; Frenking, G. *J. Am. Chem. Soc.* **1994**, *116*, 4937-4946.

22 Norrby, P. O.; Kolb, H. C.; Sharpless, K. B. *Organometallics* **1994**, *13*, 344-347.

23 Norrby, P.-O.; Becker, H.; Sharpless, K. B. *J. Am. Chem. Soc.* **1996**, *118*, 35-42.

24 Becke3LYP/6-31G*; ECP for osmium.

25 Singleton, D. A.; Thomas, A. A. *J. Am. Chem. Soc.* **1995**, *117*, 9357-9358.

26 Ujaque, G.; Maseras, F.; Lledós, A. *J. Am. Chem. Soc.* **1999**, *121*, 1317-1323.

27 Ujaque, G.; Maseras, F.; Lledós, A. *J. Org. Chem.* **1997**, *62*, 7892-7894.

28 Norrby, P.-O.; Rasmussen, T.; Haller, J.; Strassner, T.; Houk, K. N. *J. Am. Chem. Soc.* **1999**, *121*, 10186-10192.

29 Moitessier, N.; Maigret, B.; Chretien, F.; Chapleur, Y. *Eur. J. Org. Chem.* **2000**, 995-1005.

30 Maseras, F. *Top. Organomet. Chem.* **1999**, *4*, 165-191.

31 Sharpless, K. B.; Amberg, W.; Beller, M.; Chen, H.; Hartung, J.; Kawanami, Y.; Lubben, D.; Manoury, E.; Ogino, Y.; et al. *J. Org. Chem.* **1991**, *56*, 4585-4588.

32 Kolb, H. C.; Andersson, P. G.; Sharpless, K. B. *J. Am. Chem. Soc.* **1994**, *116*, 1278-1291.

33 Sharpless, K. B.; Amberg, W.; Bennani, Y. L.; Crispino, G. A.; Hartung, J.; Jeong, K. S.; Kwong, H. L.; Morikawa, K.; Wang, Z. M.; et al. *J. Org. Chem.* **1992**, *57*, 2768-2771.

34 Houk, K. N.; Strassner, T. *J. Org. Chem.* **1999**, *64*, 800-802.

35 Lee, D. G.; Brownridge, J. R. *J. Am. Chem. Soc.* **1974**, *96*, 5517-5523.

36 Lee, D. G.; Brown, K. C. *J. Am. Chem. Soc.* **1982**, *104*, 5076-5081.

37 Wolfe, S.; Ingold, C. F. *J. Am. Chem. Soc.* **1981**, *103*, 940-941.

38 Simandi, L. I.; Jaky, M. *J. Am. Chem. Soc.* **1976**, *98*, 1995-1997.

39 Lee, D. G.; Nagarajan, K. *Can. J. Chem.* **1985**, *63*, 1018-1023.

40 Strassner, T. 2002, submitted.

41 Simandi, L. I.; Jaky, M.; Freeman, F.; Fuselier, C. O.; Karchefski, E. M. *Inorg. Chim. Acta* **1978**, *31*, L457-L459.

42 Freeman, F. *Rev. React. Species Chem. React.* **1973**, *1*, 37.

43 Freeman, F.; Kappos, J. C. *J. Org. Chem.* **1986**, *51*, 1654-1657.

44 Schaefer, A.; Huber, C.; Ahlrichs, R. *J. Chem. Phys.* **1994**, *100*, 5829-5835.

45 Frenking, G.; Antes, I.; Boehme, C.; Dapprich, S.; Ehlers, A. W.; Jonas, V.; Neuhaus, A.; Otto, M.; Stegmann, R.; Veldkamp, A.; Vyboishchikov, S. F. In *Reviews in Computational Chemistry*; Lipkowitz, K. B., Boyd, D. B., Eds.; Wiley VCH: New York, 1996; Vol. 8, pp 63 - 144.

46 Strassner, T.; Busold, M. *J. Org. Chem.* **2001**, *66*, 672-676.

47 Wistuba, T.; Limberg, C. *Chem. Eur. J.* **2001**, *7*, 4674-4685.

48 Herrmann, W. A.; Fischer, R. W.; Marz, D. W. *Angew. Chem.* **1991**, *103*, 1706-1709 .

49 Romao, C. C.; Kuehn, F. E.; Herrmann, W. A. *Chem. Rev.* **1997**, *97*, 3197-3246.

50 Herrmann, W. A.; Kuehn, F. E. *Acc. Chem. Res.* **1997**, *30*, 169-180.

51 Pietsch, M. A.; Russo, T. V.; Murphy, R. B.; Martin, R. L.; Rappé, A. K. *Organometallics* **1998**, *17*, 2716-2719.

52 Gisdakis, P.; Antonczak, S.; Koestlmeier, S.; Herrmann, W. A.; Roesch, N. *Angew. Chem., Int. Ed. Eng.* **1998**, *37*, 2211-2214.

53   Gisdakis, P.; Roesch, N.; Bencze, E.; Mink, J.; Goncalves, I. S.; Kuehn, F. E. *Eur. J. Inorg. Chem.* **2001**, 981-991.

54   Gisdakis, P.; Roesch, N. *Eur. J. Org. Chem.* **2001**, 719-723.

55   Roesch, N.; Gisdakis, P.; Yudanov, I. V.; Di Valentin, C. *Peroxide Chem.* **2000**, 601-619.

56   Gisdakis, P.; Yudanov, I. V.; Roesch, N. *Inorg. Chem.* **2001**, *40*, 3755-3765.

57   Di Valentin, C.; Gandolfi, R.; Gisdakis, P.; Roesch, N. *J. Am. Chem. Soc.* **2001**, *123*, 2365-2376.

58   Koestlmeier, S.; Haeberlen, O. D.; Roesch, N.; Herrmann, W. A.; Solouki, B.; Bock, H. *Organometallics* **1996**, *15*, 1872-1878.

59   Gable, K. P.; Phan, T. N. *J. Am. Chem. Soc.* **1994**, *116*, 833-839.

60   Deubel, D. V.; Frenking, G. *J. Am. Chem. Soc.* **1999**, *121*, 2021-2031.

61   Deubel, D. V.; Schlecht, S.; Frenking, G. *J. Am. Chem. Soc.* **2001**, *123*, 10085-10094.

62   Gisdakis, P.; Roesch, N. *J. Am. Chem. Soc.* **2001**, *123*, 697-701.

63   Makhija, R. C.; Stairs, R. A. *Can. J. Chem.* **1969**, *47*, 2293-2299.

64   Makhija, R. C.; Stairs, R. A. *Can. J. Chem.* **1968**, *46*, 1255-1260.

65   Stairs, R. A.; Diaper, D. G.; Gatzke, A. L. *Can. J. Chem.* **1963**, *41*, 1059.

66   Rappé, A. K.; Goddard, W. A., III *J. Am. Chem. Soc.* **1982**, *104*, 448-456.

67   Rappé, A. K.; Goddard, W. A., III *J. Am. Chem. Soc.* **1982**, *104*, 3287-3294.

68   Rappé, A. K.; Goddard, W. A., III *J. Am. Chem. Soc.* **1980**, *102*, 5114-5115.

69   Torrent, M.; Deng, L.; Duran, M.; Solà, M.; Ziegler, T. *Can. J. Chem.* **1999**, *77*, 1476-1491.

70   Torrent, M.; Deng, L.; Ziegler, T. *Inorg. Chem.* **1998**, *37*, 1307-1314.

Chapter 11

The Dötz Reaction: A Chromium Fischer Carbene-Mediated Benzannulation Reaction

Miquel Solà,[1,*] Miquel Duran,[1] and Maricel Torrent[2]

[1] Institut de Química Computacional and Departament de Química, Universitat de Girona, E-17071 Girona, Catalonia, Spain;. [2] Medicinal Chemistry Dept., Merck Research Laboratories, Merck & Co., West Point PA.

Abstract: A complete revision of the most widely accepted mechanistic schemes for the Dötz reaction is proposed. According to our calculations the addition of the alkyne molecule to the carbene complex takes place before CO loss in the initial steps of the reaction. Further, our study shows that a novel proposal involving a chromahexatriene intermediate entails lower energy barriers and more stable intermediates than the previous reaction mechanisms postulated by Dötz and Casey. The novel findings query revision of the classically assumed paths and put forward that additional experimental and theoretical studies are necessary to definitely unravel the reaction mechanism of this intringuing reaction.

Key words: Dötz reaction, benzannulation, Fischer carbene complexes, reaction mechanism, density functional theory

1.   INTRODUCTION

The use of group 6 heteroatom stabilized carbene complexes (Fischer carbene complexes) in organic synthesis is relatively recent and, in spite of that, it has already produced impressive synthetic results [1]. These versatile organometallic reagents have an extensive chemistry, and they are probably one of the few systems that undergo cycloadditions of almost any kind. For instance, [1+2], [2+2], [3+2], [3+3], [4+1], [4+2], [4+3], [6+3] cycloadditions and multicomponent cycloadditions such as [1+1+2], [1+2+2], [3+2+1], [4+2+1] or even [4+2+1-2] and [2+2+1+1] to Fischer carbenes have been reported [2].

F. Maseras and A. Lledós (eds.), Computational Modeling of Homogeneous Catalysis, 269–287.
© 2002 Kluwer Academic Publishers. Printed in the Netherlands.

Among the synthetically useful reactions of Fischer carbenes, the benzannulations are certainly the most prominent. In particular, the so-called Dötz reaction, first reported by Dötz in 1975 [3], is an efficient synthetic method that starting from aryl- or alkenyl-substituted alkoxycarbene complexes of chromium affords *p*-alkoxyphenol derivatives by successive insertion of the alkyne and one CO ligand in an α,β-unsaturated carbene, and subsequent electrocyclic ring closure (see Figure 1).

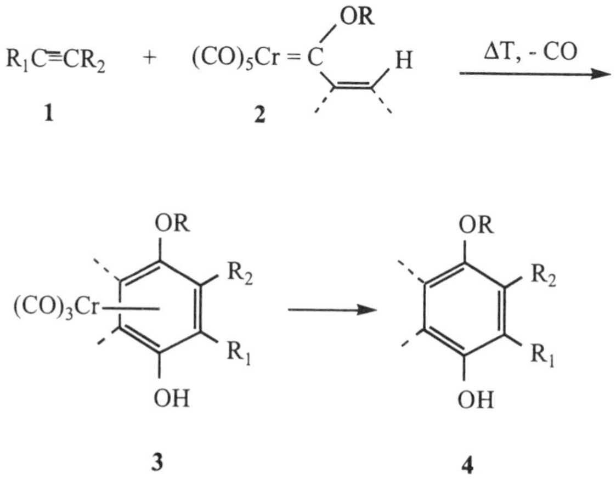

$$R_1C{\equiv}CR_2 \quad + \quad (CO)_5Cr=C \overset{\displaystyle OR}{\underset{\displaystyle H}{}} \quad \xrightarrow{\Delta T, -CO}$$

1   **2**

$$(CO)_3Cr \cdots \begin{array}{c} OR \\ R_2 \\ R_1 \\ OH \end{array} \longrightarrow \begin{array}{c} OR \\ R_2 \\ R_1 \\ OH \end{array}$$

3   **4**

Figure 1. Schematic representation of the Dötz reaction.

This formally [3+2+1] cycloaddition reaction proceeds with remarkable regioselectivity when unsymmetrical alkynes **1** are used. In this case, it is observed that the larger substituent (e.g. R_1 in Figure 1) is incorporated *ortho* to the phenol function [4], with only few exceptions [5, 6].

Generally phenol formation is the major reaction path; however, relatively minor modifications to the structure of the carbene complex, the alkyne, or the reaction conditions can dramatically alter the outcome of the reaction [7]. Depending on reaction conditions and starting reactants roughly a dozen different products have been so far isolated, in addition to phenol derivatives [7-12]. In particular, there is an important difference between the products of alkyne insertion into amino or alkoxycarbene complexes. The electron richer aminocarbene complexes give indanones **8** as the major product due to failure to incorporate a carbon monoxide ligand from the metal, while the latter tend to favor phenol products **7** (see Figure 2).

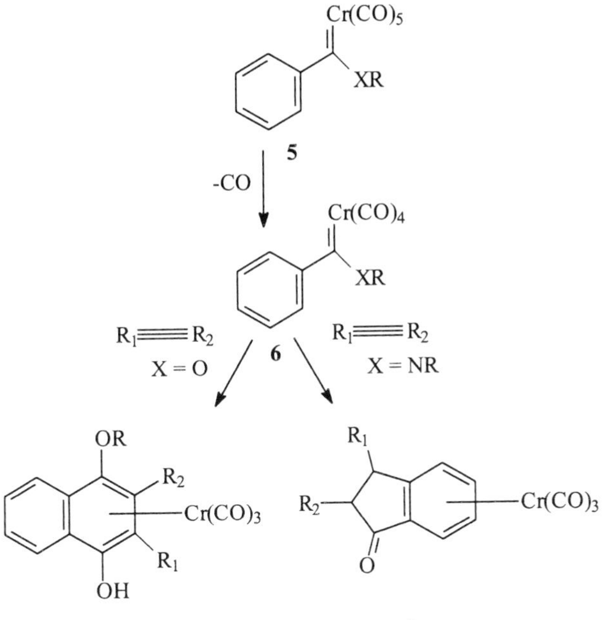

Figure 2. Benzannulation vs. cyclopentannulation in Fischer carbene complexes.

The compatibility with different functional groups, the remarkable regio- and stereoselectivity, and the development of asymmetric procedures have made benzannulation an attractive methodology for the synthesis of natural products with densely functionalized quinoid or fused phenolic substructures [13-20]. Some pertinent examples are the syntheses of vitamins K and E [17], and the production of anthracyclinones or naphtoquinone antibiotics [13, 14a, 15, 21].

Despite the undeniable synthetic value of the benzannulation reaction of aryl and alkenyl Fischer carbene complexes, the details of its mechanism at the molecular level remain to be ascertained. Indeed, although a relatively large number of theoretical studies have been directed to the study of the molecular and electronic structure of Fischer carbene complexes [22], few studies have been devoted to the analysis of the reaction mechanisms of processes involving this kind of complexes [23-30]. The aim of this work is to present a summary of our theoretical research on the reaction mechanism of the Dötz reaction between ethyne and vinyl-substituted hydroxycarbene species to yield *p*-hydroxyphenol.

1.1   Mechanistic background

Before we turn to the results of our work, let us briefly summarize the relevant mechanistic information available from experimental studies and describe the mechanistic proposals presented in the literature. Figure 3 reflects the mechanistic pathways postulated for the Dötz reaction.

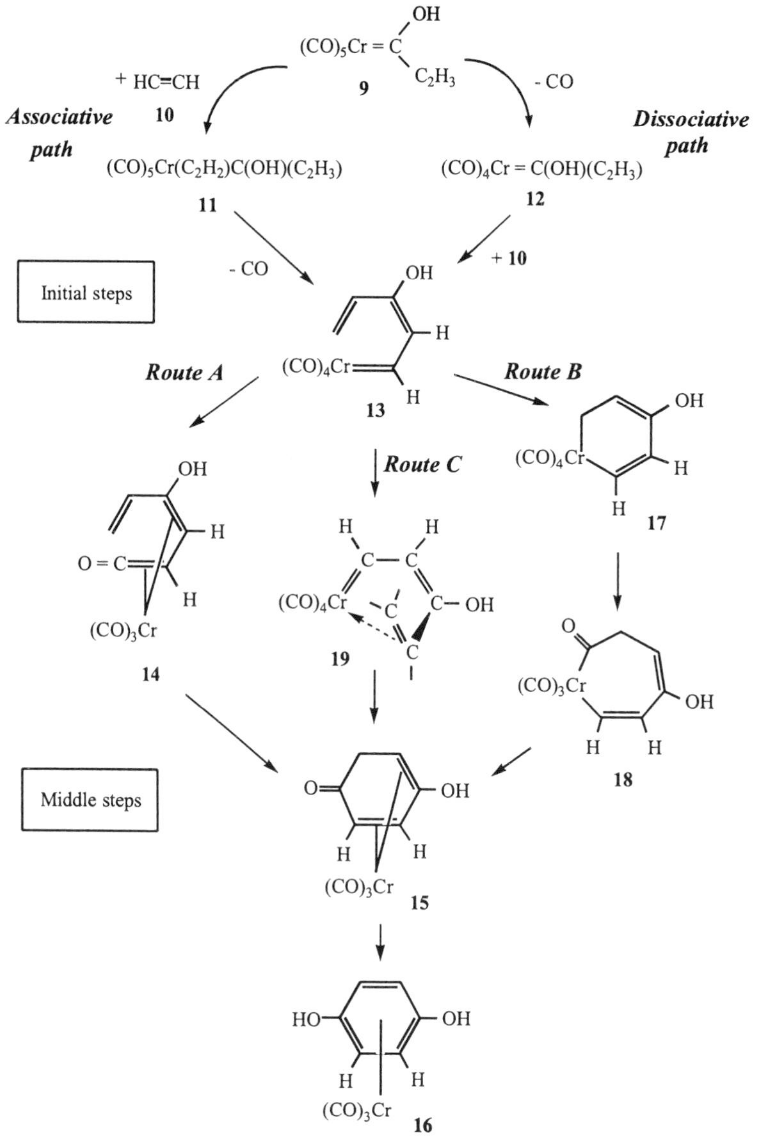

Figure 3. Postulated (A,B) and non-canonical (C) routes for the benzannulation reaction.

The most widely accepted mechanism [1b] for the Dötz benzannulation reaction assumes that the reaction starts with a reversible *cis*-CO dissociation from **9** followed by alkyne coordination, yielding η^3-vinylallylidene complex **13**. Two diverging routes have been postulated from this point. According to Dötz [1b], intermediate **13** evolves to η^3-vinylketene intermediate **14** by CO insertion into the chromium-carbene bond, followed by electrocyclic ring closure to yield η^4-cyclohexadienone **15**, which can finally turn into the direct precursor of the phenol derivative **16** after enolization and aromatization (route A). As suggested by Casey [31], another possibility is that metallation of the arene ring occurs in complex **13** to form **17**, which subsequently undergoes carbonylation to yield the chromacycloheptadienone complex **18** (route B). Reductive elimination from **18** gives **15**, where both scenarios merge again.

It has proven difficult to validate these mechanistic suggestions since the rate-limiting step typically involves CO ligand loss to open a coordination site for the alkyne to bind. Once bound, rapid ring closure occurs to give the observed products or the metal-complexed cyclohexadienone **15** [7h, 32]. No single reaction has been followed completely through the individual steps depicted in Figure 3, although there are some examples of isolation and characterization of presumed model intermediates. For instance, Barluenga and coworkers [8] have recently published the first example of a Dötz reaction in which the decarbonylation step is carried out in the absence of the alkyne. In this way, the authors have been able to isolate and characterize an η^3-vinylcarbene complex analogous to **12**. Other works [1h, 18a] have provided examples of this first intermediate in the benzannulation reactions. Barluenga *et al.* [8] also reported the isolation and characterization of a metallatriene intermediate corresponding to complex **13**. Subsequent thermal decomposition of this complex produced either cyclopentadienes or phenol derivatives. Examples of η^4-vinylketene intermediates **14** containing Rh(I), Co(I), and Ir(I) and also containing Cr(0) have been reported, together with the reaction of η^4-vinylketene cobalt complexes with alkynes to form phenols [33, 34]. However, although experimental evidence has been provided for the formation of η^4-vinylketene complexes, such species have never been observed during a benzannulation reaction itself [34]. This can only mean that either (*i*) η^4-vinylketene complexes are actual intermediates, but they are not observed because of their easy, almost barrierless conversion to **15**, or (*ii*) they are not observed because the actually operating mechanism does not take place through η^4-vinylketene complexes, other intermediates being involved instead.

Attempts to determine whether the reaction proceeds through route A or B have been made by Garret *et al.* [35] and by Wulff and coworkers [36]. The approach followed by the former authors was to generate a coordinatively unsaturated carbene complex at a temperature sufficiently low to retard subsequent reactions so that they could be studied step by step [35]. In the work of Wulff and coworkers [36], the criterion taken for the mechanistic discussion was consistency to the observed product distribution, together with other experimental data. It was found that the vinylketene route seemed to explain more satisfactorily all data; however, at that time the authors concluded that there was not enough evidence to definitively rule out any of the studied mechanisms [36]. In both cases [35, 36], the authors were unable to distinguish which mechanistic proposal is the operative one.

The Dötz reaction mechanism has received further support from kinetic and theoretical studies. An early kinetic investigation [37] and the observation that the reaction of the metal carbene with the alkyne is supressed in the presence of external carbon monoxide [38] indicated that the rate-determining step is a reversible decarbonylation of the original carbene complex. Additional evidence for the Dötz mechanistic hyphotesis has been provided by extended Hückel molecular orbital [23, 24] and quantum chemical calculations [25].

Despite these important studies, most steps of the reaction mechanism are still only postulated. Therefore, we have decided to undertake a theoretical investigation of the Dötz reaction by discussing whether the reaction proceeds via a dissociative or an associative pathway in the initial steps of the process. We have also analyzed the central part of the reaction, the key issue being whether the reaction proceeds through a vinylketene intermediate (route A) or, instead, via a metallacycloheptadienone (route B). As will be seen, we came across a novel third pathway (route C) that turns out to be the best alternative from thermodynamic and kinetic points of view

2.   RESULTS AND DISCUSSION

This section describes the main results obtained in our studies of the Dötz reaction mechanism [26-29, 39]. The section is divided as follows: First, the results for the initial part of the reaction (9→13) are presented. The central discussion will be whether the alkyne binds the carbene complex after or before CO loss. Then, the results for routes A, B and C (Figure 3) are discussed. In particular, we will examine the suitability of the novel route C involving a chromahexatriene intermediate.

2.1   Initial part of the reaction

Two different mechanistic proposals have been investigated theoretically for the initial steps of this reaction (Figure 3): a dissociative mechanism and an associative route. As mentioned above, early studies have shown that reactions of chromium carbene complexes with alkynes are suppressed in the presence of external CO [38]. This makes most probable that carbon monoxide ligand dissociates at the first step (dissociative route). This suggestion has been later reinforced by kinetic data [37]. An alternative mechanistic scenario is the insertion of an alkyne ligand into the metal coordination sphere of the metal carbene complex prior to CO elimination (associative route). Such a mechanism is not unknown in the field of TM chemistry, e.g. in carbonylmetal complexes with a 17-valence electron shell such as $V(CO)_6$.

2.1.1   Dissociative mechanism

The dissociative route has been explored in two independent DFT studies [25, 27] using ethyne as the alkyne model. The investigation by Gleichmann, Dötz, and Hess (GDH) [25] shows that the dissociation reaction

$$(OC)_5Cr=C(X)(CH=CH_2) \rightarrow (OC)_4Cr=C(X)(CH=CH_2) + CO \quad (1)$$

with X = OH or NH_2 is highly endothermic (125-202 kJ mol^{-1} for X = OH; 115-217 kJ mol^{-1} for X = NH_2), and proceeds without barrier. Similar conclusions were reported by Torrent, Duran, and Solà (TDS) [27] using the same functional and a slightly better basis set (144.7-188.6 kJ mol^{-1} for X = OH). Both studies also found that, as expected, dissociation of the *trans*-CO ligand requires more energy than dissociation of any of the four *cis*-CO ligands [25, 27]. An interesting point, though, is that depending on the particular *cis*-CO ligand removed the resulting species, **12**, can be either suitable to interact with an alkyne molecule or unable to undergo the cyclization reaction. TDS found that the most stable and suitable dissociation product was none of the two species earlier proposed by GDH [25] (a tetracarbonyl carbene complex, **12a**, and an η^3-allylidene complex, **12b**) but a third one having an agostic interaction between the metal and the H atom on the C_α to the $C_{carbene}$, **12c** (Figure 4).

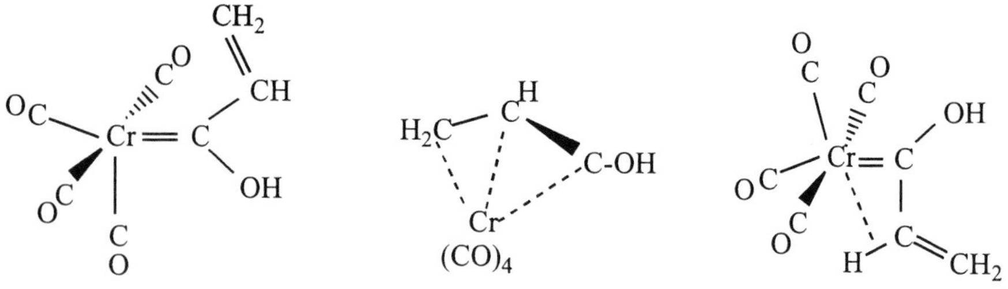

Figure 4. Different conformations for the CO dissociation product.

The next step of the dissociative route is coordination of ethyne to the dissociation product **12** followed by ethyne-carbene insertion to form the metallatriene intermediate **13**. This process is highly exothermic, the release of energy starting from **12c** being of 175.7 kJ mol^{-1} [27].

2.1.2   Associative mechanism

We found that the decarbonylation of **9** requires at least 144.7 kJ mol^{-1} [27]. Because experiments are carried out under mild conditions (moderate temperatures [3, 4]: 45-55 °C), it is not straightforward to understand how CO loss can take place at this stage. Obviously, the $(CO)_4Cr(C(OH)C_2H_3)$ intermediate can be stabilized by solvent molecules. However this does not necessarily change the kinetics of the reaction significantly, since prior to formation of the $(CO)_4$(solvent)$Cr(C(OH)C_2H_3)$ complex, it is necessary that CO leaves a vacant coordination site in the $(CO)_5Cr(C(OH)C_2H_3)$ complex, and as a consequence the same 144.7 kJ mol^{-1} are again required.

An alternative mechanistic scenario for the initial steps of this reaction (associative route) was the subject of a study by TDS [27, 39]. It considers the possibility that the cycloaddition with alkynes takes place initially by direct reaction with the coordinatively saturated chromium carbene complex **9**.

Starting from the reactants **9** and **10** (Figure 5), we have found a weak interacting complex **20** that is only 16.5 kJ mol^{-1} stabilized with respect to reactants. The addition of ethyne to $(CO)_5Cr(C(OH)C_2H_3)$ goes through transition state **TS(20→11)** with an energy barrier of 149.1 kJ mol^{-1} to give complex **11** which is stabilized by 163.4 kJ mol^{-1} with respect to reactants [39]. From **11** the release of a CO molecule to form complex **13a** (see Figure 6) requires 132.4 kJ mol^{-1}. The lower CO dissociation energy in **11** as compared to **9** may well be attributed to the fact that not only is 3-hydroxy-2,4-pentadienylidene a better σ-donor than 1-hydroxy-2-propenylidene, but it also turns into a better π-acceptor as well [27]. Both features make the CO ligands more labile in **11** than in **9**.

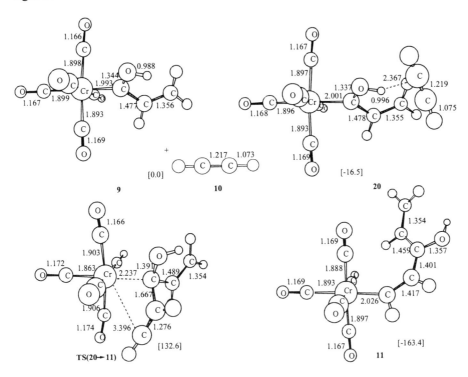

Figure 5. Geometries and energies relative to reactants of all the stationary points related to the associative route. Selected bond distances are given in Å and energies in kJ mol^{-1}.

As a whole, the energy barrier from reactants that must be surpassed for the associative pathway (132.6 kJ mol^{-1}) is lower than the dissociation energy of CO from $(CO)_5Cr(C(OH)C_2H_3)$ to give **12** (144.7 kJ mol^{-1}) and therefore TDS concluded that the associative route is kinetically favorable. It is important to remark that since we have not taken into account the effect of the solvent, the proposed associative mechanism must be seen as an alternative to explain how the reaction can proceed when it is performed at gas-phase or in noncoordinating solvents.

Such conclusions have been initially regarded with reluctance by some experimentalists [40]. Despite its potential interest, the new alternative does not fit the large body of experimental observations available so far. As pointed out by Fischer and Hofmann [40], kinetic studies are not consistent with the associative mechanism and are clearly in favor of a dissociative path. However, in a recent kinetic study, Waters, Bos, and Wulff (WBW) [41] have provided the first example of a bimolecular reaction of a heteroatom-stabilized carbene complex with an alkyne in the absence of CO pressure, supporting TDS's theoretical results [27]. The unprecedented bimolecular mechanism observed by WBW for the reaction of o-methoxyphenyl chromium carbene complex with 1-phenyl-1-propyne impelled these authors to investigate whether this was a general phenomenon. Although the kinetics of chromium carbene complexes with substituted acetylenes had been early investigated by Fischer and Dötz [37], all of their studies were done at 5 x 10^{-3} M carbene complex and at 90. °C, whereas the most typical conditions for running a benzannulation reaction are 45 °C and 0.1 - 0.5 M carbene complex concentration, which is significantly more concentrated than Fischer's kinetic studies. WBW have found [41] that, under typical reaction conditions (0.1 - 0.5 M carbene and 45 °C), the reaction is bimolecular being first-order in carbene complex and alkyne. Since the calculated energy barriers associated to the dissociative and associative paths differ by only 8.1 kJ mol^{-1}, it is not unexpected that different reaction conditions may lead to unimolecular or bimolecular behavior as found experimentally [41]. As new data become available, also new theoretical studies are necessary. In particular, from a computational point of view, further research on the initial part of the reaction including solvent effects is mandatory to get a more precise answer on the validity of these two possible reaction pathways.

2.2   Central part of the reaction

This section is divided as follows: First, the results for the vinylketene route (section 3.2.1) and for the chromacycloheptadienone route (section 3.2.2) are presented, followed by a brief comparison of the two studied canonical paths. Once pondered the most classically assumed mechanisms, we discuss and compare (section 3.2.3) the suitability of a novel route involving a chromahexatriene intermediate as an alternative to redefine the standard pathway believed to be operative.

2.2.1   The vinylketene route

The starting point of the central part of the reaction is assumed to be the η^1-vinylcarbene complex **13a**.

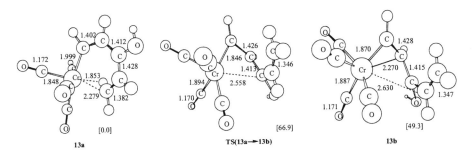

Figure 6. Optimized geometries of η^1-vinylcarbene complex **13a**, η^3-vinylcarbene complexe **13b**, and the transition state connecting them. Selected bond distances are given in Å and energies are given with respect to **13a** in kJ mol^{-1}.

In order to follow the vinylketene route suggested by Dötz (route A), **13a** has to convert into **14**. Such a conversion takes place in two steps. First, **13a** isomerizes into **13b** through **TS(13a→13b)** (Figure 6), and then turns into **14** via **TS(13b→14)** (Figure 7). No TS connecting directly **13a** and **14** was found [29]. Conversion from **13a** to **13b** involves rotation and folding of the organic chain, which can be expected to occur without many hindrances given the unrestricted mobility of the carbene ligand. The corresponding barrier is 66.9 kJ mol^{-1}. Once **13b** is formed from its precursor **13a**, a CO migration takes place in the step **13b → 14** followed by a ring closure in the step **14 → 15**. The activation barrier for the conversion **13b → 14** (28.9 kJ mol^{-1}) corresponds roughly to the energy required for a CO ligand to migrate into the Cr-C double bond [29]. In the next step, conversion from **14** to **15** is notably exothermic (101.7 kJ mol^{-1}) and kinetically very favorable, with a smooth activation barrier of only 1.2 kJ mol^{-1}.

The last step of the reaction is the keto-enol tautomerization from η^4-cyclohexadienone intermediates (**15**) to aromatic products (**16**). Such a step is accompanied with a considerable gain in energy: about 80 kJ mol^{-1} for vinylcarbenes [29], (where a phenol system is formed by the tautomerization step), and about 175 kJ mol^{-1} for phenylcarbenes [25] (where a naphtol system is produced). The energy barrier for such step should be lower than 40 kJ mol^{-1} according to previous calculations on similar systems [42].

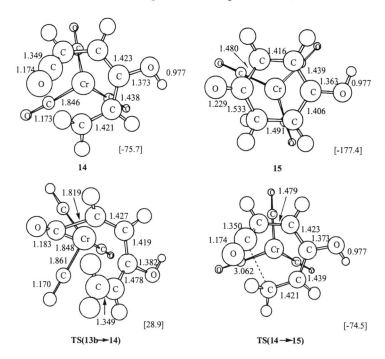

Figure 7. Optimized geometries for intermediates and transition states involved in route A. Selected bond distances are given in Å and energies are given with respect to complex **13b** in kJ mol^{-1}.

2.2.2   The chromacycloheptadienone route

According to Figure 3, the main intermediates postulated for the conversion of **13** to **15** through route B are chromacyclohexadiene **17** and chromacycloheptadienone **18**. We found no minimum structure [29] having the geometrical traits shown by the six-membered species labeled **17**, *i.e.*, a species with two short Cr-C distances. Different input geometries led systematically to complex **13a** directly. Opening of the chain and stabilization through agostic interaction (as in **13a**) were invariably the preferred trends observed along any optimization process. We concluded that **17** does not exist as a real stationary point, and that chromacycloheptadienone **18** (see Figure 8) does directly derive from complex **13a**. This is consistent with a recent isotopic study by Hughes *et al.* [43] where metallacyclohexadiene complexes similar to **17** were suggested to be excluded as intermediates along the reaction pathway of the Dötz and related reactions.

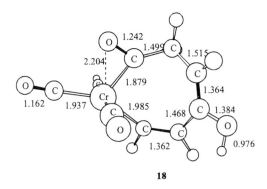

18

Figure 8. Optimized geometrical parameters for the main intermediate involved in route B: chromacycloheptadienone complex **18**. Bond distances are given in Å.

Conversion of **13a** into **18** was found to be a notably endothermic process (104.6 kJ mol⁻¹). In the next step, reduction of the strain in the seven-membered ring by turning it into a six-membered one (step **18** → **15**) results in a substantial release of energy (-233.0 kJ mol⁻¹). No TSs have been reported for route B [29]; thermodynamic data are conclusive enough to see that route A is more favorable than route B. The reaction energy required for the formation of intermediate **18** is larger than any of the activation energies involved in route A [29]. Our calculations indicate that the route initially suggested by Casey for the benzannulation reaction has few (or null) chances to compete against route A.

2.2.3   The chromahexatriene route

Route C in Figure 3 is a novel mechanistic proposal [28] for the central part of the Dötz reaction that invokes formation of a chromahexatriene intermediate **19** through rearrangement of the branch point species **13**, followed by insertion of a CO ligand to yield **15** in the subsequent step.

The structural arrangements involved in this sequence are summarized in Figure 9. Starting with **13a** (which is the most stable conformer of complex **13**), the ring chain evolves to complex **19** that is characterized by a d_π interaction between chromium and the terminal olefinic bond. A crucial feature of complex **19** as compared to complex **13a** is the shortening of the distance between the terminal C atom not connected to chromium and the CO ligand to be transferred (C-CO = 3.517 Å in **13a**, and 2.559 Å in **19**), which facilitates CO insertion in the subsequent step (C-CO = 2.262 Å in TS(**19**→**15**), and 1.533 Å in **15**). Interestingly, the particular orientation of the carbene ligand in **19** makes the migration of CO very accessible. During the process each of the two π* orbitals of CO can interact simultaneously with the π orbitals of the Cr=C and the terminal olefinic bonds. Therefore, CO migration in **19** is favored over CO migration in **13**, and, consequently, the energy profile for route C has the advantage of being globally smoother.

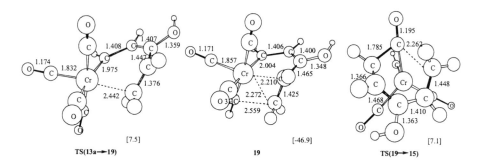

Figure 9. Optimized geometries for intermediates and transition states involved in route C. Selected bond distances are given in Å and energies are given with respect to complex **13a** in kJ mol^{-1}.

Support for the existence of **19** comes from a recent investigation by Barluenga *et al.* [44], where a chromahexatriene similar to **19** was isolated and characterized by ^{1}H and ^{13}C NMR spectroscopy. Interestingly, chromahexatriene **19** is not an endpoint of the reaction because it connects to intermediate complex **15** through the **TS(19→15)** shown in Figure 9. The fact that the reaction does not terminate at **19** is consistent again with the experimental results reported by Barluenga *et al.* [44]. These authors found that the synthesized chromahexatriene was not stable in solution at room temperature, and decomposed to yield the most common final product in the Dötz reaction with aminocarbenes [44].

3.   FINAL REMARKS

The results of the present work are summarized in Figure 10. As far as the initial part of the reaction is concerned, we have found that the associative mechanism is slightly more favored than the dissociative one. However, the small energy difference found between the two mechanisms and the possible effect of the solvent, which has not been included in the present study, precludes formulation of a definitive conclusion on the most effective reaction pathway. Most likely depending on the reaction conditions and the initial reactants the two mechanisms can be operative. In fact, experimental evidence in favor of both the dissociative [37, 38] and associative [41] mechanisms has been provided.

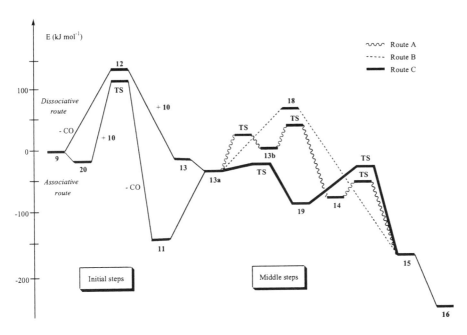

Figure 10. Energy profile for the whole benzannulation reaction: comparison of the profiles for the two studied mechanisms in the initial part of the reaction and the three analyzed mechanisms of the central part of the reaction.

Two main conclusions can be drawn from the studies on the middle steps of the reaction. First, the results shown herein indicate that, from a thermodynamic point of view, Casey's mechanistic proposal for the central part of the benzannulation reaction should be rejected. The conversion of vinylcarbenes into chromacycloheptadienones is computed to be a highly endothermic process, more energy-demanding than any of the steps postulated by Dötz in the vinylketene route. Second, our study shows that the novel proposal involving a chromahexatriene intermediate emerges as a concurrent alternative in the context of the two previously assumed mechanisms. Comparison of the new route to the one suggested by Dötz reveals that, when the classical order of the two central steps is reformulated by postponing CO migration (as in route C), the reaction proceeds through a lower-energy barrier path. It suggests that, instead of assuming CO migration as the first step from the branching point followed by structural rearrangement (route A), the mechanism could be best formulated as inverting these two processes. This obviously means replacing the vinylketene intermediate by the one capable of undergoing CO migration in the second step, *i.e.*, the chromahexatriene intermediate **19**, for which experimental support exists [44].

From a mechanistic point of view, all data reported so far indicate that altogether the associative mechanism and route C stands out as potentially suitable alternative to redefine the classical mechanism. However, a word of caution applies here. Given the similarity in energies for the investigated routes, the use of different substrates may alter the relative order of the barriers and, therefore, modify the mechanistic behavior of the reactants, making the reaction proceed through other pathways. New studies will contribute to widen the range of concurrent, competing alternatives, for this intriguing benzannulation reaction whose experimental simplicity (a one-pot reaction) contrasts with its mechanistic complexity.

APPENDIX: COMPUTATIONAL DETAILS

Encouraged by the success of the density functional theory (DFT) in recent gas-phase studies involving chromium complexes [22d-i, 45], we decided to carry out our calculations using this method and the Gaussian program [46, 47]. All geometry optimizations and energy differences were computed including nonlocal corrections (Becke's nonlocal exchange correction [48] and Perdew's nonlocal correlation correction [49]). The 6-31G** basis set [50] was employed for C, O, and H atoms. For the chromium atom we utilized a basis set as described by Wachters [51] (14s9p5d)/[8s4p2d], using the d-expanded contraction scheme (62111111/3312/311). The stationary points were located with the Berny algorithm [52] using redundant internal coordinates, and characterized by the correct number of negative eigenvalues of their analytic Hessian matrix; this number must be zero for minima and one for any true TS. We also verified that the imaginary frequency exhibits the expected motion. Only closed-shell states were considered. Kinetical relativistic effects are unimportant for an accurate calculation of chromium complexes [53] and were neglected in our calculations.

ACKNOWLEDGEMENTS

Support for this work under Grant PB98-0457-C02-01 from the Dirección General de Enseñanza Superior e Investigación Científica y Técnica (MECD-Spain) is acknowledged. M.S. is indebted to the Departament d'Universitats, Recerca i Societat de la Informació of the Generalitat de Catalunya for financial support through the Distinguished University Research Promotion, 2001.

REFERENCES AND NOTES

1 For reviews see: (a) Dötz, K. H.; Fischer, H.; Hofmann, P.; Kreissl, F. R.; Schubert, U.; Weiss, K. *Transition Metal Carbene Complexes*; VCH: Weinheim, 1983. (b) Dötz, K. H. *Angew. Chem.* **1984**, *96*, 573; *Angew. Chem., Int. Ed. Engl.* **1984**, *23*, 587. (c) Hegedus, L. S. *Comprehensive Organometallic Chemistry II*; Abel, E. W.; Stone, F. G. A.; Wilkinson, G., Eds.; Pergamon Press: Oxford, 1995; Vol. 12, p. 549. (d) Wulff, W. D. *Comprehensive Organic Synthesis*; Trost, B. M.; Fleming, I., Eds.; Pergamon Press: Oxford, 1991; Vol. 5, p. 1065. (e) Wulff, W.D. *Comprehensive Organometallic Chemistry II*; Abel, E.W.; Stone, F. G. A.; Wilkinson, G., Eds.; Pergamon Press: Oxford, 1995; Vol. 12, p. 469. (f) Harvey, D. F.; Sigano, D. M. *Chem. Rev.* **1996**, *96*, 271. (g) de Meijere, A. *Pure Appl. Chem.* **1996**, *68*, 61. (h) Barluenga, J. *Pure Appl. Chem.* **1996**, *68*, 543. (i) Aumann, R.; Nienaber, H. *Adv. Organomet. Chem.* **1997**, *41*, 163. (j) Dötz, K. H.; Tomuschat, P. *Chem. Soc. Rev.* **1999**, *28*, 187. (k) Sierra, M. A. *Chem. Rev.* **2000**, *100*, 3591. (l) Davies, M. W.; Johnson, C. N.; Harrity, J. P. A. *J. Org. Chem.* **2001**, *66*, 3525. (m) Barluenga, J.; Suárez-Sobrino, A. L.; Tomás, M.; García-Granda, S.; Santiago-García, R. *J. Am. Chem. Soc.* **2001**, *123*, 10494.

2 (a) Schmalz, H.-G. *Angew. Chem.* **1994**, *106*, 311; *Angew. Chem., Int. Ed. Engl.* **1994**, *33*, 303. (b) Frühauf, H.-W. *Chem. Rev.* **1997**, *97*, 523. (c) Barluenga, J.; Martínez, S.; Suárez-Sobrino, A. L.; Tomás, M. *J. Am. Chem. Soc.* **2001**, *123*, 11113.

3 Dötz, K. H. *Angew. Chem.* **1975**, *87*, 672; *Angew. Chem., Int. Ed. Engl.* **1975**, *14*, 644.

4 (a) Dötz, K. H.; Dietz, R. *Chem. Ber.* **1978**, *111*, 2517. (b) Dötz, K. H.; Pruskil, I. *Chem. Ber.* **1980**, *113*, 2876. (c) Wulff, W. D.; Tang, P. C.; McCallum, J. S. *J. Am. Chem. Soc.* **1981**, *103*, 7677. (d) Dötz, K. H.; Mühlemeier, J.; Schubert, U.; Orama, O. *J. Organomet. Chem.* **1983**, *247*, 187. (e) Wulff, W. D.; Chan, K. S.; Tang, P. C. *J. Org. Chem.* **1984**, *49*, 2293. (f) Yamashita, A.; Toy, A. *Tetrahedron Lett.* **1986**, *27*, 3471.

5 Benzannulation with stannylalkynes shows the reversed regioselectivity: Chamberlain, S.; Waters, M. L.; Wulff, W. D. *J. Am. Chem. Soc.* **1994**, *116*, 3113.

6 Wang, H.; Wulff, W. D.; Rheingold, A. L. *J. Am. Chem. Soc.* **2000**, *122*, 9862.

7 (a) Dötz, K. H. *J. Organomet. Chem.* **1977**, *140*, 177. (b) Dötz, K. H.; Pruskil, I. *Chem. Ber.* **1978**, *111*, 2059. (c) Wulff, W. D.; Gilbertson, S. R.; Springer, J. P. *J. Am. Chem. Soc.* **1986**, *107*, 5823. (d) Semmelhack, M. F.; Park, J. *Organometallics* **1986**, *5*, 2550. (e) Garret, K. E.; Sheridan, J. B.; Pourreau, D. B.; Weng, W. C.; Geoffroy, G. L.; Staley, D. L.; Rheingold, A. L. *J. Am. Chem. Soc.* **1989**, *111*, 8383. (f) Yamashita, A. *Tetrahedron Lett.* **1986**, *27*, 5915. (g) Dötz, K. H.; Larbig, H. *J. Organomet. Chem.* **1991**, *405*, C38. (h) Wulff, W. D.; Bax, B. A.; Brandvold, B. A.; Chan, K. S.; Gilbert, A. M.; Hsung, R. P.; Mitchell, J.; Clardy, J. *Organometallics* **1994**, *13*, 102. (i) Aumann, R.; Jasper, B.; Fröhlich, R. *Organometallics* **1995**, *14*, 231. (j) Parlier, A.; Rudler, M.; Rudler, H.; Goumont, R.; Daran, J.-C.; Vaissermann, J. *Organometallics* **1995**, *14*, 2760. (k) Dötz, K. H.; Gerhardt, A. *J. Organometal. Chem.* **1999**, *578*, 223. (l) Tomuschat, P.; Kröner, L.; Steckhan, E.; Nieger, M.; Dötz, K. H. *Chem. Eur. J.* **1999**, *5*, 700.

8 Barluenga, J.; Aznar, F.; Gutiérrez, I.; Martín, A., García-Granda, S.; Llorca-Baragaño, M. *J. Am. Chem. Soc.* **2000**, *122*, 1314.

9 Barluenga, J.; Aznar, F.; Palomero, M. A. *Angew. Chem., Int. Ed. Engl.* **2000**, *39*, 4346.

10 Barluenga, J.; Tomás, M.; Ballesteros, A.; Santamaría, J.; Brillet, C.; García-Granda, S.; Piñera-Nicolás, A.; Vázquez, J. T. *J. Am. Chem. Soc.* **1999**, *121*, 4516.

11 Barluenga, J.; Tomás, M.; Rubio, E.; López-Pelegrín, J. A.; García-Granda, S.; Pérez Priede, M. *J. Am. Chem. Soc.* **1999**, *121*, 3065.

12　(a) Yamashita, A. *Tetrahedron Lett.* **1986**, *27*, 5915. (b) Wulff, W. D.; Box, B. M.; Brandvold, T. A.; Chan, K. S.; Gilbert, A. M.; Hsung, R. P. *Organometallics* **1994**, *13*, 102. (c) Wulff, W. D.; Gilbert, A. M.; Hsung, R. P.; Rahm, A. *J. Org. Chem.* **1995**, *60*, 4566. (d) Bos, M. E.; Wulff, W. D.; Miller, R. A.; Brandvold, T. A.; Chamberlin, S. *J. Am. Chem. Soc.* **1991**, *113*, 9293. (e) Waters, M. L.; Branvold, T. A.; Isaacs, L.; Wulff, W. D.; Rheingold, A. L. *Organometallics* **1998**, *17*, 4298.

13　(a) Semmelhack, M. F.; Bozell, J. J.; Keller, L.; Sato, T.; Spiess, E. J.; Wulff, W. D.; Zask, A. *Tetrahedron* **1985**, *41*, 5803. (c) Wulff, W. D.; Tang, P. C. *J. Am. Chem. Soc.* **1984**, *106*, 434. (c) Yamashita, A. *J. Am. Chem. Soc.* **1985**, *107*, 5823. (d) Dötz, K. H.; Popall, M. *J. Organomet. Chem.* **1985**, *291*, C1. (e) Dötz, K.H.; Popall, M. *Tetrahedron* **1985**, *41*, 5797. (f) Dötz, K. H.; Popall, M.; Müller, G.; Ackermann, K. *J. Organomet. Chem.* **1990**, *383*, 93. (g) Paetsch, D.; Dötz, K. H. *Tetrahedron Lett.* **1999**, *40*, 487.

14　(a) Semmelhack, M. F.; Bozell, J. J.; Sato, T.; Wulff, W. D.; Spiess, E.; Zask, A. *J. Am. Chem. Soc.* **1982**, *104*, 5850. (b) Wulff, W.; Su, J.; Tang, P.-C.; Xu, Y.-C. *Synthesis* **1999**, 415. (c) Gilbert, A. M.; Miller, R.; Wulff, W. *Tetrahedron* **1999**, *55*, 1607. (d) Miller, R. A.; Gilbert, A. M.; Xue, S.; Wulff, W. D. *Synthesis* **1999**, 80.

15　(a) Dötz, K. H.; Popall, M. *Angew. Chem.* **1987**, *99*, 1220; *Angew. Chem., Int. Ed. Engl.* **1987**, *26*, 1158. (b) Wulff, W. D.; Xu, Y.-C. *J. Am. Chem. Soc.* **1988**, *110*, 2312.

16　(a) Boger, D. L.; Jacobson, I. C. *J. Org. Chem.* **1991**, *56*, 2115. (b) Boger, D. L.; Hüter, O.; Mbiya, K.; Zhang, M. *J. Am. Chem. Soc.* **1995**, *117*, 11839.

17　(a) Dötz, K. H.; Pruskil, I. *J. Organomet. Chem.* **1981**, *209*, C4. (b) Dötz, K. H.; Pruskil, I.; Mühlemeier, J. *Chem. Ber.* **1982**, *115*, 1278. (c) Dötz, K. H.; Kuhn, W. *Angew. Chem.* **1983**, *95*, 755; *Angew. Chem., Int. Ed. Engl.* **1983**, *22*, 732.

18　(a) Hohmann, F.; Siemoneit, S.; Nieger, M.; Kotila, S.; Dötz, K. H. *Chem. Eur. J.* **1997**, *3*, 853. (b) Dötz, K. H.; Stinner, C. *Tetrahedron: Asymmetry* **1997**, *8*, 1751. (c) Gordon, D.M.; Danishefsky, S. H.; Schulte, G.M. *J. Org. Chem.* **1992**, *57*, 7052. (d) King, J.; Quayle, P.; Malone, J. F. *Tetrahedron Lett.* **1990**, *31*, 5221. (e) Parker, K. A.; Coburn, C. A. *J. Org. Chem.* **1991**, *56*, 1666. (f) Boger, D. L.; Jacobson, I. C. *J. Org. Chem.* **1990**, *55*, 1919. (g) Semmelhack, M. F.; Jeong, N. *Tetrahedron Lett.* **1990**, *31*, 650. (h) Semmelhack, M. F.; Jeong, N.; Lee, G. R. *Tetrahedron Lett.* **1990**, *31*, 609. (i) Yamashita, A.; Toy, A.; Scahill, T. A. *J. Org. Chem.* **1989**, *54*, 3625. (j) Wulff, W.D.; McCallum, J. S.; Kung, F. A. *J. Am. Chem. Soc.* **1988**, *110*, 7419.

19　Fogel, L.; Hsung, R. P.; Wulff, W. D.; Sommer, R. D.; Rheingold, A. L. *J. Am. Chem. Soc.* **2001**, *123*, 5580.

20　Wulff, W. D. *Organometallics* **1998**, *17*, 3116.

21　(a) Dötz, K. H.; Popall, M.; Müller, G.; Ackermann, K. *Angew. Chem.* **1986**, *98*, 909; *Angew. Chem., Int. Ed. Engl.* **1986**, *25*, 911. (b) Dötz, K. H.; Sturm, W.; Popall, M.; Riede, J. *J. Organomet. Chem.* **1985**, *277*, 257.

22　(a) Nakatsuji, H.; Ushio, J.; Han, S.; Yonezawa, T. *J. Am. Chem. Soc.* **1983**, *105*, 426. (b) Taylor, T. E.; Hall, M. B. *J. Am. Chem. Soc.* **1984**, *106*, 1576. (c) Ushio, J.; Nakatsuji, H.; Yonezawa, T. *J. Am. Chem. Soc.* **1984**, *106*, 5892. (d) Jacobsen, H.; Ziegler, T. *Organometallics* **1995**, *14*, 224. (e) Jacobsen, H.; Ziegler, T. *Inorg. Chem.* **1996**, *35*, 775. (f) Wang, C.-C.; Wang, Y.; Lin, K.-J.; Chou, L.-K.; Chan, K.-S. *J. Phys. Chem. A* **1997**, *101*, 8887. (g) Frenking, G.; Pidun, U. *J. Chem. Soc., Dalton Trans.* **1997**, 1653. (h) Frölich, N.; Pidun, U.; Stahl, M.; Frenking, G. *Organometallics* **1997**, *16*, 442. (i) Vyboishchikov, S. F.; Frenking, G. *Chem. Eur. J.* **1998**, *4*, 1428.

23　Hofmann, P.; Hämmerle, M. *Angew. Chem.* **1989**, *101*, 940; *Angew. Chem., Int. Ed. Engl.* **1989**, *28*, 908.

24　Hofmann, P.; Hämmerle, M.; Unfried, G. *New. J. Chem.* **1991**, *15*, 769.

25 Gleichmann, M. M.; Dötz, K. H.; Hess, B. A. *J. Am. Chem. Soc.* **1996**, *118*, 10551.

26 Torrent, M. Thesis, Universitat de Girona, 1998; chapters 9 and 10.

27 Torrent, M.; Duran, M.; Solà, M. *Organometallics* **1998**, *17*, 1492.

28 Torrent, M.; Duran, M.; Solà, M. *Chem. Commun.* **1998**, 999.

29 Torrent, M.; Duran, M.; Solà, M. *J. Am. Chem. Soc.* **1999**, *121*, 1309.

30 Arrieta, A.; Cossío, F. P.; Fernández, I.; Gómez-Gallego, M.; Lecea, B.; Mancheño, M. J.; Sierra, M. A. *J. Am. Chem. Soc.* **2000**, *122*, 11509.

31 Casey, C. P. *Reactive Intermediates*; Jones, M., Jr.; Moss, R. A., Eds.; Wiley: New York, 1981; Vol. 2, p. 155.

32 For complex **15** see: Tang, P.-C.; Wulff, W. D. *J. Am. Chem. Soc.* **1984**, *106*, 1132.

33 (a) Huffman, M. A.; Liebeskind, L. S.; Pennington, W. T. *Organometallics* **1992**, *11*, 255. (b) Grojthan, D. B.; Bikzhanova, G. A.; Collins, L. S. B.; Concolino, T.; Lam, K.-C.; Rheingold, A. L. *J. Am. Chem. Soc.* **2000**, *122*, 5222. (c) Wulff, W. D.; Gilberstson, S. R.; Springer, J. P. *J. Am. Chem. Soc.* **1986**, *108*, 520. (d) Richards, C. J.; Thomas, S. E. *J. Chem. Soc., Chem. Commun.* **1990**, 307.

34 (a) Anderson, B.A.; Wulff, W.D. *J. Am. Chem. Soc.* **1990**, *112*, 8615. (b) Chelain, E.; Parlier, A.; Rudler, H.; Daran, J.-C.; Vaissermann, J. *J. Organomet. Chem.* **1991**, *419*, C5.

35 Garret, K. E.; Sheridan, J. B.; Pourreau, D. B.; Feng, W. C.; Geoffroy, G. L.; Staley, D. L.; Rheingold, A. L. *J. Am. Chem. Soc.* **1989**, *111*, 8383.

36 (a) Chan, K. S.; Petersen, G. A.; Brandvold, T. A.; Faron, K. L.; Challener, C. A.; Hydahl, C.; Wulff, W. D. *J. Organomet. Chem.* **1987**, *334*, 9. (b) Bos, M. E.; Wulff, W. D.; Miller, R. A.; Chamberlin, S.; Brandvold, T. A. *J. Am. Chem. Soc.* **1991**, *113*, 9293.

37 Fischer, H.; Mühlemeier, J.; Märkl, R.; Dötz, K. H. *Chem. Ber.* **1982**, *115*, 1355.

38 Dötz, K. H.; Dietz, R. *Chem. Ber.* **1977**, *110*, 1555.

39 Solà, M.; Torrent, M.; Duran, M., to be published.

40 Fischer, H.; Hofmann, P. *Organometallics* **1999**, *18*, 2590.

41 Waters, M. L.; Bos, M. E.; Wulff, W. D. *J. Am. Chem. Soc.* **1999**, *121*, 6403.

42 (a) Guallar, V.; Moreno, M.; Lluch, J. M.; Amat-Guerri, F.; Douhal, A. *J. Phys. Chem.* **1996**, *100*, 19789.

43 Hughes, R. P.; Trujillo, H. A.; Gauri, A. J. *Organometallics* **1995**, *14*, 4319.

44 Barluenga, J.; Aznar, F.; Martín, A.; García-Granda, S.; Pérez-Carreño, E. *J. Am. Chem. Soc.* **1994**, *116*, 11191.

45 (a) Torrent, M.; Gili, P.; Duran, M.; Solà, M. *J. Chem. Phys.* **1996**, *104*, 9499. (b) Torrent, M.; Duran, M.; Solà, M., In *Advances in Molecular Similarity*; Carbó, R., Mezey, P. G., Eds.; JAI Press: Greenwich, CT, 1996, pp 167-186.

46 Frisch, M. J et al. Gaussian 94 (Revision A.1); Gaussian, Inc. Pittsburg PA, 1995.

47 Frisch, M. J et al.Gaussian98, (Revision A.7); Gaussian Inc.: Pittsburgh, PA 1998.

48 Becke, A. D. *Phys. Rev.* **1988**, *A38*, 3098.

49 Perdew, J. P. *Phys. Rev.* **1986**, *B33*, 8822; **1986**, *B34*, 7406E.

50 (a) Binkley, J. S.; Pople, J. A.; Hehre, W. J. *J. Am. Chem. Soc.* **1980**, *102*, 939; (b) Gordon, M. S.; Binkley, J. S.; Pople, J. A.; Pietro, W. J.; Hehre, W. J. *J. Am. Chem. Soc.* **1982**, *104*, 2797; (c) Pietro, W. J.; Francl, M. M.; Hehre, W. J.; Defrees, D. J.; Pople, J.A.; Binkley, J. S. *J. Am. Chem. Soc.* **1982**, *104*, 5039.

51 Wachters, A. J. H. *J. Chem. Phys.* **1970**, *52*, 1033.

52 Peng, C.; Ayala, P. Y.; Schlegel, H. B.; Frisch, M. J. *J. Comp. Chem.* **1996**, *17*, 49.

53 Jacobsen, H.; Schreckenbach, G.; Ziegler, T. *J. Phys. Chem.* **1994**, *98*, 11406.

Chapter 12

Mechanism of Olefin Epoxidation by Transition Metal Peroxo Compounds

Notker Rösch,[1] Cristiana Di Valentin,[1,2] and Ilya V. Yudanov[1,3]

[1] *Institut für Physikalische und Theoretische Chemie, Technische Universität München, 85747 Garching, Germany;* [2] *Dipartimento di Scienza dei Materiali, Università degli Studi di Milano-Bicocca, via Cozzi 53, 20125 Milano, Italy;* [3] *Boreskov Institute of Catalysis, Siberian Branch of the Russian Academy of Sciences, 630090 Novosibirsk, Russia*

Abstract: Density functional calculations on model systems show that olefin epoxidation by peroxo complexes of early transition metals (Mo, W, Re) in general proceeds by direct transfer via spiro-type transition structures, rather than via insertion. Oxygen transfer by hydroperoxo complexes is at least competitive, if not preferred for Mo, while it is clearly preferred for Ti. A simple MO analysis allows to rationalize many trends (e.g. effects of the metal center and of additional base ligands) found for the computational results.

Key words: Olefin epoxidation, allylic alcohol epoxidation, insertion, direct oxygen transfer, early transition metals, peroxo and hydroperoxo intermediates, effects of base ligands, activation barriers, DFT-based calculations

1.   INTRODUCTION

During the last three decades, peroxo compounds of early transition metals (TMs) in their highest oxidation state, like Ti^{IV}, V^V, Mo^{VI}, W^{VI}, and Re^{VII}, attracted much interest due to their activity in oxygen transfer processes which are important for many chemical and biological applications. Olefin epoxidation is of particular significance since epoxides are key starting compounds for a large variety of chemicals and polymers [1]. Yet, details of the mechanism of olefin epoxidation by TM peroxides are still under discussion.

F. Maseras and A. Lledós (eds.), Computational Modeling of Homogeneous Catalysis, 289–324.

1.1 Evolution of experimental views

Several general reviews describe the state of the art of peroxide epoxidation catalyzed by TM compounds at about a decade ago [2-4]. Later on, specialized reviews dealt with particular peroxides of Cr, Mo, and W [5], V [6], and with epoxidation reactions catalyzed by methyltrioxorhenium (MTO) [7] that involve Re peroxo complexes as species responsible for the oxygen transfer.

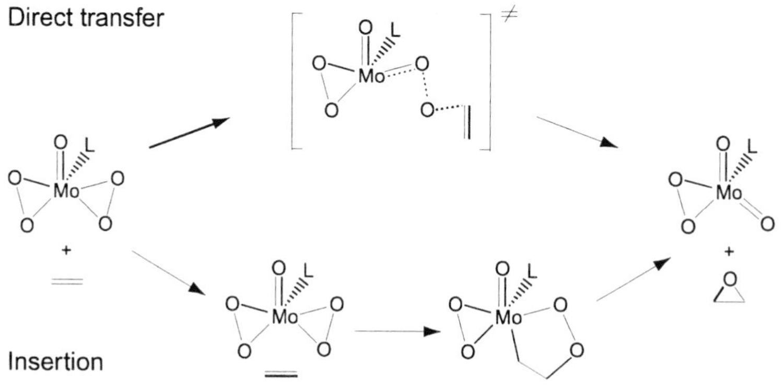

Figure 1. Insertion and direct transfer mechanisms of ethene epoxidation by TM peroxo compounds, exemplified for a Mo complex.

The mechanism of oxygen transfer from a metal peroxo complex to an olefin double bond, resulting in the formation of an epoxide, has been discussed extensively. In their pioneering work [8] on the kinetics of the stoichiometric epoxidation by the di(peroxo) Mo^{VI} complex $MoO(O_2)_2$·hmpt (hmpt = hexamethylphosphoric triamide), Mimoun et al. suggested a multi-step mechanism. They assumed pre-coordination of the olefin to the Mo center with subsequent insertion into a metal-peroxo bond leading to a five-membered metallacycle intermediate that involves two carbon atoms, the metal center and a peroxo group (Figure 1). In the final stage of this process the epoxide molecule is extruded from the metallacycle. Such intermediates are known for reactions of peroxo complexes of late TMs, like Pd and Pt [9]; however, they have never been detected for d^0 TM peroxides. While Mimoun assumed the coordination of the olefin to the metal center to proceed as substitution of the phosphane oxide ligand [9] Arakawa et al. advocated coordination of the olefin as additional ligand [10].

With regard to epoxidation activity, the peroxo complex $MoO(O_2)_2$·hmpt exhibits close similarity to peracids. Based on this observation, Sharpless et al. suggested [11] a concerted (direct) mechanism as an alternative to the insertion mechanism. That mechanism, which is assumed to proceed via a transition state (TS) of a spiro-like structure (Figure 1), does not require pre-coordination of the olefin since it implies a direct attack of the peroxo group on the olefin. Despite of extensive further studies, this controversy between the two mechanisms was not resolved for a long time by means of experimental investigations alone [12-15].

In addition to the general mechanism of oxygen transfer, the nature of active peroxo species has been discussed for many particular cases. Recently, Herrmann and co-workers identified MTO, CH_3ReO_3, as well as its derivatives as efficient catalysts for olefin epoxidation [16-18]. They were able to isolate a di(peroxo) complex of rhenium, $CH_3ReO(O_2)_2(H_2O)$, which is stabilized by a water ligand, and to characterize it structurally [19]. Based on these investigations, they proposed a reaction mechanism which involves mono- and di(peroxo) Re compounds, formed via interaction of MTO with H_2O_2, to be the active intermediates in epoxidation [19-21]. Yet, final experimental clarification is still lacking whether mono- [20] or di(peroxo) [19] complexes are the catalytically active intermediates.

In epoxidation reactions catalyzed by Ti^{IV} compounds, hydroperoxo or alkylperoxo moieties TiOOR (R = H, alkyl) are generally accepted as oxygen donors, while $Ti(\eta^2-O_2)$ peroxo groups with two symmetrically coordinated oxygen centers at Ti are considered to be inert. For instance, a TiOO*t*Bu species is assumed to be responsible for epoxidation in the Sharpless process [22] and a TiOOH species for epoxidation on Ti silicalite [23]. However, in early reports also active $Ti(O_2)$ species have been suggested for the latter system [24, 25]. Actually, a number of stable and experimentally well characterized Ti^{IV} η^2-peroxo complexes are *not* active in olefin epoxidation. Thus, despite similar electronic and geometric structures, the η^2-peroxo complexes of Ti form a striking contrast to the closely related mono- and di(peroxo) complexes of Re as well as to the di(peroxo) complexes of Mo and W which are known to epoxidize alkenes. A clear rationalization why η^2-peroxo complexes of Ti are not active in epoxidation is so far lacking.

Another important aspect is the influence of strongly coordinating basic ligands on the epoxidation activity of the peroxo group. For Mo and W complexes $MO(O_2)_2L$ (M = Mo, W), the deactivating effect of strongly coordinating solvents was considered in close connection to the general reaction mechanism [8-10, 13, 15]. According to Mimoun et al. [8, 9] a solvent molecule occupies the free coordination position at the metal center and in this way blocks the pre-coordination initializing the insertion process. In the direct mechanism, the solvent effect is attributed to reduced electrophilicity of the peroxo group [26] due to induction of electron density from the additional solvent ligand via the metal center [12-15].

The proper choice of the base ligand is also important for MTO-catalyzed processes. It was detected early on that the use of Lewis base adducts of MTO significantly decreases the formation of undesired diols due to the reduced Lewis acidity of the catalyst system [27], however, while base ligands increase the selectivity, they also decrease the epoxidation rate. Another attempt to overcome the formation of diols relies on the urea/H_2O_2 complex instead of aqueous hydrogen peroxide [28, 29]. Later on, it was discovered that phase transfer systems (water phase/organic phase) and an excess of pyridine as Lewis base does not only hamper the formation of diols, but also improves the reaction rate compared to MTO as catalyst precursor [30-32]. Recently it was shown that 3-cyanopyridine and especially pyrazole as Lewis bases are even more effective than pyridine [33, 34] while pyridine N-oxides are less efficient [35, 36]. From *in situ* measurements it was concluded that both electronic and steric factors of the aromatic Lewis base involved play a significant role during the formation of the catalytically active species. The presence of pyridines lowers the activity of hydronium ions, thus reducing the rate of epoxide ring opening [37].

As mentioned above, di(peroxo) molybdenum complexes, like $MoO(O_2)_2$·hmpt, are deactivated by strongly coordinating solvents via addition of solvent molecule as seventh ligand of the complex [8]. However, seven-coordinated complexes $(L-L)MoO(O_2)_2$ with bidentate L-L base ligands recently were shown to catalyze epoxidation by *tert*-butyl hydroperoxide as oxidant [38]. This finding led to the hypothesis that MoOO*t*Bu and MoOOH groups are formed via interaction of *t*BuOOH with one of the peroxo groups of the complex [38]. This is an interesting suggestion since to date TM alkyl- and hydroperoxo species were experimentally found only for Ti^{IV} and V^V (see Section 3.3).

1.2   Theoretical studies

1.2.1   Semiempirical and early ab initio calculations

Several theoretical investigations described the interaction of a d^0 metal center with a peroxide in order to understand oxygen transfer from a TM center to an alkene [2, 39-42] or a sulfide [43]. These studies, using semiempirical methods or the ab initio Hartree-Fock SCF method, were limited to an orbital analysis of the ground state of metal peroxides.

Extended Hückel theory (EHT) was applied to study the decomposition of the five-membered metallacycle intermediate proposed by Mimoun for the epoxidation by Mo bisperoxo complexes [44, 45]. Another EHT study [40] proposed the coordination of ethene to the metal center of an $MoO(O_2)_2$ complex as the first step, followed by a slipping motion of ethene toward one of the peroxygen centers and the formation of a three-membered TS rather than the five-membered metallacycle intermediate. Using an NDDO approach, the electrophilic properties of six- and seven-coordinated di(peroxo) complexes of Mo were compared via calculated electron affinities [46]. Although a higher electron affinity was computed for the six-coordinated species favoring the mechanism that involves a direct electrophilic attack of the peroxo group at the olefin, no detailed information on the reaction mechanism was reported in this semiempirical approach.

1.2.2   Calculations based on Density Functional Theory

During the last decade quantum chemical calculations using density functional (DF) methods proved to be valuable for investigating olefin epoxidation by TM complexes. While early DF studies focused on structural aspects of peroxo complexes [47-50], the mechanism of olefin epoxidation, including the location of TSs and activation barriers of different mechanisms of oxygen transfer, were studied in more recent works. In this way, epoxidation of olefins by peroxo complexes of Ti [51-60], Cr, Mo, W [61-66] and Re [67, 68] was investigated. Recently, the epoxidation of allylic alcohol by rhenium peroxo complexes was studied computationally with special focus on the role of hydrogen bonding between an alcohol substrate and a peroxo complex [69]. Structure and energetics of hypothetic Mo [70, 71] and Re [71] hydroperoxo intermediates as well as their epoxidation activity [71] were also studied computationally.

An important finding is that all peroxo compounds with d^0 configuration of the TM center exhibit essentially the same epoxidation mechanism [51, 61, 67-72] which is also valid for organic peroxo compounds such as dioxiranes and peracids [73-79]. The calculations revealed that direct nucleophilic attack of the olefin at an electrophilic peroxo oxygen center (via a TS of spiro structure) is preferred because of significantly lower activation barriers compared to the multi-step insertion mechanism [51, 61-67]. A recent computational study of epoxidation by Mo peroxo complexes showed that the metallacycle intermediate of the insertion mechanism leads to an aldehyde instead of an epoxide product [62].

Despite of the common reaction mechanism, peroxo complexes exhibit very different reactivities – as shown by the calculated activation energies – depending on the particular structure (nature of the metal center, peroxo or hydroperoxo functionalities, type and number of ligands). We proposed a model [72, 80] that is able to qualitatively rationalize differences in the epoxidation activities of a series of structurally similar TM peroxo compounds $CH_3Re(O_2)_2O \cdot L$ with various Lewis base ligands L. In this model the calculated activation barriers of direct oxygen transfer from a peroxo group to an olefin are correlated with the energies of pertinent orbitals, the π(C-C) HOMO of the olefin and the σ*(O-O) LUMO of the peroxo group [80, 81]. When applied to organic peroxides, this rather simple model performs especially well [82-84].

1.2.3   Calculations on organic epoxidation and solvent effects

The similarity of olefin epoxidation by TM peroxo and hydroperoxo complexes with epoxidation by dioxirane derivatives R_2CO_2 and percarboxylic acids RCO(OOH) was confirmed by computational studies [73-79]. This similarity holds in particular for the spiro-type transition structure.

Most computational investigations on TM peroxo and hydroperoxo complexes were limited to reactions in the gas phase. Since organic oxidants are simpler to treat computationally than the TM systems, it is easier to investigate more complex aspects of the epoxidation reaction, like the solvent effects. Short-range (via explicit solvent molecule coordination) and long-range electrostatic solvent effects on the activation barriers have been investigated for dioxirane and acetic percarboxylic acid [82]. The elaborated relationship between the solvent-induced change of the activation energy and the value of the dielectric constant of the solvent was also applied to estimate the solvent effect on the activity of $CH_3Re(O)(O_2)_2 \cdot L$ (L = H_2O, pyridine, pyrazole) [82].

2.   COMPUTATIONAL STRATEGY

All electronic structure calculations of our group to be discussed in this review were performed with the hybrid B3-LYP [85] density functional scheme [86] using LANL2DZ effective core potentials for the TM metal centers [87]. For all other centers, a 6-311G(d,p) basis set was employed [88]; this basis set affords an adequate description also in cases of hydrogen bonds and proton transfer among ligands. Geometry optimizations were carried out without any symmetry restrictions. Finally, one (Ti) or two f exponents (Cr, Mo, W, Re) were added to the TM basis set to evaluate energies in single-point fashion [59, 61, 62, 67]. This computational strategy has been successfully applied in a series of oxygen transfer investigations [59, 61, 62, 67, 69, 71, 72, 80, 89-91]. It has been validated for Re-O bond energies of the complexes LReO$_3$ (including L = Cp* and CH$_3$); calculated and experimentally derived values agree very well for L = Cp* (113 and 116 kcal/mol, respectively).

Since the investigations often focused on trends and their rationalization, we refrained from correcting stabilization energies and reaction barriers for enthalpy and solvent effects. Other studies had shown that such corrections do not affect trends [67, 89]. Charges of atoms were determined by a natural bond orbital analysis (NBO) [92]. For more computational details we refer the reader to the publications reviewed here [59, 61, 62, 67, 69, 71, 72, 80, 82].

3. MODELING OF EPOXIDATION BY TRANSITION METAL COMPLEXES

3.1 Cr, Mo, W: The effect of the metal center on the reactivity of peroxo complexes

A considerable amount of experimental information is available for the peroxo complexes of the group VI metals Cr, Mo, and W in their highest oxidation state [5]. The complex $MoO(O_2)_2$·hmpt was the first identified to epoxidize alkenes in nonpolar solvents stoichiometrically, and it became the complex of choice in various studies [5]. There is evidence that the tungsten analogue $WO(O_2)_2$·hmpt exhibits a higher epoxidation activity [13]. However, no activity as oxygen transfer agent was reported for analogous Cr^{VI} peroxo compounds [5]. Various di(peroxo) complexes $MoO(O_2)_2L_1L_2$ (M = Cr, Mo, W) with different combinations of base ligands L_1 and L_2 have been experimentally characterized [5]. Therefore, these complexes provide an opportunity for a comparative investigation of how the nature of the metal atom affects the activation/deactivation of peroxo groups toward olefin epoxidation.

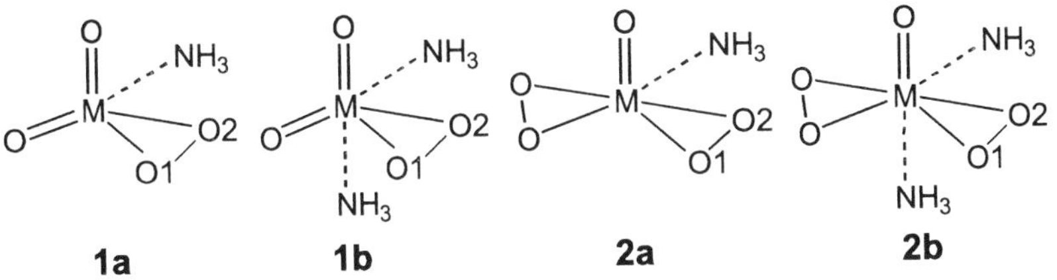

Figure 2. Model peroxo complexes of M = Cr, Mo, and W.

3.1.1 Model complexes

We consider the di(oxo)-mono(peroxo) and oxo-di(peroxo) model complexes shown in Figure 2, with one (**a**) or two (**b**) base ligands L. The metal center M is varied, M = Cr, Mo, and W. NH_3 is chosen as monodentate model base ligand L. While there are no experimental structures available corresponding to the mono(peroxo) species **1a** and **1b**, related complexes should occur as a result of oxygen transfer from di(peroxo) species (**2a** and **2b**) to the nucleophilic substrate. For di(peroxo) complexes of Cr, Mo and W X-ray structures are available [93-95]. A comparison of pertinent calculated structural parameters to experimental data shows that the present models perform satisfactorily for peroxo complexes with amino ligands, but they are less accurate for complexes with oxygen containing base ligands such as H_2O or hmpt.

Calculated structural parameters and a population analysis allow an instructive comparison of such peroxo complexes: (i) except for **1b**, the two oxygen centers of a peroxo group differ in M-O distances and atomic charges and hence are not equivalent; (ii) the M-O distances of corresponding mono- and di(peroxo) complexes are rather similar, but for Cr complexes they are by ~0.1 Å shorter than in the Mo and W species, consistent with smaller ionic radius of Cr^{VI}; (iii) the O-O distance increases from 1.40 Å for Cr to 1.45 Å for Mo and 1.48 Å for W, in agreement with experimental data [96]. Comparing these bond lengths to that of the isolated H_2O_2 molecule (1.45 Å), it becomes obvious that the peroxo bond of the W complex is activated, while for the Cr complex a deactivation of the peroxo group results, again in line with the calculated activation barriers (see below). Thus, the geometry of the metal peroxo moiety depends crucially on the metal center.

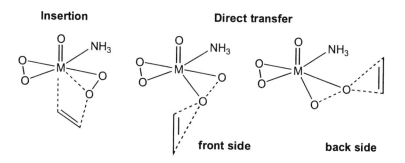

Figure 3. Schematic representation of the TS structures of the various mechanisms of olefin epoxidation.

3.1.2   Mechanism and energetics

To analyze the epoxidation activity of these peroxo complexes, we characterized for each complex TSs of oxygen transfer to the model olefin ethene. We compared the two mechanisms mostly discussed in the literature, namely insertion [2, 8, 9, 97] and a direct attack of the peroxo group on the olefin [11] (see Section 1). Direct oxygen transfer can be envisaged to occur either from the "front" (O1) or the "back" (O2). All located transition structures of direct transfer were of spiro type where the olefin C-C bond is almost perpendicular to the plane formed by the metal-peroxo moiety (Figure 3). The calculated activation barriers are collected in Table 1.

The direct attack of the front-oxygen peroxo center yields the lowest activation barrier for all species considered. Due to repulsion of ethene from the complexes we failed [61] to localize intermediates with the olefin precoordinated to the metal center, proposed as a necessary first step of the epoxidation reaction via the insertion mechanism. Recently, Deubel et al. were able to find a local minimum corresponding to ethene coordinated to the complex $MoO(O_2)_2 \cdot OPH_3$; however, the formation of such an intermediate from isolated reagents was calculated to be endothermic [63, 64]. The activation barriers for ethene insertion into an M-O bond leading to the five-membered metallacycle intermediate are at least ~5 kcal/mol higher than those of a direct front-side attack [61]. Moreover, the metallacycle intermediate leads to an aldehyde instead of an epoxide [63]. Based on these calculated data, the insertion mechanism of ethene epoxidation by d^0 TM peroxides can be ruled out.

Table 1. Activation barriers (in kcal/mol) calculated for various mechanisms of ethene epoxidation by model peroxo complexes of Cr, Mo and W (see Fig. 1).

		Cr	Mo	W
Front	1a	22.8	18.8	13.7
	1b	40.1	34.1	31.5
	2a	19.3	14.1	10.7
	2b	–[a]	19.7	16.9
Back	1a	28.9	22.8	19.8
	2a	34.5	28.9	25.9
	2b	–[a]	27.5	24.2
Insertion	1a	28.1	26.4	20.5
	2a	27.3	22.3	18.2

[a] One NH_3 ligand is expelled during the search of the local energy minimum.

For the favorable reaction mechanism, the direct front-side attack, three findings are worth noting. (i) The activation barriers for analogous species decrease along the series Cr > Mo > W. (ii) The barriers for the di(peroxo) species are lower than those of the corresponding mono(peroxo) complexes. (iii) Coordination of a second base ligand significantly increases the activation barrier.

The calculated transition structures share important features. The distance between the attacking oxygen atom and the metal center is only slightly stretched, by 0.05–0.1 Å, relative to the starting complex; the other M-O bond concomitantly contracts by ~0.1 Å. The largest change occurs for the peroxo group where the O-O bond length of the TS is stretched by 0.3–0.4 Å. The O-O distances of the intermediates cover a moderate range, from 1.40 Å to 1.48 Å along the metal series. The elongation of the peroxo bond in the transition structure is rather uniform, 0.34–0.35 Å, for the complexes **2a** of Cr, Mo, and W.

In all cases the transfer of the oxygen atom from the metal peroxo moiety to ethene is exothermic.

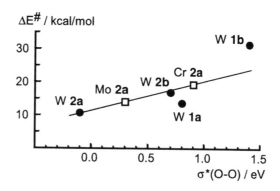

Figure 4. Calculated activation energies $\Delta E^{\#}$ vs. σ^*(O-O) orbital energies (averaged in case of **2a** and **2b**). Linear regression line derived from the model complexes **2a**.

3.1.3   Factors affecting the reactivity

Just as with organic oxidants, olefin epoxidation is governed by the electrophilicity of the d^0 metal peroxo complex [2, 97]. For example, peroxo and alkylperoxo complexes of TMs react faster with electron-rich (alkyl substituted) olefins. Sulfoxidation of thianthrene 5-oxide has previously been used as mechanistic probe of oxygen transfer reactions [98], among them dimethyldioxirane as oxidant [26]. In the same fashion, the electrophilic nature of peroxo oxygen centers of V, Mo, and W complexes was convincingly demonstrated [99, 100]. From a computational point of view, the electrophilic character of the reaction is manifested through electron density transfer in the TS of ~0.2 e from ethene to the peroxo oxygen centers [59, 67, 80]. This electronic charge transfer results from the interaction of the π(C-C) HOMO of ethene and the unoccupied O-O antibonding orbital σ^*(O-O) of the peroxo moiety. Thus, the energy of the σ^*(O-O) level plays a significant role in determining the activity of the peroxo complexes in oxygen transfer [80]. Along the triad Cr/Mo/W, the energy of the σ^*(O-O) MO decreases linearly with the calculated activation barrier (Figure 4).

A peroxo group, formally an O_2^{2-} ligand, interacts with the metal center via donation from its filled orbitals to the formally empty d orbitals of the metal center. The highest occupied orbitals of the peroxo group, π^*(O-O), are strongly involved in the interaction with the empty d orbitals of the metal center (Figure 5). This reduces the O-O antibonding character of π^* orbitals and leads to a strengthening of the O-O bond. It is important to understand how the interaction of the metal center and the peroxo group varies for different metal centers and to rationalize the calculated elongation of the O-O bonds in peroxo complexes as a consequence of the weakening of the interaction between the π^*(O-O) MO and the d orbitals of the metal center.

(i) Energy and radial extension of the d(M) orbitals increase in group VI, but they are rather similar for Mo and W due to the "lanthanide contraction". Therefore, Cr exhibits the lowest lying acceptor d orbitals [101] available for donation from π^*(O-O) orbitals. (ii) The polarizing power of a metal cation which determines the covalent character of the M-L bond varies inversely with its ionic radius. Indeed, for Cr one finds the smallest charge separation in the M-O bond, indicative for the highest electron density at the metal center or, in other words, the largest donation from π^*(O-O). A population analysis also yields decreasing contributions of the metal center to the metal-peroxo bond in the order Cr > Mo > W. (iii) The stability of the highest oxidation state increases in the series Cr, Mo, W. In fact, Cr^{VI} is a strong oxidizing agent and therefore a very good electron acceptor [102]. The strong donation from the occupied antibonding levels of the peroxo moiety to the Cr complex deactivates the peroxo bond and reduces the reactivity towards epoxidation. The influence of the metal center on the epoxidation activity is rather strong. The MO analysis rationalizes both activation as in the case of W and deactivation as in the case of Cr.

In summary, one can identify three factors that mainly affect the epoxidation activity of a TM peroxo complex: (i) the strength of the M-O and O-O interactions, (ii) the electrophilicity of the peroxo oxygen centers and the olefin, and (iii) the interaction between the π(C-C) HOMO of the olefin and the peroxo σ^*(O-O) orbital in the LUMO group of the metal complex to which we will also refer as the relevant unoccupied MO, RUMO (Figure 5).

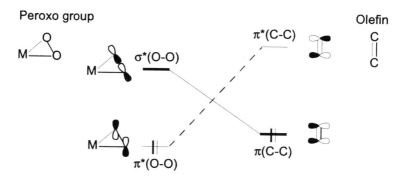

Figure 5. Frontier orbital interaction between a transition metal peroxo group and an olefin.

3.2   Re: Epoxidation of ethene and more complex olefins

3.2.1   Model complexes, simple alkenes, and mechanism

An important improvement in the catalysis of olefin epoxidation arose with the discovery of methyltrioxorhenium (MTO) and its derivatives as efficient catalysts for olefin epoxidation by Herrmann and coworkers [16-18]. Since then a broad variety of substituted olefins has been successfully used as substrates [103] and the reaction mechanism was studied theoretically [67, 68, 80].

We start by considering various rhenium oxo and peroxo complexes (Figure 6): MTO (**3**) as well as the corresponding monoperoxo (**4**) and di(peroxo) (**5**) complexes. Since the epoxidation reaction takes place in aqueous environment in the presence of H_2O_2 [19-21], we compare the free complexes **3a** to **5a** with their water ligated analogues $CH_3ReO_3 \cdot H_2O$ (**3b**), $CH_3Re(O)_2(O_2) \cdot H_2O$ (**4b**), and $CH_3ReO(O_2)_2 \cdot H_2O$ (**5b**).

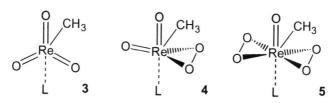

Figure 6. Methyltrioxorhenium (MTO) (**3**) and the corresponding mono- (**4**) and di(peroxo) (**5**) complexes. The coordination of the base ligand L is also indicated.

For the sake of simplicity the aqua ligand is added in cis position to the methyl group, as suggested by the X-ray structure of the di(peroxo) complex [19]. Agreement between calculated and experimental geometries of **5b** is very satisfactory: bond lengths deviate, in general, at most by 0.04 Å. The peroxo groups are asymmetrically bound to the rhenium atom which correlates with differing epoxidation activities of the peroxo ligands with respect to its "front" or "back" side, i.e. for olefin attack at the peroxo oxygen distant from or close to the methyl ligand, respectively [67, 78], as found for the peroxo complexes of Mo and W (see Section 3.1).

For the model olefin ethene, we again investigated various epoxidation mechanisms (Figure 7) [67]. As before for the group VI metals, insertion was found to exhibit significantly higher activation barriers.

Calculated activation energies $\Delta E^{\neq}$ for ethene epoxidation are presented in Figure 7. Comparing the two direct transfer reactions for each starting complex, we note that for complexes **5a** and **5b** attack for the "front" is clearly favored over a "back-side" transfer (by about 7 kcal/mol); for **4a** the analogous processes exhibit comparable barriers. For **4b** the water ligand is extruded during the optimization and we were unable to find a TS. Analyzing the energies of the different TSs with respect to the direct precursors we note that the aqua ligand in **4b** and **5b** enhances the barriers of all the three types of mechanisms (with the exception of back-side attack on **4b**). For instance, for the "front" attack, compare **5a**, 12.4 kcal/mol, and **5b**, 16.2 kcal/mol. Apparently a front attack on the unligated complex **5a** has the lowest calculated barrier (12.4 kcal/mol), but **5a** is significantly less stable than **5b** (by about 16 kcal/mol). The back-side attack on **4b** exhibits a rather similar activation energy (16.2 kcal/mol) as the front-side attack on **5b**; since **4b** is only slightly less stable than **5b**, the two processes can be competitive.

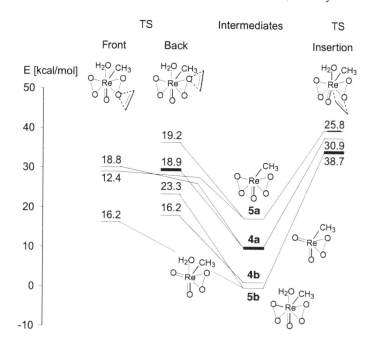

Figure 7. Energies of intermediates and TS of ethene epoxidation by various rhenium peroxo complexes. Barrier heights (in kcal/mol) given relative to the corresponding direct precursors.

The calculated energy barriers are notably larger than the activation enthalpy of 10.2±0.4 kcal/mol measured for the epoxidation of 4-methoxy-styrene with a di(peroxo) complex **5b** [21]. Instead of undertaking the computationally demanding task of finding a TS for this complex olefin, we resort to an MO analysis. Invoking the frontier orbital model (Figure 5), one expects the reactivity of the metal complex to depend to a significant extent on the interaction between the olefin HOMO π(C-C) and the peroxide RUMO σ*(O-O). Alkyl substituents at the olefin double bond raise the π(C-C) energy through electron donation, reducing therefore the energy gap between the frontier orbitals. A higher energy of the olefin HOMO reflects a stronger nucleophilic character of the olefin and one expects a concomitant lower activation barrier of the epoxidation reaction. Indeed, the calculated energy barriers for epoxidation of ethene and its methyl substituted derivatives by the base-free complex **3a** correlate linearly with the energy of the olefin HOMO π(C-C) (Figure 8). By extrapolating this regression line to the calculated energy of the HOMO of 4-methoxystyrene (5.75 eV), we deduce an activation barrier of ~4.5 kcal/mol. If we take the increase of the barrier by water adduct formation from **5a** to **5b** as indication of the base effect on the front-spiro transition structure of ethene epoxidation (~4 kcal/mol, Figure 7), we estimate the calculated epoxidation barrier of 4-

methoxystyrene at ~8.5 kcal /mol, in reasonable accordance with experimental value [21]. But for a full comparison with experiment, one should keep in mind other solvent effects on the reaction barrier, as investigated for epoxidation by organic oxidants [82].

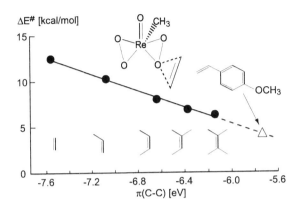

Figure 8. Calculated energy barriers $\Delta E^{\#}$ for epoxidation of substituted olefins by the base-free complex **5** as a function of the energy of the olefin HOMO π(C-C). Also, the estimated barrier height of 4-methoxystryrene is marked based on the corresponding HOMO energy.

3.2.2 Effect of a base ligand

Adduct formation with a base ligand L stabilizes a di(peroxo) complex $CH_3ReO(O_2)_2 \cdot L$ thermodynamically, but it also reduces its activity in olefin epoxidation [27]. To probe this effect computationally we investigated several base adducts with L = NH_3 (**5c**), NMe_3 (**5d**), pyridine N-oxide (**5f**), and pyrazole (**5g**). In Figure 9, we compare the stabilization of ligated complexes **5x** relative to the base-free reference compound $CH_3ReO(O_2)_2$ (**5a**) and the corresponding barrier heights of ethene epoxidation for a front-side attack. The interaction energies E(Re-L) of **5a** and the six base ligands studied vary over a moderate range. NH_3 and pyrazole exhibit strong binding, ~20 kcal/mol, followed by pyridine-N-oxide and pyridine with very similar interaction energies of ~18 kcal/mol. H_2O and NMe_3 feature noticeably smaller metal-ligand binding energies [80].

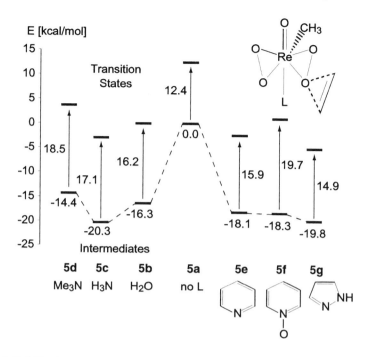

Figure 9. Stabilization energies of base adducts $CH_3ReO(O_2)_2\cdot L$ relative to the energy of the base-free di(peroxo) complex **5a** and activation barriers of the corresponding epoxidation TSs for front-side attack of ethene as model olefin.

The base effect on the barrier of epoxidation can be rationalized with the help of the frontier orbital model (Figure 5). A higher epoxidation barrier is found when the pertinent orbital of the metal-peroxo moiety, namely the RUMO $\sigma^*(O\text{-}O)$, is raised in energy by the coordination of a base (similar to Figure 4) [80]. Upon adduct formation, the metal center withdraws less electron density from the peroxo group and hence reduces its electrophilicity for an attack of an olefin. This charge rearrangement has been corroborated by a population analysis and the change of the O1s core level energy [80].

The overall effect of a base ligand on the reactivity of rhenium di(peroxo) complex **5a** is a combination of stabilization of the base adduct and partial deactivation of its peroxo group (Figure 9). Obviously, the base-free system features the lowest activation barrier relative to its intermediate precursor, yet it also exhibits the highest TS (by absolute energy) since it lacks the stabilization afforded by the base ligand. In general, the stabilization of the intermediate (~14–20 kcal/mol) surpasses by far the increase of the energy barrier to the TS. Thus, the TS can lie even at energies below that of the base-free complex **5a**, ranging from -2.2 kcal/mol for pyridine to -3.2 kcal/mol for NH_3, and -4.9 kcal/mol for pyrazole (Figure 9). Among the model bases investigated, NMe₃ and pyridine-N-oxide form noteworthy exceptions to the latter trend.

From a chemical point of view, the proper reference for the various TSs is not the base-free complex **5a**, but the almost iso-energetic TS of **5b**. The strong stabilization afforded by the water ligand is essentially compensated by a strongly enhanced activation barrier. Thus, **5b** exhibits an epoxidation TS at a higher energy (by absolute value) than the TSs of **5e** and **5g**. Since water is always present under catalytic condition, one of its effects is now obvious. A favorable base extrudes the water ligand from the complex by providing a stronger stabilization of the precursor than water, but at the same time the base ligand must not deactivate the complex too much. Pyridine fulfils both conditions, but it can easily be oxidized to pyridine N-oxide which provides a similar stabilization but induces a much higher epoxidation barrier (Figure 9). Among the ligands studied, pyrazole affects the catalytic reaction in optimal fashion since it affords the largest stabilization of the intermediate and leads to the smallest increase of the activation energy. This analysis rationalizes the experimentally determined catalytic activity of MTO/H₂O₂ in styrene epoxidation, where addition of pyrazole as base to the reaction system yields the highest epoxidation activity followed by the base pyridine [27, 30, 32, 104].

3.2.3   Allylic alcohol epoxidation

Allylic alcohols are interesting substrates for epoxidation because they produce epoxides with a hydroxyl group as additional functional group that is able to play an important role in the subsequent synthesis of complex molecules [105]. This synthesis aspect certainly benefits from the hydroxy-group directed selectivity of oxygen delivery.

Compared to ethene or other simple alkenes, several new variants of the mechanism have to be considered for the epoxidation of allylic alcohols, in particular, competition among three conceivable reaction routes: i) coordination of the allylic alcohol (as lone pair donor) to the rhenium complex followed by intramolecular epoxidation, ii) formation of a metal alcoholate (derived from addition of the allylic alcohol to the complex with formation of a metal-OR bond) followed by a stereocontrolled intramolecular oxygen delivery by the peroxo group, and iii) an intermolecular oxygen transfer process assisted, already at the start of the reaction and even more so in the TS, by a hydrogen bonding interaction in which the rhenium complex acts as hydrogen bond acceptor and HOR as hydrogen bond donor. Hydrogen bonding could severely affect the TS geometries observed in the epoxidation of simple alkenes.

An extensive set of experimental data on regioselectivity, face selectivity [106-109] and kinetics [110] of allylic alcohol epoxidation by the MTO system is available. On the basis of these results experimentalists have suggested a variety of transition structures (Figure 10) [28, 108, 110].

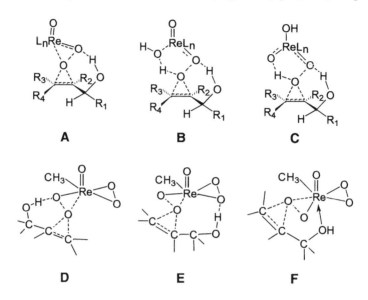

Figure 10. Proposed model transition structures.

The proposed TS models **A-F** do not only leave many mechanistic questions open, they also look somewhat too schematic. In view of the increasing synthetic relevance of MTO catalyzed epoxidations, reliable prototype transition structures are required for rationalizing and, if possible predicting, reaction rates as well as the selectivity of these reactions.

We performed a computational study [69] to assess which interaction (H bonding, metal-alcoholate formation, or metal-alcohol coordination between the allylic hydoxyl moiety and the Re complex) affects the TS and to determine which oxygen of the Re peroxo moiety acts as H-bond acceptor in the case of an H-bonded TS. A summary of the results with propenol as model allylic alchohol is presented in the following.

3.2.3.1   Alcohol coordination and metal-alcoholate formation and the corresponding transition structures

When the water ligand of **5b** is replaced by propenol as axial donor or in case of equatorial coordination of propenol at the metal center, intermediates are formed which could undergo intramolecular epoxidation through TSs **S-6** and **S-7b**, respectively (Figure 11). Here, the prefix **S** indicates a transition structure (saddle point), later on in this section we will use the prefix **I** to designate intermediates. As done previously, we indicate a complex with a water ligand coordinated in the axial position by the suffix **b**. The TSs **S-6** and **S-7b** have very high activation barriers (20-27 kcal/mol) which are considerably larger than those calculated for hydrated H-bonded TSs. These barrier heights reflect the strain that occurs when the spiro-like transition structure **S-6** is attained or the weak stabilization of **S-7b** by the equatorial coordination. Thus, these TSs represent feasible, but not competitive entries to epoxidation pathways in accord with the qualitative proposition by Espenson et al [110].

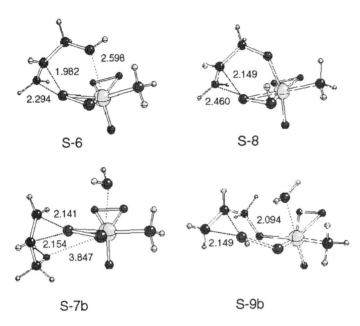

Figure 11. Transition structures **S-6** to **S-9b** for the epoxidation of propenol by the Re di(peroxo) complex **5**. Selected distances in Å.

The metal-alcoholate mechanism is well established for allylic alcohol epoxidation in the presence of Ti and V catalysts. [41, 51, 52, 111-113]. In principle, it can provide a viable pathway also for catalysis by a Re complex. In fact, allylic alcohols may add, at least formally, to either an oxo-Re or peroxo-Re moiety (e.g. of **5a** or **5b**) in a process which is referred to as metal-alcoholate binding; this mechanism gives rise to metal-alcoholate intermediates. We identified four intermediates of alcohol addition to di(peroxo) complexes; two resulting transition states, **S-8** and **S-9b**, are shown in Figure 11. All metal-alcoholate intermediates lie significantly higher in energy (by 10-22 kcal/mol) than **5b** + propenol, except the hydroperoxo complex **I-9b** with the alcoholate in equatorial position. The latter complex is 6.2 kcal/mol more stable than the reactants and hence might be competitive in the epoxidation reaction. However, intramolecular epoxidation in **I-9b** can take place only via the high-lying TS **S-9b** (~30 kcal/mol relative to **5b** + propenol), which has an energy very similar to that of TS **S-8**.

The corresponding intermediates either have high energies and/or are associated with large activation energies (20-35 kcal/mol) for intramolecular epoxidation; the metal-alcoholate TSs lie by more than 15 kcal/mol higher in energy than solvated or unsolvated H-bonded TSs (see below). These results provide the first compelling and unambiguous evidence that alcoholate-metal complexes do not contribute to product formation and can be confidently neglected when discussing Re-peroxo catalyzed epoxidation reactions of allylic alcohols. Our calculations fully confirm conclusions by Adam et al. based on experimental data [107, 114].

These findings for Re peroxo complexes are in striking contrast with Ti and V catalyzed reactions [41, 51, 52, 111, 113] in which the metal-alcoholate bond drives the allylic OH directivity. We recall that the formation of alcoholate intermediates was also rejected for epoxidations of allylic alcohols with Mo and W peroxo compounds while H-bonding (between OH and the reacting peroxo fragment) was considered consistent with kinetic data for these complexes [115].

3.2.3.2   Hydrogen-bonded transition structures

To discuss H-bonded transition structures it is instructive to start with propenol conformers. The two most stable conformers (defined as "out" and "in", according to the relative position of the C-O and C=C bonds) out of the five fast equilibrating conformers [77, 79] are close in energy. In both conformers the hydroxyl group is properly oriented for easy involvement in H-bonding with the oxygen centers of the attacking rhenium peroxo complex. Actually, the CH_2OH conformation of H-bonded syn TSs resembles that of the starting propenol conformers and, accordingly, TSs are given the descriptor out and in, respectively. By the way, the structure of each TS has been characterized by three further descriptors [69], reflecting the facial selectivity (syn/anti) and the side of the approach (endo/exo, cis/trans).

A complete investigation of all possible syn H-bonded TSs for propenol epoxidation with **4**, **4b**, **5** and **5b** (also considering the conformational freedom of the CH_2OH group) would have been too complicated, expensive, and probably useless. Consequently, we chose to investigate all those syn pathways involving the di(peroxo) complexes **5** and **5b** that are meaningful for a reliable evaluation of the H-bonding interaction in the reaction under study, and just few examples of epoxidation by the mono(peroxo) complexes **4** and **4b**.

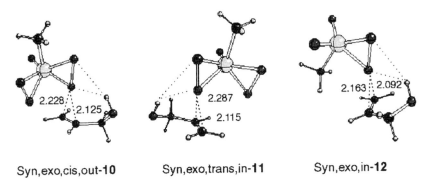

Syn,exo,cis,out-**10**   Syn,exo,trans,in-**11**   Syn,exo,in-**12**

Figure 12. Syn transition structures **10–12** for the epoxidation of propenol by the unsolvated di(peroxo) Re complex **5** and the mono(peroxo) Re complex **4**. Bond lengths in Å.

All syn TSs for the reaction of unsolvated di(peroxo) Re complex **5** with propenol are more stable, by ~1–4 kcal/mol, than the anti,exo transition structures; some of them even show significantly lower activation energies than the corresponding TSs of propene epoxidation (~10 kcal/mol). There are four TSs in which OH forms a hydrogen bond to oxygen centers of the reacting peroxo moiety, as in the case of **10** and **11** (Figure 12). The presence of H-bonding, manifested by the reduction of the OH stretching frequency, favors syn TSs by several kcal/mol relative to anti TSs. Two such TSs (e.g. **10**) have the OH group in outside position and are similar to the qualitative transition structures advocated by Adam et al [107, 114]. The other two TSs (e.g. **11**) exhibit an OH group with inside conformation. This conformation has not been considered previously [107, 114], probably because it does not permit efficient H-bonding. The most reasonable description of the hydrogen bonds in the TSs is that of "bridged" hydrogen bonds involving to some extent both peroxo oxygens. In line with Espenson's proposal [110] there are also TSs with hydrogen bonding to the non-reacting peroxo moiety. As judged by the strong reductions of the OH stretching frequency and the relatively short OH–O distances, these two TSs have the strongest hydrogen bonds. Finally, given that the oxo center exhibits the largest net negative charge, it does not come as a surprise that also this oxygen center can act as H-bond acceptor.

Based on these calculations, one can conclude that TSs for epoxidation of alkenes and allylic alcohols with peroxy acids, dioxiranes, and Re-peroxo complexes share a spiro geometry in which the plane of the attacking peroxo (peroxy) moiety is oriented almost perpendicularly to the plane of the oxirane to be formed.

All unsolvated TSs lie in a narrow range of 2 kcal/mol relative to **5** + propenol. From the point of view of a qualitative analysis, it is disappointing that we were unable to identify any structural characteristic that might definitely favor or disfavor one TS with respect to the others. Would the reaction occur via unsolvated TSs, all eight of them are expected to contribute to some extent to the formation of the final epoxide.

It is widely accepted that in the case of the di(peroxo) Re complex the actual epoxidizing species is a base adduct, e.g. with a water molecule, **5b** [20, 21, 80, 107, 110] (see Section 3.2.2). Hydrated TSs such as **11b** resemble very closely the corresponding non-hydrated TS, except for the added water molecule coordinated opposite to the oxo ligand. The absolute energies of TSs originating from complexes with a additional water ligand (type **b**) differ more widely than the energies of "unsolvated" TSs. Therefore, hydration definitely seems to tip the balance in favor of TSs with hydrogen bonding to the reacting peroxo bridge (such as **10b**) whereas those with H- bonding to oxo functionality are no longer competitive [69].

The proposal by Espenson et al. [110] of a dominating TS, with hydrogen bonding to the non-reacting peroxo bridge (**E**, Figure 10), for the Re di(peroxo) catalyzed epoxidation of allylic alcohols is not supported by our calculations. Model **E** is certainly reasonable for qualitative TSs, but it is neither the only possible structure nor does it represent the dominant pathway to the final epoxide.

As discussed above, the activation barrier of "solvated" complexes (type **b**) is larger, since the starting complex is more stabilized than the TS. Notwithstanding this barrier increase, solvated TSs reside at much lower (absolute) energies than their unsolvated counterparts. However, hydrated TSs are entropically disfavored (by ~9 kcal/mol in the gas phase) with respect to the corresponding water-free TSs. These observations prevent a definitive decision whether **5** or **5b** is the active epoxidant species in solution.

With regard to the competition between mono(peroxo) and di(peroxo) Re complexes, our calculation confirm the deduction from experimental kinetics that the epoxidation rate of allylic alcohols by the mono(peroxo) Re complex (e.g. TS **12**) is negligible compared to that by the di(peroxo) complex [69]. It is interesting to note that this observation is unique and does not hold for any other substrate studied so far.

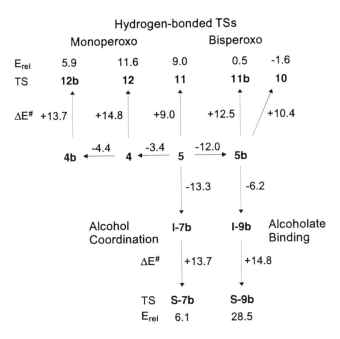

Figure13. Energies E$_{rel}$, of epoxidation TSs relative to **5** + H$_2$O + propenol. Figures with plus signs near arrows report the corresponding activation energies ΔE$^{\neq}$. Other figures with minus signs near arrows indicate the formation energies of the various complexes: **4**, the hydrated complexes **4b** and **5b** as well as the intermediates **I-7b** and **I-9b**. All energies in kcal/mol.

The most stable TSs for each mechanism are given in Figure 13. Calculations clearly suggest that the routes starting both from coordination of propenol and from formation of a metal alcoholate are viable reaction channels. However, along these routes the reacting systems have to surmount high-lying saddle points so that these pathways do not contribute appreciably to epoxide formation. The di(peroxo) rhenium complex (likely in hydrated form) is the active species in the catalyzed hydrogen peroxide epoxidation of propenol, via spiro-like TSs with hydrogen bonding to the attacking peroxo fragment.

3.3   Ti, Mo, Re: peroxo vs. hydroperoxo complexes

While numerous peroxo complexes of Ti, V, Cr/Mo/W and Re have been experimentally isolated and well characterized, there are only few examples of alkyl peroxo complexes whose X-ray structures are known, namely (di(picolinato)VO(OOtBu)(H$_2$O)   [116]   and   [((η^2-*tert*-butylperoxo) titanatrane)$_2$· 3 dichloromethane] [117]. Evidence for the existence of other

Ti alkylperoxo complexes based on a NMR analysis has been also reported [118].

3.3.1   Activity of Ti η^2-peroxo and hydroperoxo species

It is instructive to compare the properties of metal peroxo and alkyl (or hydro) peroxo groups for the case of Ti because experimental structures of both types are known [117, 119-121] and Ti compounds are catalysts for such important processes as Sharpless epoxidation [22] and epoxidation over Ti-silicalites [122], where alkyl and hydro peroxo intermediates, respectively, are assumed to act as oxygen donors. Actually, the known Ti(η^2-O$_2$) complexes are not active in epoxidation. [121-124] However, there is evidence [123] that (TPP)Ti(O$_2$) (TPP = tetraphenylporphyrin) becomes active in epoxidation of cyclohexene when transformed to the *cis*-hydroxo(alkyl peroxo) complex (TPP)Ti(OH)(OOR) although the latter has never been isolated.

To elucidate the factors that control the activity of Ti^{IV} complexes we have performed a comparative computational study of ethene epoxidation by both types of peroxo groups, Ti(η^2-O$_2$) and TiOOR [59]. We paid special attention to the coordination number of the Ti center. Other theoretical investigations focused on epoxidation by TiOOH species [51-54, 57, 58] and considered oxygen transfer from H$_2$O$_2$ coordinated to a Ti center [55]; a Ti-O-O-Si moiety has also been investigated as epoxidizing agent [60].

Energy profiles of ethene epoxidation starting from various Ti(O$_2$) and TiOOH intermediates are sketched in Figure 14 where the energy of the intermediate (HO)$_2$Ti(O$_2$)·(NH$_3$)$_2$ (**13c**) + C$_2$H$_4$ is taken as reference. The relative energies of the intermediates (HO)$_2$Ti(O$_2$) (**13a**) and (HO)$_3$TiOOH (**14a**) are determined from the formal water addition **13a** + H$_2$O → **14a**. The simplest model of η^2-peroxo complexes, (HO)$_2$Ti(O$_2$) (**13a**), exhibits a significant affinity to base ligands. The binding energy of NH$_3$ in the first base adduct (HO)$_2$Ti(O$_2$)·NH$_3$ (**13b**) is -32.3 kcal/mol (Figure 14); addition of a second NH$_3$ molecule to **13b** leads to a further energy gain of -16.5 kcal/mol. That species **13c** features the lowest energy (Figure 14); it can be considered as a model of the complex Ti(TPP)(O$_2$) [123] with the porphyrin ring of the latter substituted by two NH$_3$ and two hydroxide ligands. We represented Ti hydroperoxo species by the model complexes **14a** and **14b** (Figure 14). Addition of a NH$_3$ ligand to **14a**, resulting in the model complex **14b**, yields an energy gain of -13.2 kcal/mol. It does not come as a surprise that no local symmetry is preserved in the Ti(OOH) moiety. The distances between Ti and oxygen centers of the hydroperoxo group of **14a** differ significantly; the distance Ti-Oα (oxygen bound directly to the metal center) and

Ti-Oβ (oxygen in β position with regard to the metal center) are calculated at 1.89 Å and 2.26 Å, respectively.

The calculated TSs of ethene epoxidation by the various Ti(O$_2$) and TiOOH systems have a number of features in common and, in general, resemble the structures of TSs of direct oxygen transfer for the peroxo complexes of Mo, W, and Re [61, 67]. Inspection of Figure 14 reveals that the intermediates **13c** and **14b** feature the lowest energies and similar thermodynamic stability. However, epoxidation by **14b** is favored for both thermodynamic (higher reaction energy) and kinetic reasons (lower activation barrier). Evidently, the kinetic argument plays the key role: epoxidation by the lowest Ti(O$_2$) intermediate **13c** can be ruled out since the corresponding TS exhibits an activation barrier of more than 27 kcal/mol; this is more than twice larger than the activation barrier of the TiOOH intermediate **14b**. The low-coordinated intermediate **13a** was found to exhibit a relatively low barrier of oxygen transfer, 11.0 kcal/mol; however, it is thermodynamically not stable since it readily forms base adducts.

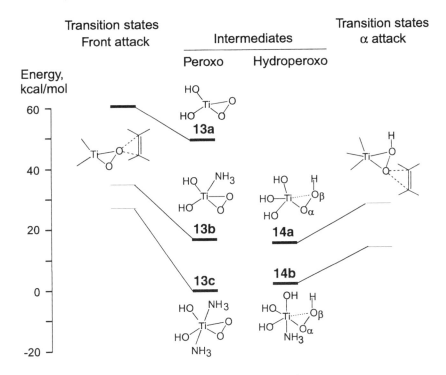

Figure 14. Energies (in kcal/mol) of intermediates with Ti(η^2-O$_2$) or TiOOH groups and the corresponding TSs of ethene epoxidation, relative to the energy of **13c** + C$_2$H$_4$.

Note that in Figure 14 we consider only attack at the center Oα of the TiOOH group. An attack of the other oxygen center, Oβ (with the hydrogen attached), results in a higher activation energy since in this case, the epoxidation reaction is accompanied by a proton transfer from Oβ to Oα [59]. For instance, the activation barrier of ethene epoxidation by **14a** via α attack is calculated at 12.7 kcal/mol compared to 25.7 kcal/mol via β attack. At variance with the Ti(O$_2$) complexes, the reaction barrier of TiOOH intermediates hardly changes upon addition of a base ligand at the metal center; the TS of the highly coordinated species **14b** exhibits even a slightly smaller activation energy, 12.4 kcal/mol, than that corresponding to **14a**. Activation barriers for ethene epoxidation by a TiOOH group have also been obtained in several computational studies related to the activity of Ti-silicalites [51, 53, 57].

The high activation barriers of coordinatively saturated Ti(O$_2$) species agrees with the observed chemical stability of Ti^{IV} peroxo complexes [119, 121]. The inactivity of (TPP)Ti(O$_2$) in epoxidation has been attributed to steric repulsion between the alkene and the porphyrin ring [2]. However, the deactivating effect of the donor ligands on the activity of the Ti(O$_2$) moiety can be rationalized with a MO analysis (Section 3.1) [59]: (i) additional base ligands reduce the electrophilicity of the peroxo group and raise the energy of the σ^*(O-O) orbital which is crucial for breaking the O-O bond; (ii) concomitantly, the populations of the bonding levels of the peroxo group increase due to reduced donation to metal d orbitals. Subsequently, the O-O bond becomes stronger with higher metal coordination, as indicated by the calculated bond shortening [59].

The apparently small effect of a ligand on the properties of the hydro-peroxo complexes **14a** and **14b**, in particular on the oxygen transfer barrier, can be rationalized by the weaker metal-peroxo interaction in these complexes. This follows, first of all, from the fact that only one oxygen atom, Oα, is involved in a strong interaction with the Ti center; the distance Ti-Oβ is much longer (by ~0.3 Å) and thus the TM-O interaction is significantly weaker. In this sense, Ti hydroperoxo species exhibit an "end-on" coordination mode (at variance with the original designation [117]), in contrast to the almost ideal η^2 (or "side-on") coordination of peroxo species. Moreover, the Ti-Oα interaction is weaker than the corresponding Ti-O interaction in peroxo species. This can be inferred from the Ti-Oα distances in **14a** and **14b**, 1.89–1.92 Å, which are notably longer than the corresponding Ti-O distances of the peroxo complexes, 1.82–1.86 Å [59]. At variance with the bonding of the Ti(O$_2$) complexes, the π channel of the Ti-Oα interaction almost vanishes and the NBO population analysis assigns the corresponding orbital to a lone pair located at the center Oα. Substitution of methyl for hydrogen of the hydroperoxo group increases the calculated activation barrier by about 3 kcal/mol, from 12.7 kcal/mol in **14a** to 15.8 kcal/mol in $(HO)_3TiOOCH_3$ [59]. This small increase of the activation barrier is in line with the weak electron-donating character of the methyl substituent; it reduces the electrophilic character of the peroxo group and raises the $\sigma^*(O\text{-}O)$ level.

3.3.2 Mo and Re hydroperoxo intermediates

Recently, Thiel et al. investigated the mechanism of olefin epoxidation by seven-coordinated molybdenum peroxo complexes of the type $MoO(O_2)_2(L\text{-}L)$ (L-L = pyrazolylpyridine) [38, 125, 126]. They reported that the transferred oxygen does not originate from any of the $\eta^2\text{-}O_2$ peroxo groups. They proposed a new mechanism for olefin epoxidation by Mo peroxo complexes, where an alkylperoxide ROOH as oxidizing agent coordinates to the Mo^{VI} center, undergoes a proton transfer to one of the peroxo ligands, and thus is activated for oxygen transfer forming a MoOOR alkylperoxo moiety. This proposal was supported by experiments on kinetics and proton transfer [125, 126]. In a computational study [70] of this mechanism, using $MoO(O_2)_2(NH_3)_2$ as model complex and CH_3OOH as model oxidizing agent, various isomeric species of the type $MoO(O_2)(OOH)(OOCH_3)(NH_3)_2$ were found with a lower energy (by 11–16 kcal/mol) than the reactants, although a large part of the stabilization should be attributed to various interligand H-bonds. Note that replacement of an NH_3 ligands by CH_3OOH and the subsequent formation of different isomers $MoO(O_2)(OOH)(OOCH_3)\cdot NH_3$ were found to be endothermic relative to $MoO(O_2)_2(NH_3)_2 + CH_3OOH$.

We considered [71] the epoxidation activity of Mo^{VI} hydroperoxo intermediates formed via "opening" an $\eta^2\text{-}O_2$ peroxo group:

$$MoO(O_2)_2L_{eq}L_{ax} + H_2O \rightarrow MoO(O_2)(OOH)(OH)L_{eq}L_{ax} \qquad (1)$$

We used different equatorial base ligands $L_{eq} = NH_3$, ONH_3 and $L_{ax} = NH_3$ or none as axial ligand. There is striking structural similarity with the di(peroxo) complexes of Re^{VII} that arise in the MTO/H_2O_2 catalytic system $(L_{ax} = H_2O$ or none):

$$CH_3ReO(O_2)_2L_{ax} + H_2O \rightarrow CH_3ReO(O_2)(OOH)(OH)L_{ax}, \qquad (2)$$

Formal water addition (1) is exothermic for all Mo di(peroxo) complexes considered. Process (2) is exothermic for $CH_3ReO(O_2)_2$ (**5a**, Figure 7) and endothermic for $CH_3ReO(O_2)_2\cdot H_2O$ (**5b**) [71]. The activation barrier of the ring opening that transforms $MoO(O_2)_2(NH_3)_2$ into $MoO(O_2)(OOH)(NH_3)_2$ has been calculated at 11.3 kcal/mol [71].

We now discuss the reaction profiles (Figure 15) starting from the di(peroxo) complex $MoO(O_2)_2(NH_3)_2$ (**2b**) in more detail; a discussion of the analogous Re systems has been given elsewhere [71]. The hydroperoxo species feature several isomers and corresponding TSs since a $M\text{-}O_\alpha\text{-}O_\beta\text{-}H$ group has more conformational freedom than a peroxo group. The olefin may interact with an α oxygen center of the metal hydroperoxo moiety or with a β oxygen center. In the case of an α attack, the OH group remaining after the O-O bond is broken has to be transferred back to the metal center. We have found such a process conceivable only for hydroperoxo intermediates (**15a**) without an axial base ligand L_{ax}; for the corresponding base adducts the back transfer of the remaining OH group to the metal center is sterically strongly hindered because the coordination sphere is too crowded. For an attack at $O\beta$, we have confined our investigations to intramolecular proton transfer where the acceptor group is part of the same complex, in analogy to transition structures of olefin epoxidation by peracids [73-78]. Thus, three reaction pathways were considered for β attack of hydroperoxo species where an oxo, a hydroxo, or a peroxo group acts as acceptor of the proton; the corresponding transition states are referred to as $\beta O15x$, and βO_215x, $\beta OH15x$, respectively (with $x = $ **a**, **b**), see Figure 15.

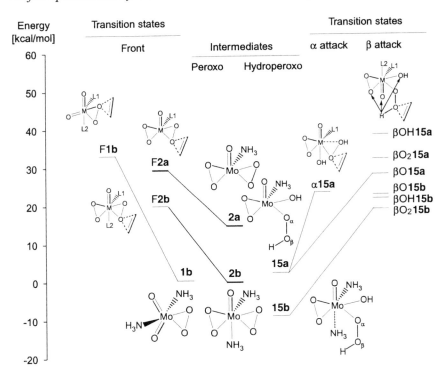

Figure 15. Energies (in kcal/mol) of intermediates and TSs of ethene epoxidation by various molybdenum peroxo and hydroperoxo complexes, relative to the energy of $MoO(O_2)_2\cdot(NH_3)_2$ (**2b**) + C_2H_4.

As for Ti hydroperoxo species [59], attack of Oα requires less activation energy than attack at Oβ. The Mo hydroperoxo species **15a** (Figure 15) exhibits slightly higher activation barriers than the initial di(peroxo) complexes **2a** ($MoO(O_2)_2L_{eq}$): 20.5 and 16.5 kcal/mol for hydroperoxo species with L_{eq} = NH₃ and ONH₃, respectively, in comparison to 14.1 and 15.5 kcal/mol for the corresponding di(peroxo) complexes. However, at the absolute energy scale the TSs for α attack lie lower due to the lower energy of hydroperoxo intermediates **15a** (Figure 15).

The lowest TS βO₂**15b** lies at almost the same energy as the TS F**2b** of front-side attack of **2b** because of the higher stability of the preceding intermediate **15b** while the activation barrier is much higher (27.5 kcal/mol) than that of **2b**. The relative stability of **2b** and **15b** may be sensitive to environmental effects that were not taken into account in the models. Note also that TS α**15a** of α attack lies only 4.4 kcal/mol higher than TS F**3b**.

Experimental studies, from the pioneering work of Mimoun [8] to the most recent investigations [127], provide unambiguous evidence that oxygen transfer from seven-coordinated Mo bisperoxo complexes such as **2b** is significantly slowed down or even inhibited. Thus, the calculated activation barrier of about 20 kcal/mol for **2b** seems to be beyond the threshold where Mo peroxo complexes are still reactive; therefore, the lowest intermediate, the hydroperoxo complex **15b**, can be as well considered as inert in epoxidation. In our model, the most probable pathway is a direct front-side attack of **2a** (with an activation barrier of 14.1 kcal/mol) – provided that the reaction conditions do not prevent coordination of a second strong base to this species [127]. Nevertheless, the hydroperoxo epoxidation pathway via **15a** seems to be competitive [71].

The computational results show that transition structures derived from hydroperoxo Re complexes lie slightly higher in energy than those obtained for the corresponding peroxo complexes, nevertheless their involvement in the epoxidation reaction cannot be excluded. However, for neither Mo^{VI} nor Re^{VII} evidence (let alone preference) for hydroperoxo reaction pathways is as clear as for Ti^{IV} complexes. Of course, more complex mechanisms involving intermolecular proton transfer and/or hydrogen bonded intermediates may change this picture to some extent.

4. SUMMARY

Density functional calculations reveal that epoxidation of olefins by peroxo complexes with TM d^0 electronic configuration preferentially proceeds as direct attack of the nucleophilic olefin on an electrophilic peroxo oxygen center via a TS of spiro structure (Sharpless mechanism). For the insertion mechanism much higher activation barriers have been calculated. Moreover, decomposition of the five-membered metallacycle intermediate occurring in the insertion mechanism leads rather to an aldehyde than to an epoxide [63].

The most reactive di(peroxo) complexes (i.e. those with the lowest activation barriers) are those of the third-row transition elements, W and Re (Table 2). For complexes of the same structural type, the calculated activation barriers of direct oxygen transfer to ethene decrease along Cr > Mo > W with W complexes being the most active. These computational results are in line with the experimentally reported higher activity of W species compared to Mo. Also, di(peroxo) complexes are more active than their monoperoxo congeners. Coordination of a Lewis base ligand to the metal center generally increases the activation barrier of olefin epoxidation because the base ligand donates electron density via the metal center to the peroxo group, hence reducing its electrophilicity.

Table 2. Calculated activation energies (kcal/mol) for ethene epoxidation by TM complexes.[a]

Complex		Activation energy	Ref.
Mono(peroxo)	$(HO)_2Ti(O_2)\cdot(NH_3)_2$	27.4	59
	$CrO_2(O_2)\cdot(NH_3)_2$	40.1	61
	$MoO_2(O_2)\cdot(NH_3)_2$	34.1	61
	$WO_2(O_2)\cdot(NH_3)_2$	31.5	61
	$CH_3ReO_2(O_2)\cdot H_2O$	16.2	71, 80
	$(HO)_2MoO(O_2)\cdot H_2O\cdot NH_3$[b]	23.1	62
	$Cl_2MoO(O_2)\cdot H_2O\cdot NH_3$[b]	22.6	62
	$MoO(O_2)(di(picolinato)\cdot H_2O$[b]	21.7	62
Di(peroxo)	$CrO(O_2)_2\cdot NH_3$	19.3	61
	$MoO(O_2)_2\cdot(NH_3)_2$	19.7	61
	$WO(O_2)_2\cdot(NH_3)_2$	16.9	61
	$CH_3ReO(O_2)_2\cdot NH_3$	17.1	80
	$CH_3ReO(O_2)_2\cdot H_2O$	16.2	71, 80
	$CH_3ReO(O_2)_2\cdot pyridine$	15.9	80
	$CH_3ReO(O_2)_2\cdot pyrazole$	14.9	80
Hydroperoxo[c]	$(HO)_3TiOOH\cdot NH_3$	12.4[d]	59
	$(HO)_3TiOOCH_3\cdot NH_3$	15.8[d]	59
	$MoO(O_2)(OOH)\cdot NH_3$	20.5[d]	71
	$MoO(O_2)(OOH)\cdot ONH_3$	16.5[d]	71
	$CH_3ReO(O_2)(OOH)$	15.8[d]	71
	$MoO(O_2)(OOH)\cdot(NH_3)_2$	27.5[e]	71
	$CH_3ReO(O_2)(OOH)\cdot H_2O$	17.8[e]	71

[a] B3LYP calculations as described in Section 2. [b] Structures optimized with a 6-31G(d) basis for the main group elements. [c] Also alkylperoxo complexes. [d] Attack of $O\alpha$ of the hydroperoxo (alkylperoxo) group. [e] Attack of $O\beta$ of the hydroperoxo group.

For Ti, hydroperoxo complexes exhibit lower activation barriers than the corresponding peroxo species. Peroxo complexes of Ti and Cr can be considered as inert in epoxidation. Hydroperoxo species may be competitive with di(peroxo) compounds in the case of Mo. For the system MTO/H_2O_2 the di(peroxo) complex $CH_3ReO(O_2)_2\cdot H_2O$ was found to be most stable and to yield the lowest TS for epoxidation of ethene, in line with experimental findings. However, olefin epoxidation by Re monoperoxo and hydroperoxo complexes cannot be excluded.

The frontier orbital interaction between the olefin HOMO $\pi(C-C)$ and the orbitals with $\sigma^*(O-O)$ character in the LUMO group of the metal peroxo moiety controls the activation of O-O bond. Electron donating alkyl substituents at the olefin double bond raise the energy of the HOMO, with the epoxidation barrier dropping concomitantly. On the other hand, a base coordinated at the metal center pushes the $\sigma^*(O-O)$ LUMO to higher energies and thus entails a higher barrier for epoxidation.

Since the catalytic system MTO/H_2O_2 was shown to be an excellent catalyst for alkene epoxidation, it is also applied in the epoxidation of allylic alcohols. The characteristics of these reactions are very similar to those of alkene epoxidation, even though quite strong hydrogen bonding between the allylic alcohol hydroxy functionality and the oxygen centers of the peroxo moiety occur along the reaction path. Reaction routes via coordination or ligation of the hydroxy group to the metal center of the peroxo complex have too high calculated activation barriers and, therefore, are not expected to contribute appreciably to epoxide formation.

ACKNOWLEDGMENT

With his creativity and energy, P. Gisdakis made essential contributions to the work described in this review. We also acknowledge the contributions of R. Gandolfi. N.R. would like to thank W. A. Herrmann for introducing him to the exciting field of olefin epoxidation. I.Y. thanks the program of the Russian Academy of Sciences supporting young scientists. This work has been supported by the Deutsche Forschungsgemeinschaft, the Bayerische Forschungsstiftung (FORKAT), the Fonds der Chemischen Industrie, and INTAS-RFBR (grant IR-97-1071).

REFERENCES

1   Cornils, B.; Herrmann, W. A. (Eds.) *Applied Homogeneous Catalysis with Organometallic Compounds*, VCH: Weinheim, 1996.
2   Jørgensen, K. A. *Chem. Rev.* **1989**, *89*, 431.
3   Sheldon, R. A. *Catalytic Oxydations with Hydrogen Peroxides as Oxidant*; Kluwer: Rotterdam, 1992.
4   Jacobsen, E. N. In *Catalytic Asymmetric Synthesis*, I. Ojima, Ed.; VCH: New York, 1 993; p. 159.
5   Dickman, M. H.; Pope, M. T. *Chem. Rev.* **1994**, *94*, 569.
6   Batler, A.; Clague, M. J.; Meister, G. E. *Chem. Rev.* **1994**, *94*, 625.
7   Romão, C. C.; Kühn, F. E.; Herrmann, W. A. *Chem. Rev.* **1997**, *97*, 3197.
8   Mimoun, H.; De Roch, I. S.; Sajus, L. *Tetrahedron* **1970**, *26*, 37.
9   Mimoun, H. *Angew. Chem. Int. Ed. Engl.* **1982**, *21*, 734.
10   Arakawa, H.; Moro-Oka, Y.; Ozaki, A. *Bull. Chem. Soc. Japan* **1974**, *47*, 2958.
11   Sharpless, K. B.; Townsend, J. M.; Williams, D. R. *J. Am. Chem. Soc.* **1972**, *94*, 295.
12   Arcoria, A.; Ballistreri, F. P.; Tomaselli, G. A.; Di Furia, F.; Modena, G. *J. Mol. Catal.* **1983**, *18*, 177.
13   Amato, G.; Arcoria, A.; Ballistreri, F. P.; Tomaselli, G. A.; Bortolini, O.; Conte, V.; Di Furia, F.; Modena, G.; Valle, G. *J. Mol. Catal.*, **1986**, *37*, 165.
14   Camprestini, S.; Conte, V.; Di Furia, F.; Modena, G.; Bortolini, O. *J. Org. Chem.* **1988**, *53*, 5721.
15   Talsi, E. P.; Shalyaev, K. V.; Zamaraev, K. I. *J. Mol. Catal.* **1993**, *83* 347.

16   Herrmann, W. A.; Fischer, R. W.; Marz, D. W. *Angew. Chem., Int. Ed. Engl.* **1991**, *30*, 1638.

17   Herrmann, W. A. *J. Organomet. Chem.* **1995**, *500*, 149.

18   Herrmann, W. A.; Kühn, F. E. *Acc. Chem. Res.* **1997**, *30*, 169.

19   Herrmann, W. A.; Fischer, R. W.; Scherer, W.; Rauch, M. U. *Angew. Chem., Int. Ed. Engl.* **1993**, *32*, 1157.

20   Al-Ajlouni, A. M.; Espenson, J. H. *J. Am. Chem. Soc.* **1995**, *117*, 9243.

21   Al-Ajlouni, A. M.; Espenson, J. H. *J. Org. Chem.* **1996**, *61*, 3969.

22   Finn, M. G.; Sharpless, K. B. *J. Am. Chem. Soc.* **1991**, *113*, 113.

23   Clerici, M. G.; Ingallina, P. *J. Catal.* **1993**, *140*, 71.

24   Notari, B. *Stud. Surf. Sci. Catal.* **1988**, *37*, 413.

25   Huybrechts, D. R. C.; De Bruycker, L.; Jacobs, P. A. *Nature* **1990**, *345*, 240.

26   Adam, W.; Golsch, D. *Chem. Ber.* **1994**, *127*, 1111–1113.

27   Herrmann, W. A.; Fischer, R. W.; Rauch, M. U.; Scherer, W. *J. Mol. Catal.* **1994**, *86*, 243.

28   Adam, W.; Mitchel, C. M. *Angew. Chem., Int. Ed. Engl.* **1996**, *35*, 533.

29   Boehlow, T. R.; Spilling, C. D. *Tetrahedron Lett.* **1996**, *37*, 2717.

30   Herrmann, W. A.; Kühn, F. E.; Mattner, M. R.; Artus, G. R. J.; Geisberger, M. R.; Correia, J. D. G. *J. Organomet. Chem.* **1997**, *538*, 203.

31   Rudolph, J.; Reddy, K. L.; Chiang, J. P.; Sharpless, K. B. *J. Am. Chem. Soc.* **1997**, *119*, 6189.

32   Yudin, A. K.; Sharpless, K. B. *J. Am. Chem. Soc.* **1997**, *119*, 11536.

33   Copéret, C.; Adolfsson, H.; Sharpless, K. B. *J. Chem. Soc., Chem. Commun.* **1997**, 1915.

34   Herrmann, W. A.; Kratzer, R. M.; Ding, H.; Glas, H.; Thiel, W. R. *J. Organomet. Chem.* **1998**, *555*, 293.

35   Herrmann, W. A.; Ding, H.; Kratzer, R. M.; Kühn, F. E.; Haider, J. J.; Fischer, R. W.; *J. Organomet. Chem.* **1997**, *549*, 319.

36   Herrmann, W. A.; Correia, J. D. G.; Rauch, M. U.; Artus, G. R. J.; Kühn, F. E. *J. Organomet. Chem.* **1997**, *118*, 33.

37   Wang, W. D.; Espenson, J. H. *J. Am. Chem. Soc.* **1998**, *120*, 11335.

38   Thiel, W. R.; Priermeier, T. *Angew. Chem. Int. Ed. Engl.* **1995**, *34*, 1737.

39   Bach, R. D.; Wolber, G. J.; Coddens, B. A. *J. Am. Chem. Soc.* **1984**, *106*, 6098.

40   Jørgensen, K. A.; Hoffmann, R. *Acta Chemica Scandinavica B* **1986**, *40*, 411.

41   Jørgensen, K. A.; Wheeler, R.A.; Hoffmann, R. *J. Am. Chem. Soc.* **1987**, *109*, 3240.

42   Jørgensen, K. A.; Swanstrøm, P. *Acta Chem. Scandinavica* **1992**, *46*, 82.

43   Jørgensen, K. A. *J. Chem. Soc. Perkin Trans.* **1994**, *2*, 117.

44   Purcell, K. F. *J. Organomet. Chem.* **1983**, *252*, 181.

45   Purcell, K. F. *Organometallics* **1985**, *4*, 509.

46   Filatov, M. J.; Shalyaev, K. V.; Talsi, E. P. *J. Mol. Cat.* **1994**, *87*, L5.

47   Szyperski, T.; Schwerdtfeger, P. *Angew. Chem. Int. Ed. Engl.* **1989**, *28*, 1228.

48   Köstlmeier, S.; Pacchioni, G.; Herrmann, W. A.; Rösch, N. *J. Organomet. Chem.* **1996**, *514*, 111.

49   Köstlmeier, S.; Häberlen, O. D.; Rösch, N.; Herrmann, W. A.; Solouki, B.; Bock, H. *Organometallics* **1996**, *15*, 1872.

50   Bagno, A.; Conte, V.; Di Furia, F.; Moro, S. *J. Phys. Chem. A* **1997**, *101*, 4637.

51   Wu, Y. D.; Lai, D. K. W. *J. Org. Chem.* **1995**, *60*, 673.

52   Wu, Y. D.; Lai, D. K. W. *J. Am. Chem. Soc.* **1995**, *117*, 11327.

53   Neurock, M.; Manzer, L. E. *Chem. Commun.* **1996**, 1133.

54   Karlsen, E.; Schöffel, K. *Catal. Today* **1996**, *32*, 107.
55   Vayssilov, G. N.; van Santen, R. A. *J. Catal.* **1998**, *175*, 170.
56   Zhidomirov, G. M.; Yakovlev, A. L.; Milov, M. A.; Kachurovskaya, N. A.; Yudanov, I. V. *Catal. Today* **1999**, *51*, 1.
57   Tantanak, D.; Vincent, M. A.; Hillier, I. H. *Chem. Commun.* **1998**, 1031.
58   Sinclair, P. E.; Catlow, C. R. A. *J. Phys. Chem.* **1999** *103*, 1084.
59   Yudanov, I. V.; Gisdakis, P.; Di Valentin, C.; Rösch, N. *Eur. J. Inorg. Chem.* **1999**, 2135.
60   Munakata, H.; Oumi, Y.; Miyamoto, A. *J. Phys. Chem. B* **2001**, *105*, 3493.
61   Di Valentin, C.; Gisdakis, P.; Yudanov, I. V.; Rösch, N. *J. Org. Chem.* **2000**, *65*, 2996.
62   Yudanov, I. V.; Di Valentin, C.; Gisdakis, P. Rösch, N. *J. Mol. Catal. A* **2000**, *158*, 189.
63   Deubel, D. V.; Sundermeyer, J.; Frenking, G. *J. Am. Chem. Soc.* **2000**, *122*, 10101.
64   Deubel, D. V.; Sundermeyer, J.; Frenking, G. *Inorg. Chem.* **2000**, *39*, 2314.
65   Deubel, D.V.; Sundermeyer, J.; Frenking, G. *Eur. J. Inorg. Chem.* **2001** 1819.
66   Deubel, D.V. *J. Phys. Chem. A* **2001**, *105*, 4765.
67   Gisdakis, P.; Antonzcak, S.; Köstlmeier, S.; Herrmann, W. A.; Rösch, N. *Angew. Chemie Int. Ed. Engl.* **1998**, *37*, 2211.
68   Wu, Y. D.; Sun, J. *J. Org. Chem.* **1998**, *63*, 1752.
69   Di Valentin, C.; Gandolfi, R.; Gisdakis, P.; Rösch, N. *J. Am. Chem. Soc.* **2001**, *123*, 2365.
70   Hroch, A.; Gemmecker, G.; Thiel, W. R. *Eur. J. Inorg. Chem.* **2000**, 1107.
71   Gisdakis, P.; Yudanov, I. V.; Rösch, N.; *Inorg. Chem.* **2001**, *40*, 3755.
72   Rösch, N.; Gisdakis, P.; Yudanov, I. V.; Di Valentin, C. In *Peroxide Chemistry: Mechanistic and Preparative Aspects of Oxygen Transfer*, W. Adam (Ed.); Wiley-VCH: Weinheim, 2000; p. 601.
73   Houk, K. N.; Liu, J.; DeMello, N. C.; Condroski, K. R. *J. Am. Chem. Soc.* **1997**, *119*, 10147.
74   Singleton, D. A.; Merrigan, S. R.; Liu, J.; Houk, K. N. *J. Am. Chem. Soc.* **1997**, *119*, 3385.
75   Lucero, M. J.; Houk, K. N. *J. Org. Chem.* **1998**, *63*, 6973.
76   Bach, R. D.; Canepa, C.; Winter, J. E.; Blanchette, P. E. *J. Org. Chem.* **1997**, *62*, 5191.
77   Bach, R. D.; Estévez, C. M.; Winter, J. E.; Glukhovtsev, M. N. *J. Am, Chem, Soc.* **1998**, *120*, 680.
78   Bach, R. D.; Glukhovtsev, M. N.; Gonzalez, C. *J. Am. Chem. Soc.* **1998**, *120*, 9902.
79   Freccero, M.; Gandolfi, R.; Sarzi-Amadè, M.; Rastelli, A. *J. Org. Chem.* **1999**, *64*, 3853.
80   Kühn, F. E.; Santos, A. M.; Roesky, P. W.; Herdtweck, E.; Scherer, W.; Gisdakis, P.; Yudanov, I. V.; Di Valentin, C.; Rösch, N. *Chem. Eur. J.* **1999**, *5*, 3603.
81   For a discussion of the chemical interpretation of Kohn-Sham orbital energies, see Görling, A.; Trickey, S. B.; Gisdakis, P.; Rösch, N. In *Topics in Organometallic Chemistry*, Vol. 4, .Brown, J.; Hofmann, P. (Ed.); Springer: Heidelberg, 1999; p. 109.
82   Gisdakis, P.; Rösch, N. *Eur. J. Org. Chem.* **2001**, 719.
83   Gisdakis, P.; Rösch, N. *J. Phys. Org. Chem.* **2001**, 14, 328.
84   Kim, C.; Traylor, T. G.; Perrin, C. L. *J. Am. Chem. Soc.* **1998**, *120*, 9513.
85   a) Becke, A. D. *J. Chem. Phys.* **1993**, *98*, 5648. b) Lee, C.; Yang, W.; Parr, R. G. *Phys. Rev. B* **1988**, *37*, 785.
86   Frisch, M. J. et al. *Gaussian 94,* Gaussian, Inc.: Pittsburgh PA, USA, **1995**.
87   Hay, P. J.; Wadt, W. R.; *J. Chem. Phys.* **1985**, *82*, 299.

88   a) Krishnan, R.; Binkley, J.; Seeger, R.; Pople, J. *J. Chem. Phys.* **1980**, *72*, 650; b)
      McLean, A.; Chandler, G. *J. Chem. Phys.* **1980**, *72*, 5639.
89   Gisdakis, P.; Antonczak, S.; Rösch, N. *Organometallics* **1999**, *18*, 5044.
90   Gisdakis, P.; Rösch, N. *J. Am. Chem. Soc.* **2001**, *123*, 697.
91   Gisdakis, P.; Rösch, N.; Bencze, É.; Mink, J.; Gonçalves, I. S.; Kühn, F. E. *Eur. J.*
      Inorg. Chem. **2001**, 981.
92   Reed, A. E.; Curtiss, L. A.; Weinhold, F. *Chem. Rev.* **1988**, *88*, 899.
93   Stomberg, R. *Ark. Kemi*, **1964**, *22*, 29–47.
94   Stomberg, R.; Ainalem, I.-B. *Acta Chem. Scand.* **1968**, *22*, 1439.
95   Schlemper, E. O.; Schrauzer, G. N.; Hughes, L. A. *Polyhedron* **1984**, *3*, 377.
96   Westland, A. D.; Haque, F.; Bouchard, J.-M. *Inorg. Chem.* **1980**, 19, 2255.
97   Mimoun, H. In *Chemistry of Peroxides*; Patai, S., Ed.; Wiley: Chichester, 1983, p. 463.
98   Adam, W.; Hass, W.; Lohray, B. B. *J. Am. Chem. Soc.* **1991**, *113*, 6202.
99   Bonchio, M.; Conte, V.; Conciliis, M. A. D.; Furia, F. D.; Ballistreri, F. P.; Tomaselli,
      G. A.; Toscano, R. M. *J. Org. Chem.* **1995**, *60*, 4475.
100   Adam, W.; Golsch, D.; Sundermeyer, J.; Wahl G. *Chem. Ber.* **1996**, *129*, 1177.
101   Neuhaus, A.; Veldkamp, A.; Frenking G. *Inorg. Chem.* **1994**, *33*, 5278.
102   Lee, J. D. in *Concise Inorganic Chemistry,* Chapman & Hall: London, 1991, p.713.
103   Herrmann, W.A.; Kühn, F.E.; Lobmaier, G.M. in *Aqueous-Phase Organometallic*
      Catalysis: Concept and Applications, Cornils, B.; Herrmann, W.A. (Eds.); Wiley-
      VCH: Weinheim, 1998, p. 529.
104   Rudolph, J.; Reddy, K. L.; Chiang, J. P.; Sharpless, K. B. *J. Am. Chem. Soc.* **1997**, *119*,
      6189.
105   Hanson, R. M. *Chem. Rev.* **1991**, *92*, 437.
106   Adam, W.; Smerz, A. K. *J. Org. Chem.* **1996**, *61*, 3506.
107   Adam, W.; Kumar, R.; Reddy, T. I.; Renz, M. *Angew. Chem. Int. Ed. Engl.* **1996**, *35*,
      880.
108   Adam, W.; Mitchell, C. M.; Saha-Möller, C. R. *J. Org. Chem.* **1999**, *64*, 3699.
109   Adam, W.; Wirth, T. *Acc. Chem. Res.* **1999**, *32*, 703.
110   Tetzlaff, H. R.; Espenson, J. H. *Inorg. Chem.* **1999**, *38*, 881.
111   Sharpless, K. B.; Michaelson, R. C. *J. Am. Chem. Soc.* **1973**, *95*, 6136.
112   Rossiter, B. E.; Verhoeven, T. R.; Sharpless, R. B. *Tetrahedron Letters* **1979**, *49*, 4733.
113   Hoveyda, A. H.; Evans, D. A.; Fu, G. C. *Chem. Rev.* **1993**, 1307.
114   Adam, W.; Mitchell, C. M. *Angew. Chem. Int. Ed. Engl.* **1996**, *35*, 533.
115   Arcoria, A.; Ballistreri, F.; Tomaselli, G.; DiFuria F.; Modena, G. *J. Org. Chem.* **1986**,
      51, 2374.
116   Mimoun, H.; Chaumette, P.; Mignard, M.; Saussine, L. *Nouv. J. Chim.* **1983**, *7*, 467.
117   Boche, G.; Möbus, K.; Harms, K. K.; Marsch, M. *J. Am. Chem. Soc.* **1996**, *118*, 2770.
118   Talsi, E. P.; Shalyaev, K. V. *J. Mol. Catal.* **1996**, *105*, 131.
119   Manohar, H.; Schwarzenbach, D. *Helv. Chim. Acta* **1974**, *57*, 1086.
120   Guilard, R.; Latour, J.-M.; Lecomte, C.; Marchon, J.-C.; Protas, J.; Ripoll, D. *Inorg.*
      Chem. **1978**, *17*, 1228.
121   Mimoun, H.; Postel, M.; Casabianca, F.; Fischer, J.; Mitschler, A. *Inorg. Chem.* **1982**,
      21, 1303.
122   Clerici, M. G.; Bellussi, G.; Romano, U. *J. Catal.* **1991**, *129*, 1; *ibid.* **1991**, *129*, 159.
123   Ledon, H. J., Varescon, F. *Inorg. Chem.* **1984**, *23*, 2735.
124   Talsi, E. P.; Babushkin, D. E. *J. Mol. Catal. A* **1996**, *106*, 179.
125   Thiel, W. R. *Chem. Ber.* **1996**, *129*, 575.
126   Thiel, W. R. *J. Mol Catal. A* **1997**, *117*, 449.

127　Wahl, G.; Kleinhenz, D.; Schorm, A.; Sundermeyer, J.; Stowasser, R.; Rummey, C.; Bringmann, G.; Fickert, C.; Kiefer, W. *Chem. Eur. J.* **1999**, *5*, 3237.

Chapter 13

The N≡N Triple Bond Activation by Transition Metal Complexes

Djamaladdin G. Musaev,[1,*] Harold Basch,[1,2] and Keiji Morokuma[1]

[1] *Cherry L. Emerson Center for Scientific Computation, and Department of Chemistry, Emory University, Atlanta, Georgia, 30322, USA;* [2] *Department of Chemistry, Bar-Ilan University, Ramat-Gan, 52900, Israel*

Abstract: The activation of the N≡N triple bond requires the coordination of N_2 molecule to transition metal centers. It is predicted that the stronger $M-N_2$ bond the easier N≡N triple bond utilization, which could occur via various ways including protonation, nucleophilic addition, hydrogenation and coordination of another transition metal center. As example, we report the density functional (B3LYP) studies of the reaction mechanism of model complex **A1**, $[p_2n_2]Zr(\mu-\eta^2-N_2)Zr[p_2n_2]$, where $[p_2n_2]= (PH_3)_2(NH_2)_2$ with a H_2 molecule. It was shown that reaction with the first H_2 molecule proceeds via 21 kcal/mol barrier at the "metathesis-like" transition state, **A2**, and produces the diazenido-μ-hydride complex, **A7(B1)**. Complex **A7(B1)** is the only experimentally observed product of the reaction **A1** + H_2 reaction, and separated by nearly 55-60 kcal/mol barriers from the energetically more (by about 40-50 kcal/mol) favorable hydrazono **A13**, $[p_2n_2]Zr(\mu-NH_2)(\mu-N)Zr[p_2n_2]$ and hydrado **A17**, $[p_2n_2]Zr(\mu-NH)_2Zr[p_2n_2]$, complexes. The addition of the second H_2 molecule to complex **A1** (the addition of the first H_2 to **A1**) take place with a 19.5 kcal/mol barrier, which is 1.2 kcal/mol smaller than that for the first H_2 addition reaction. Since the addition of the first H_2 molecule to **A1** is known to occur at laboratory conditions, one predicts that the addition of the second hydrogen molecule to **A1** should also be feasible. Furthermore, the complex **A17**, the thermodynamically most stable but kinetically not accessible product of the first H_2 addition reaction to **A1** could be obtained with the aid of the second reacting H_2 molecule. We predict that addition of the second (even third) hydrogen molecule to complex $[p_2n_2]Zr(\mu-\eta^2-N_2)Zr[p_2n_2]$, **A1** should be feasible under appropriate laboratory conditions. We encourage experimentalists to check our theoretical prediction.

Key words: Activation of the N≡N triple bond, Utilization of the N≡N bond, Nitrogen fixation

F. Maseras and A. Lledós (eds.), Computational Modeling of Homogeneous Catalysis, 325–361.
© *2002 Kluwer Academic Publishers. Printed in the Netherlands.*

1.   INTRODUCTION

Reduced nitrogen is an essential component of nucleic acids and proteins. All organisms require this nutrient for growth. The major part of all the nitrogen required in human nutrition is still obtain by biological nitrogen fixation [1]. Nature has found ways to convert the inert dinitrogen molecule to a useable form like ammonia under mild conditions using a small but diverse group of diazotrophic microorganisms that contain the enzyme nitrogenase. Nitrogenase consists of two component metalloproteins, the Fe-protein and the molybdenum iron (MoFe) protein, and converts dinitrogen to ammonia by a sequence of electron- and proton-transfer reactions:

$$N_2 + 8H^+ + 8e^- + 16MgATP \rightarrow 2NH_3 + H_2 + 16MgADP + 16P(phosphate)$$

Currently, almost all industrial dinitrogen fixation is due to the old Haber-Bosch [2] process. This operates at high temperature and pressure, and uses a promoted metallic-Fe catalyst. The reaction assumed to occur by coordination of both dinitrogen and dihydrogen on the catalyst surface, followed by stepwise assembly of ammonia from these molecules.

During the past century, chemists have been actively looking to mimic the nitrogenase and/or Haber-Bosch process using mild conditions, and to find economical and new methods of fixing atmospheric dinitrogen, which comprises 79% of the earth's atmosphere [3]. Although the thermodynamics of nitrogen fixation are favorable, realization of ammonia synthesis from the N_2 and H_2 molecules, as well as other chemical transformations of dinitrogen molecule are complicated by the kinetic stability of the N≡N triple bond. Indeed, the standard free-energy change, $\Delta G°$, for the gas-phase reaction

$$N_2(g) + 3H_2(g) \rightarrow 2NH_3 (g)$$

is found to be −7.7 kcal/mol at 298 K and 1atm pressure [1a]. The nitrogen-nitrogen triple, double, and single bond energies are 225, 100, and 39 kcal/mol, respectively. By comparison, carbon-carbon triple, double and single bond energies are 200, 146, and 83 kcal/mol, respectively [1a]. Thus, the N≡N triple bond is more significantly stabilized over N=N double and N-N single bonds than the C≡C triple bond over C=C double and C-C single bonds. These comparisons indicate that activation of the N≡N triple bond is going to be a much more difficult process than the C≡C triple bond, while N=N double and N-N single bonds could be relatively more easily activated than C=C double and C-C single bonds.

Therefore, it is not surprising that no one has yet been able to reduce the inert dinitrogen molecule catalytically under the mild conditions employed by nitrogenase. However, these numerous studies have led to discovery the new class of processes and elementary reactions, which are believed to be occurring in the nitrogenase and Haber-Bosch processes. In the early 1960's, Vol'pin and Shur [4] demonstrated that the transition metal complexes and a strong reducing agent in a non-aqueous environment could react with N_2, yielding materials which produce ammonia upon hydrolysis. At the same time, Shilov and co-workers [5] demonstrated that mixture of a V(II) (or Mo) compound with a polyphenols fixes nitrogen in an aqueous environment. Some of these processes are genuinely catalytic, but occur within a narrow range of pH, and definite mechanisms have never been determined, like the Haber-Bosch process.

Also, it is worth mentioning (a) the process of $N \equiv N$ triple bond cleavage by Mo(III) complex [6], and (b) the reactivity of coordinated dinitrogen with electrophiles, including the protons and organic free radicals [7], and coordinated [8] and free [9] dihydrogen.

Since the first step of all of these reactions is dinitrogen coordination to either the surface of the catalyst or transition metal center of the complex, let us briefly discuss the nature and importance of the $M-N_2$ interaction, and the possible coordination modes of N_2 to transition metal centers. These issues were the subjects of many discussions in the literature [10, 11] and it is commonly agreed that the interaction of the N_2 molecule with transition metal centers facilitates the activation of the $N \equiv N$ triple bond; the stronger the $M-N_2$ interaction, the easier to break the $N \equiv N$ triple bond.

The electronic configuration of the N_2 molecule is $(1\sigma)^2(2\sigma)^2(3\sigma)^2(4\sigma)^2(5\sigma)^2(1\pi)^4$, and its interaction with transition metal centers could be described in terms of donation and back donation. Since N_2 has a filled σ orbital it could be considered as a σ-donor. However, since its highest occupied σ-orbital lies much lower in energy than that, for example, of the iso-electronic CO molecule, N_2 is expected to be a much weaker σ-donor. This weak σ-donation is expected to favor a linear, η^1-fashion, coordination of N_2 to transition metal centers. N_2 also has an empty π^* orbital for back bonding. Since the empty π^* orbital is lower in energy for N_2 than that for the CO molecule, it might be expected to be more accessible. However, this is not true because this π^* orbital is equally distributed between two nitrogen atoms. Since in most dinitrogen complexes the N_2 molecule is coordinated to the transition metal center in a η^1-fashion, one may conclude that back donation is not the largest component a $M-N_2$ interaction. Of these two M-N interactions, the back donation is more important for $N \equiv N$ triple bond activation. Since the two ends of N_2 are the same, the molecule can also act as a bridging ligand between two ligands.

The M-N-N-M moiety in these complexes could be linear, $M(\eta^1-N_2)M$, (for weak back donation interaction) or zigzag, or even rhombic, $M(\eta^2-N_2)M$, depending on the strength of the π-back donation (see Figure 1).

Below, we plan to discuss only the reactivity of the coordinated dinitrogen with dihydrogen molecules.

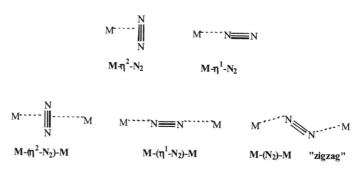

Figure 1. Possible coordination modes of the dinitrogen molecule in the mono- and dinuclear complexes

2. THE REACTION MECHANISM OF COMPLEX $[P_2N_2]Zr(\mu-\eta^2-N_2)Zr[P_2N_2]$, A1, WITH MOLECULAR HYDROGEN.

Now, let us discuss the mechanism of this fascinating reaction, namely, the reaction of the coordinated dinitrogen with molecular hydrogen via hydrogenation reported by Fryzuk et al [9]. Recently, Fryzuk and coworkers have reported [9] that the $[P_2N_2]Zr(\mu-\eta^2-N_2)Zr[P_2N_2]$, A1, complex, where $P_2N_2 = PhP(CH_2SiMe_2NSiMe_2CH_2)_2PPh$, reacts with molecular hydrogen to form $[P_2N_2]Zr(\mu-\eta^2-N_2H)Zr[P_2N_2](\mu-H)$, A2, containing an N-H bond on the bridging N_2 molecule and a metal bridging hydride. This is the first example of the reaction of a coordinated N_2 molecule with molecular hydrogen. Although the N-N bond is not split in this reaction, product A2 could be an intermediate in the process of obtaining ammonia directly from N_2 and H_2 at ambient conditions. This same study also reported that the reaction of A1 with primary silanes $RSiH_3$ results in formation of only one product, $[P_2N_2]Zr(\mu-\eta^2-N_2SiH_2R)Zr[P_2N_2](\mu-H)$.

However, these novel experimental studies left several key questions unanswered about the mechanism of the reaction of A1 with H_2. In the present paper we will report on the calculations and address some of these questions; namely, 1) what is the mechanism of the reaction of A1 with H_2, and (2) what are the structures and energies of all the equilibrium structures, intermediates and transition states on the reaction path of **A1** + H_2. In order to answer these questions we have performed [12, 13, 14] theoretical studies of the mechanisms the reaction of **A1** with H_2. Also, we will very briefly discuss the role of the auxiliary ligands of $[P_2N_2]$.

2.1   Computational Procedure

Since the explicit computation of the experimentally used complex I with the $[P_2N_2] = PhP(CH_2SiMe_2NSiMe_2CH_2)_2PPh$ tetradentate ligand, and the potential energy surface of the reaction **A1** + H_2 at a reasonably high level of theory is technically impossible, some kind of modeling of the studied complexes and reaction is needed. However, one should be aware that unreasonable modeling may lead to unrealistic results, therefore, we should be exteremely careful. Here, each $[P_2N_2] = PhP(CH_2SiMe_2NSiMe_2CH_2)_2PPh$ tetradentate ligand in the real complex will be replaced by $[p_2n_2] = (PH_3)_2(NH_2)_2$. In other words, the coordinated phosphine and nitrogen atoms of the tetradentate macrocyclic $[P_2N_2]$ on each Zr atom are conserved, while H atoms replace the $-CH_2$, $-SiMe_2$ and phenyl groups. Since the coordinating P and N atoms are not connected in the $[p_2n_2]$ model as in the actual macrocyclic $[P_2N_2]$ ligand, the structurally more rigid aspect of the extended ligand is lost. In addition, the electronic and the steric effects of the substituents are lost in the model phosphine ligands. However, the primary electronic effects of phosphine and imine ligands are retained. As each NH_2 group formally accepts one electron from Zr atom and the bridging N_2 molecule accepts four more, each Zr is formally d^0 and the electronic ground state of the model complex is a closed shell singlet state. Triplet electronic states were not examined.

The electronic structure calculations were carried out using the hybrid density functional method B3LYP [15] as implemented in the GAUSSIAN-94 package [16], in conjunction with the Stevens-Basch-Krauss (SBK) [17] effective core potential (ECP) (a relativistic ECP for Zr atom) and the standard 4-31G, CEP-31 and (8s8p6d/4s4p3d) basis sets for the H, (C, P and N), and Zr atoms, respectively.

Because the N_2 molecule in the complex is still expected to have partial multiple bond character, the d-type functions could be needed. Therefore, d-type function have been added to nitrogen atoms (using a Gaussian exponent of 0.8) of the dinitrogen molecule. Our preliminary test of the effect of phosphorus atom d-type functions on the calculated Zr-PH_3 bond lengths of model complexes **A1** and **A7** showed that the influence of d-type functions on the P atom is insignificant. Therefore, here we will not include d-type of functions for P-atoms.

However, computational studies [14] of the model $(H)_3ZrPR_3^+$ complex show that the stepwise substitution of R=H by R=Me steadily increased the Zr-PR$_3$ binding energy, which is consistent with experimental studies on the Rh and Ru complexes [18], indicating a strengthening of the Zr-P bond. Thus, the simplification of using NH$_2$ and PH$_3$ in the model [p$_2$n$_2$] ligand instead of the real macrocyclic [P$_2$N$_2$] ligand is expected to result in weaker Zr-P bonds in all the geometric structures studied here. The more rigid macrocyclic structure of the real [P$_2$N$_2$] ligand, together with the stronger Zr-N bonding, will not allow a substantial lengthening of the Zr-P bond distance. In the optimized geometric structure for complex **A1** using the [p$_2$n$_2$] ligand, the computed Zr-PH$_3$ distances were found to be too long by ~0.2Å compared to experiment [9]. We could not put substituents on the phosphines, because this would put the size of the calculations beyond current capabilities. In some of our present calculations, the weaker Zr-P bond in the model complexes and its expected effect on the calculated Zr-P distances were compensated by reoptimizing all the equilibrium geometries of the complexes studied here using a fixed Zr-P bond distance of 2.80Å. This procedure is intended to simulate the constraints of the macrocyclic ligand and their results will be discussed and interpreted alongside the unconstrained results. The external NH$_2$ groups (formally as NH$_2^-$) are expected to be more strongly bound to Zr than PH$_3$ by the electrostatic interaction and d-type functions were not used on these N atoms. The calculated Zr-NH$_2$ distances in the model complex are much closer to the X-Ray results for **A1** than the Zr-P distances.

Geometries of the reactant, intermediates, transition states (TSs) and products of the reaction [p$_2$n$_2$]Zr(μ-η^2-N$_2$)Zr[p$_2$n$_2$] (**A1**) + H$_2$ were determined by gradient optimization. Since the systems studied here were too large for the local computer resources, it was not possible to carry out second derivative calculations. In order to confirm the nature of the calculated TS, quasi-IRC (intrinsic reaction coordinate) calculations were carried out in the following manner. For one direction, the TS geometry (optimized but with a small residual gradient) was simply released for direct equilibrium (Eq) geometry optimization. Once one Eq geometry is obtained, the other Eq geometry was obtained by stepping from the TS geometry in the reverse direction, as indicated by the eigenvector of the imaginary eigenvalue of the approximate Hessian, and releasing for equilibrium optimization. In this manner, the TS and Eq geometries in the reaction path were "connected". The energies given here and discussed below do not include zero point energy correction (ZPC) or any other spectroscopic or thermodynamic terms.

This paper is organized as follows: In the next section we will discuss the structure of the initial complex, **A1**. Afterwards, a section is devoted to the study of the mechanism of the reaction of complex **A1** with one hydrogen molecule (reaction **A**). All obtained intermediates, transition states and products of this reaction will be denoted by "**A**". Another section will be concerned with the reaction mechanism of the reaction of addition of a second hydrogen molecule (reaction **B**, we label all structures related to this reaction with "**B**"). An additional section will extend our discussions to the addition of a third hydrogen molecule to the initial complex (reaction **C**), and we will indicate all structures related to this reaction with the label "**C**". In the final section, some conclusions from these studies will be presented.

2.2 The structure of the Zr(N$_2$)Zr complex, A1.

The calculated equilibrium structure of the complex **A1** is shown in Figure 2. In general, the reactant complex, **A1**, which is calculated to have a bent Zr$_2$N$_2$ unit, could have various isomeric forms with regard to the arrangement of the NH$_2$ and PH$_3$ groups relative to the bridging N-N axis. In the optimized structure shown in Figure 2, the bridging N^1-N^2 axis (of the N$_2$ molecule) is *perpendicular* (ϕ=90°) to the ligand N-N axis (of the two NH$_2$ ligands attached to each Zr atom) but is *parallel* to the P-P axis (of the two PH$_3$ ligands attached to each Zr atom). We have confirmed that this structure is energetically the most favorable under the D$_{2h}$ constraint. Another structure with the bridging N-N *parallel* (ϕ=0°) to the ligand N-N and *perpendicular* to the P-P is calculated to be 18 kcal/mol higher. This result can be understood in terms of the larger repulsion in the latter complex between the negative charges of the nitrogen atoms in NH$_2$ (-0.83e) and in the bridging N$_2$ atoms (-0.45e). An attempt to optimize a structure with the bridging N^1-N^2 axis initially bisecting the N^{ligand}-Zr-P angles (ϕ=45°) reverted to the structure **A1** presented in Figure 2.

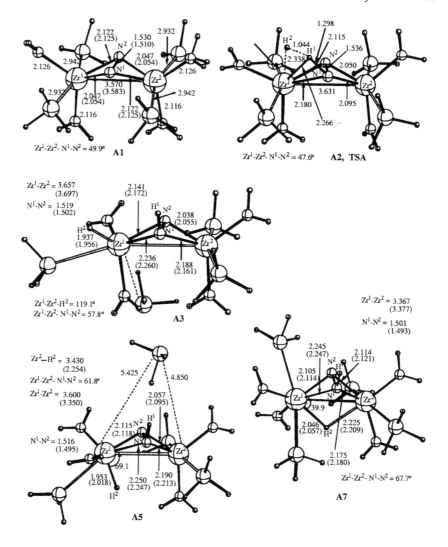

Figure 2. The calculated geometries (distances in Å and angles in deg) of the reactants, intermediates transition states and products of the reaction of $[p_2n_2]Zr(\mu\text{-}\eta^2\text{-}N_2)Zr[p_2n_2]$ with a hydrogen molecule. Numbers given in parentheses are calculated with the constraint $R(Zr\text{-}P)=2.80\text{Å}$. (Part 1 of 3)

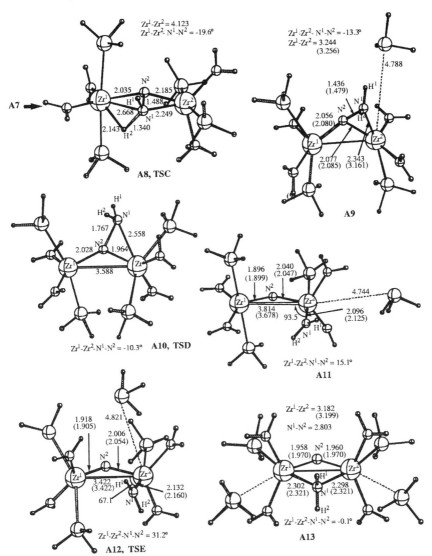

Figure 3. The calculated geometries (distances in Å and angles in deg) of the reactants, intermediates transition states and products of the reaction of [p$_2$n$_2$]Zr(μ-η^2-N$_2$)Zr[p$_2$n$_2$] with a hydrogen molecule. Numbers given in parentheses are calculated with the constraint R(Zr-P)=2.80Å. (Part 2 of 3)

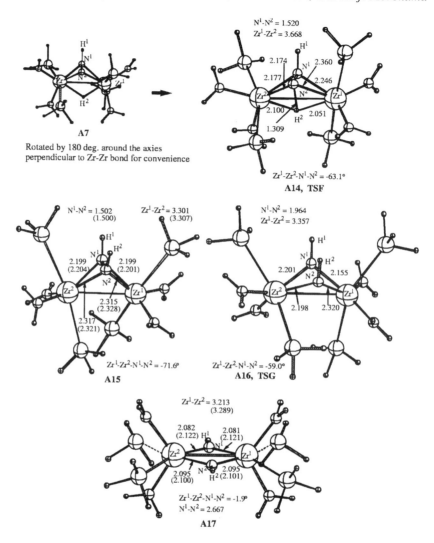

Figure 4. The calculated geometries (distances in Å and angles in deg) of the reactants, intermediates transition states and products of the reaction of [p$_2$n$_2$]Zr(μ-η^2-N$_2$)Zr[p$_2$n$_2$] with a hydrogen molecule. Numbers given in parentheses are calculated with the constraint R(Zr-P)=2.80Å. (Part 3 of 3)

In the isomers discussed above the ligands of the same type are situated *trans* to each other: N *trans* to N and P *trans* to P, which is the conformation found experimentally with the [P$_2$N$_2$] ligand. We have also calculated several possible isomers that may arise from the *cis* location of the same ligands. In these isomers, the bridging N-N axis can then bisect either the N^{ligand}-Zr-P angles or the P-Zr-P angles, or align with a N^{ligand}-Zr-P axis. In each case, the structure **A1** was found to be more stable. Thus, structure **A1** represents the most stable coordinating ligand geometry with regard to the arrangement of the external NH_2 and PH_3 ligands around each Zr atom, which is in accord with the experiment.

As noted above, **A1** is calculated to have a bent Zr_2N_2 configuration, hinged at the N-N axis, and 24.4 kcal/mol more stable than the planar Zr_2N_2 in D_{2h} symmetry. This result is contrary to the earlier theoretical studies [19] on the naked Zr_2N_2, and to the experiment on **A1** [9]. Note that the calculated bending, with the Zr^1-Zr^2-N^1-N^2 dihedral angle of 49.9° or the distance between the Zr^1-Zr^2 midpoint and the N^1-N^2 midpoint of 0.954Å, is accompanied by a substantial reduction in the N^1-N^2 (1.698Å in D_{2h} to 1.530Å in **A1**; 1.43Å), Zr-N^1 (2.077Å to 2.047Å; 2.01Å), the largest Zr-P (3.016Å to 2.942Å; 2.734Å) and Zr-Zr (3.792Å to 3.570; 3.756Å) distances (where the third number in parentheses is from the X-ray crystal data [9] for complex **A1**). The origin of the bend is probably electronic in nature, i. e. due to more extensive opportunities for orbital mixing between the Zr (4d, 5s and 5p) and molecular nitrogen (σ, σ*, π, π*) valence electrons in the lower symmetry. It is, therefore, probable that the extended macrocyclic ligands prevent the bending in the real complex **A1**. This point has been confirmed by IMOMM calculations [19] on the real complex **A1**; these calculations show that the Zr-NN-Zr core is nearly planar for the real complex **A1**, which is consistent with experimental findings. One notices in Figure 2 that the N-N distance in **A1** is 1.530Å, corresponding to a single N-N bond. In **A1**, the two π-bonds in the molecular nitrogen are completely broken, and the Zr-NN-Zr core can be considered as a metallabicycle.

2.3   Addition of the first hydrogen molecule, reaction A.

All the calculated equilibrium structures and transition states on the potential energy surfaces of the reaction **A** are shown in Figures 2 to 4. The reaction potential energy profile is shown in Figure 5.

2.3.1   Hydrogen activation by complex A1

The first step of the reaction of **A1** with the H_2 molecule is the coordination of dihydrogen to **A1**. The product of this step was interpreted as the complex $[P_2N_2]Zr(\mu-\eta^2-N_2H)(\mu-H)Zr[P_2N_2]$ on the basis of ^1H and ^{15}N NMR data [9]. However, low-temperature X-ray diffraction data [9] were interpreted in terms of a complex having a side-on bridging H_2 unit. Our calculations show that this coordination does not produce the dihydrogen complex $(H_2) \bullet \mathbf{A1}$, but leads to the diazenidohydride complex $[p_2n_2]Zr(\mu\text{-}\eta^2\text{-}N_2H)Zr[p_2p_2](\mu\text{-}H)$, $\mathbf{A3}$ (see Figure 2). The latest incoherent inelastic neutron scattering studies [20] have conclusively supported our results and showed that the dihydrogen adduct of $\mathbf{A1}$ is complex $\mathbf{A3}$. The transition state connecting the reactants with the complex $\mathbf{A3}$ is a four-center transition state, $\mathbf{A2}$, with addition of H_2 to two centers, Zr^1 and N^1, where the $H^1\text{-}H^2$ bond is broken and at the same time the terminal $Zr^1\text{-}H^2$ and the $N^1\text{-}H^1$ bonds are formed. The barrier height for the oxidative addition of H_2 to $\mathbf{A1}$ at $\mathbf{TSA}$ $\mathbf{A2}$ is calculated to be 21.6 kcal/mol. The activation barrier changed very little between the fully optimized calculation and the Zr-P constrained optimization (21.5 kcal/mol). The actual barrier will probably be lower when zero point energy differences are taken into account between $\mathbf{A1}+H_2$ and $\mathbf{A3}$. The tunneling abilities of the hydrogen atom should also be taken into account in this context. The result of all these corrections could be an effective barrier, which is smaller than the calculated value.

The product of the reaction, complex $\mathbf{A3}$, is calculated to be 13.5 and 4.9 kcal/mol exothermic relative to reactants, in the unrestricted and the Zr-P constrained geometries, respectively. The higher energy (lower stability) of the latter comes from the tendency of the two $Zr^1\text{-}P$ bonds to lengthen considerably in the unconstrained geometry to make room for the new $Zr^1\text{-}H^2$ bond. The latter bond is broken because of its eclipsed alignment with the $N^1\text{-}H^1$ bond and the higher coordination number (>6) on Zr^1. This repulsion, in turn, pushes the weakly bound phosphine groups on the same Zr atom to large, essentially dissociated Zr-P distances. The almost free motion of the phosphine groups is inconsistent with the constrained geometric structure of the real, strongly bound macrocyclic $[P_2N_2]$ ligand, which will restrict the freedom of motion of the (substituted) phosphine groups. Therefore, the constrained $\mathbf{A3}$ geometry's 4.9 kcal/mol stability is probably the more realistic value. It should be noted that the calculated Zr-P distances in $\mathbf{TSA}$, $\mathbf{A2}$, are very similar to those in $\mathbf{A1}$, so that the calculated barrier height will not be raised by constraining the Zr-P distances in $\mathbf{TSA}$, relative to $\mathbf{A1} + H_2$. In accord with its mode of formation, $\mathbf{A3}$ has the $N^1\text{-}H^1$ and $Zr^1\text{-}H^2$ bonds aligned almost parallel and protruding from the same "face" of the Zr_2N_2 core.

2.3.2   Transformation of the diazenidohydride complex A3.

The diazenidohydride complex $\mathbf{A3}$, which has the N-H and Zr-H bonds oriented parallel or on the same face of the Zr_2N_2 core, can rearrange into other conformation isomers. We have found structure $\mathbf{A5}$ in Figure 2, with the N-H and Zr-H bonds anti-parallel or on the opposite faces. In $\mathbf{A5}$, without constraint, Zr^1, which has the $Zr\text{-}H^2$ bond and is very crowded, loses one phosphine from the first coordination shell to the second coordination shell. With the Zr-P bond constraint, of course the phosphine is forced to stay in the first coordination shell. Interestingly, the interaction energy of this phosphine is weak either way, and does not have much effect on the overall energetics seen in Figure 5, **A5** is 15.6 and 12.6 kcal/mol lower than reactants **A1** + H_2 in the Zr-P unconstrained and constrained geometries, respectively. In other words, the **A3** → **A5** transformation is 2.1 and 7.7 kcal/mol exothermic at the Zr-P unconstrained and constraint geometries, respectively.

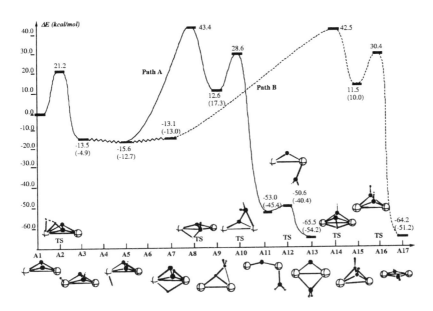

Figure 5. The calculated potential energy profile of the reaction **A1** + H_2. For clarity, the ancillary PH_3 and NH_2 ligands are omitted in the illustration. Numbers given in parentheses were obtained upon constraining the Zr-P bond distances to 2.80 Å

The search for a transition state for the **A3** → **A5** transformation is complicated. Besides the relative directions of Zr^1-H^2 and N^1-H^1 bonds, **A3** and **A5** differ by ~90° in the relative orientations of the $[p_2n_2]$ ligands on the different Zr; **A3** is N^1-H^1/Zr^1-H^2 parallel and $[p_2n_2]$ eclipsed, and **A5** is N^1-H^1/Zr^1-H^2 anti-parallel and $[p_2n_2]$ staggered. Recall that the eclipsed $[p_2n_2]$ comes from structure **A1**, which was found to be the most stable conformation theoretically and experimentally. Structure **A5** with the eclipsed $[p_2n_2]$ conformation (**A5'**) is less stable then **A5** itself, so both geometrical differences between **A3** and **A5** are real. Despite our several careful attempts, we could not find the TS connecting **A3** and **A5**. For example, an attempt involving the stepwise twisting the Zr^1-H^2 bond leads to a "transition state" with a very small negative eigenvalue, which lies only 9 kcal/mol higher than structure **A3**. This can be taken as upper limit of the barrier separating **A3** and **A5**. All attempts to search for the "real" transition state by optimizing all geometrical parameters failed and lead to the different minimum structures **A3**, **A5**, **A3'** and **A5'**, the last two stuctures are not presented in Figure 2. Therefore we have concluded that (i) the potential energy surface for the **A3** → **A5** transformation, most likely, is flat, (ii) the associated barrier is likely to be low, (iii) and the flexibility of the arrangements of the individual NH_2 and PH_3 ligands about each Zr allows multiple minima.

Another equilibrium structure **A7** was found, which had a bridging Zr^1-H^2-Zr^2 arrangement with longer Zr-H bond lengths (2.046Å, 2.225Å) than the terminal Zr^1-H^2 bond, 1.95Å, of structures **A3** and **A5**. The Zr^2-H^2-Zr^1 bridging structure **A7** can be identified with the species observed experimentally in solution [9]. In the unconstrained geometries, **A7** is less stable than **A5** by 2.5 kcal/mol and than **A3** by 0.4 kcal/mol. The energy barrier between **A5** and **A7** is expected to be very small and was not calculated. For a simplified model complex [13], $Cl_2Zr(\mu-\eta^2-N_2H)(\mu-H)ZrCl_2$, we found this barrier to be less than 1 kcal/mol.In the Zr-P constrained optimizations, however, the two structures, **A5** and **A7**, essentially merge to the very similar bridging geometry. As seen in Figure 2, a large decrease occurs in the Zr^2-H^2 distance in **A5** from 3.430Å (unconstrained) to 2.254Å (Zr-P constrained) and converges to a structure that resembles **A7**. Now structures **A7** and **A5** are nearly degenerate; the latter lies only 0.4kcal/molhigher than the former. Therefore, in the more realistic Zr-P constrained optimizations, the structures **A5** and **A7** can be considered to be a single diazenidohydride structure, the H-bridged $[p_2n_2]Zr(\mu-\eta^2-NNH)Zr[p_2n_2](\mu-H)$ structure **A7**, which agrees with the experimental finding [9]. The structure **A5** is an artifact of our unconstrained model system; one of the phosphine ligands dissociated from Zr to make space for the terminal Zr^1-H^2 bond.

If we compare the H-bridged **A7** directly to the H-terminal **A3**, we see the expected contraction of the Zr^1-Zr^2 distance due to the bridging Zr^1-H^2-Zr^2 bond which will act to hold the Zr metal atoms about 0.30Å closer in **A7** than in **A3**. The stronger Zr-Zr interaction would be expected to weaken and lengthen the Zr-N^1/N^2 bonds, but the shorter Zr-Zr distance should also act to squeeze these Zr-N distances to smaller values. The two opposing effects seem to cancel and there are no consistent trends in these Zr-N bond lengths in going from **A3** to **A7**.

2.3.3   Migration of H from Zr to N and Cleavage of the N-N Bond

From the H-bridged $[p_2n_2]Zr(\mu-\eta^2-NNH)Zr[p_2n_2](\mu-H)$ structure **A7** (or **A5**), reaction splits into two distinct branches of hydride migration, both leading to the formation of the second N-H bond and the loss of the Zr-H-Zr (or Zr-H) bond. One branch, which we call path **A**, starts from **A5** and proceeds through the H^2 migration transition state, **TSC**, **A8**, to reach a hydrazono (NH_2-N) intermediate **A9**. This is then followed by the N-N bond fission via **TSD**. The overall pathway for path A is **A5** → **TSC A8** → **A9** → **TSD A10** → **A11** → **TSE A12** → **A13**. The final structure **A13** has a completely severed N^1-N^2 bond (2.80 Å) where an N^1H_2 group and a bare N^2 atom are bridging the two Zr atoms in a symmetric geometry. The second branch, path **B**, involves migration of the bridging (Zr^1-H^2-Zr^2) hydrogen atom in **A7** to the nitrogen atom N^2 that did not carry a hydrogen atom and proceeds through a hydrazo (NH-NH) intermediate. It follows the route: **A7** → **TSF A14** → **A15** → **TSG A16** → **A17**. The N^1-N^2 bond preserved (1.50Å) in **A7** is broken (2.67Å) in the final structure **A17**, giving a planar $Zr_2(NH)_2$ arrangement. The sequence of equilibrium structures and transition states, as well as their connectivity from **A7** (or**A5**) to **A13** and from **A7**, to **A17** was obtained as described in the previous section.

Path **A** is characterized by a high initial barrier of about 60 kcal/mol at **TSC**, **A8** and a very exothermic (-65.5 kcal/mol) terminal point $[p_2n_2]Zr(\mu-NH_2)(\mu-N)Zr[p_2n_2]$ **A13**. In **A9** the NH_2 unit is already fully formed and the Zr^1-N^1 bond has been broken. The Zr^2-N^1 bond remains, but at the expense of ejecting a proximate phosphine group out to ~4.8Å. In the Zr-P constrained optimized structure for **A9**, however, the Zr^2-N^1 bond is broken, leaving a protruding NH_2 group attached only to the bridging N^2 atom. In the next step along this reaction path, the N^2-N^1 bond cleavage takes place at the **TSD** structure **A10**, leading to the intermediate **A11**. In **A11**, N^2 is located at the bridging position between two Zr atoms, while the newly formed N^1H_2 fragment is moved completely to one of the Zr atoms, Zr^2. As a result the Zr^1-N^2 bond (1.90Å) becomes shorter by 0.14Å than the Zr^2-N^2 bond (2.04Å). Later, the N^1H_2 group migrates from Zr^2, in structure **A11**, via **TSE A12** to the bridging position between two Zr atoms, leading to the formation of the N^1-Zr^1 bond in **A13**. Thus, the sequence of TS and equilibrium structures that takes **A5** to **A13** is a roundabout reaction path involving a sequence of four equilibrium and three transition state structures. The least motion, direct formation of **A13** from **A5** via a **TSC**-like structure, does not happen. The energy of **TSC** is already sufficiently high, so, as to effectively block this path. To go directly to **A13** would also require the incipient breaking of the molecular N-N bond to give an even higher saddle point energy. The actual sequence of equilibrium and transition state structures of path A apparently allows this bond breaking energy to be absorbed by other processes so that this reaction path is a cascade of energetically more stable structures (**A9** → **A11** → **A13**) and decreasing barrier heights (**TSD, TSE**) from **TSD** to **A13**. The conclusions from this path are that the direct migration of bridging hydrogen atom to a bridging NH has a high reaction barrier, which will be even higher if the N^1-N^2 bond is also broken in the process, even if the rupture eventually leads to a more stable structure.

Path **B** which leads from **A7** to **A17** has only two transition states (**TSF, A14** and **TSG, A16**) and one intermediate equilibrium structure **A15**. As with path **A**, this reaction path involves the migration of H^2 from Zr^1 to N^2 to form the metastable structure **A15**, where the N^1-N^2 bond length is essentially unchanged at ~1.50Å. The bridging nitrogen atoms in **A15** are effectively four-coordinate with the N-N bond still in place. The final step to reach the very stable **A17** structure is to stretch the N-N bond from 1.502Å in **A15** through 1.964Å (in **TSG, A16**) to the completely broken 2.667Å in **A17**. In this latter step the bent Zr_2N_2 core and the N-H bonds aligned almost perpendicular to the N-N axis, which were preserved in **A15** → **A16**, are dramatically changed, and the $Zr_2(NH)_2$ core is now completely planar with near D_{2h} symmetry. Another dramatic geometry change worthy of mention is the rotation of the incipient N^2-H^2 bond in **TSF, A14** from being anti-parallel to the existing N^1-H^1 bond (across N-N) to the parallel conformation in **A15**. This spontaneous rearrangement is apparently caused by the H^2 atom internal to the Zr_2N_2 bond being squeezed out by steric effects to the more stable external N-H position, even though this gives eclipsed N-H bonds in **A15**. As a paradigm, this barrierless rearrangement means that the H atom adding to a bridging N atom could come from a backside attack under the proper conditions. Finally, as in channel A, the large (45.6 kcal/mol) initial barrier for channel B precludes this reaction path from reactively adding a H atom to the bare N atom to form **A15**, despite the large exothermicity (-51.1 kcal/mol) for **A17** relative to **A7**.

The most important feature of the first step in both paths **A** and **B** is the very large (56-59 kcal/mol) barrier height, even though the end products are 64-65 kcal/mol more stable than reactants A1+H$_2$ or 50-51 kcal/mol more stable than **A5** and **A7**. These high barriers prevent the reaction from proceeding to form products with the second N-H bond, either on the same bridging nitrogen atom with the first N-H bond or on the second, "bare" bridging nitrogen. Thus, the dinitrogen complex $[p_2n_2]Zr(\mu\text{-}\eta^2\text{-}NNH)Zr[p_2n_2](\mu\text{-}H)$, **A7**, is the only product of the reaction of one molecule of H$_2$ with the dinitrogen complex $[p_2n_2]Zr(\mu\text{-}\eta^2\text{-}N_2)Zr[p_2n_2]$, **A1**.

In summary, we have shown that reaction of dinuclear zirconium dinitrogen complex, $[p_2n_2]Zr(\mu\text{-}\eta^2\text{-}N_2)Zr[p_2n_2]$ **A1**, (where $p_2n_2=$ $(PH_3)_2(NH_2)_2$ is a model of the experimentally studied P$_2$N$_2$ =

PhP(CH$_2$SiMe$_2$NSiMe$_2$CH$_2$)$_2$PPh), with a hydrogen molecule proceeds via: (i) the activation of H-H σ-bond via a "metathesis-like" transition state where simultaneously Zr-H and N-H bonds are formed and the H-H and one of N-N π-bonds are broken, to produce the diazenidohydride complex **A3**, [p$_2$n$_2$]Zr(μ-η^2-NNH)Zr(H)[p$_2$n$_2$], and (ii) migration of the Zr-bonded hydride ligand to a position bridging the two Zr atoms to form diazenido-μ-hydride complex **A7**, [p$_2$n$_2$]Zr(μ-η^2-NNH)Zr[p$_2$n$_2$](μ-H). The entire reaction is calculated to be exothermic by 15 kcal/mol. The rate-determining step of this reaction is found to be the activation of the H-H bond, which occurs with a 21 kcal/mol barrier.The diazenido-μ-hydride complex [p$_2$n$_2$]Zr(μ-η^2-NNH)Zr[p$_2$n$_2$](μ-H), **A7**, with only one N-H bond, which experimentally was observed in solution [9], is not the lowest energy structure in the reaction path. Complexes [p$_2$n$_2$]Zr(μ-NNH$_2$)Zr[p$_2$n$_2$], **A13** (with a bridging NH$_2$) and [p$_2$n$_2$]Zr(μ-NHNH)Zr[p$_2$n$_2$], **A17** (with two N-H units), are calculated to be more stable than the diazenido-μ-hydride complex **A7** by about 50 kcal/mol. However, these former complexes cannot be generated by the reaction of **A1** + H$_2$ at ambient laboratory conditions because of very high (nearly 55-60 kcal/mol) barriers at **A10** and **A14** separating them from **A7**. Experimentally, only **A7** is found; presumably because it is a few kcal/mol more stable than **A3** or **A5** in the Zr-P constrained calculations.

2.4 Addition of the second hydrogen molecule, reaction B.

Thus, data presented above show that only one Zr and one N center of complex **A1** are used in the initial reaction with H$_2$; one H atom is bound to the N$_2$ molecule, and the second H atom is "wasted" by forming a bond with Zr. The other N and Zr centers seem to be still available for a second H-H bond activation process. Therefore, the question can be asked whether the addition of a second molecule of H$_2$ to complex **A1** would be feasible. In order to answer to this question, we have studied the mechanism of the addition of a second molecule of hydrogen to the previously derived systems. The complexes that can serve as initial reactants for a second H$_2$ addition reaction are **A3** and **A7** described above, which are located "before" the higher barrier walls that connect to **A13** and **A17**. We have, therefore, carried out a computational experiment to explore the reaction paths for the **A3** + H$_2$ and **A7** + H$_2$ reactions.

All the calculated structures and their relative energies on the potential energy surface of the initial **B1(A7)** + H$_2$ and **B11(A3)** + H$_2$ reactions are shown in Figures 6 and 7 (where **A7** will be called **B1**), and in Figures 8 and 9 (where **A3** will be called **B11**), respectively. Below we will mainly discuss the geometries and energetics calculated with the Zr-P bonds constrained at

2.80Å. In general, the main conclusions are same for the Zr-P constrained and unconstrained calculations, and therefore, unconstrained results will be discussed only for a few specific cases.

2.4.1   Reaction of complex B1

The first step of the reaction of H_2 with **B1(A7)**, $[p_2n_2]Zr(\mu-\eta^2\text{-NNH})(\mu\text{-}H)Zr[p_2n_2]$, is coordination of a hydrogen molecule, which takes place only when H_2 approaches from above to the atoms N^2, Zr^1 and Zr^2, shown in Figure 6. This approach leads to the formation of a weakly bound molecular complex (not shown in Figure 6), (H_2)**B1**, which is only 1.2 kcal/mol more stable than reactants, H_2 + **B1**. This loose complex has a $Zr^2\text{-}H^4$ separation of ~3.4Å and a $N^2\text{-}H^4$ distance of ~2.6Å. Such a loosely bound complex would not exist when entropy is considered. Its existence/non-existence has no effect on the sequence of reaction steps described below.

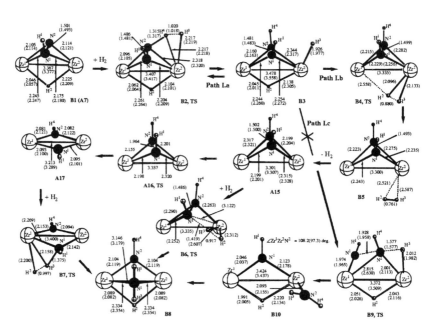

Figure 6. Calculated geometries (distances in Å and angles in deg) of the reactants, intermediates transition states and products of the reaction **B1 (A7)** + H_2. Numbers given in parentheses were obtained upon constraining the Zr-P distances to 2.80Å.

The next step of the reaction is the addition of the H^3-H^4 bond to the Zr^2-N^2 bond via transition state **B2**, shown in Figure 6. The geometric character of this TS is very similar to that of **A2**, the TS for addition of the first H_2 molecule. The active site $Zr^2 \cdots H^3 \cdots H^4 \cdots N^2$ bond distances are 2.217(2.219)Å ($Zr^2 \cdots H^3$), 1.020(1.018)Å ($H^3 \cdots H^4$), 1.315(1.317)Å ($H^4 \cdots N^2$) and 2.318(2.320)Å (Zr^2-N^2) (where the values in parentheses are calculated with the Zr-P constraint), vs. 2.266Å, 1.044Å, 1.298Å and 2.180Å, respectively, for **A2**. The similarity between the geometries of the TS for the first (**A2**) and second (**B2**) H_2 additions is reflected also in the calculated barrier heights which are ~21.2 and 19.8(19.5) kcal/mol for the reactions **A** (in Figure 5) and **B** (in Figure 7), respectively, relative to the corresponding reactants **A1** + H_2 and **B1** + H_2.

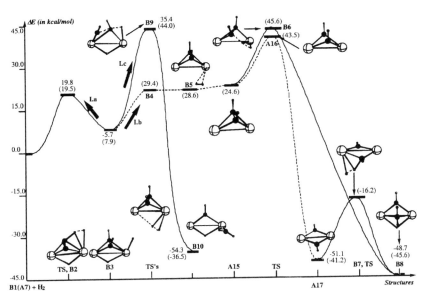

Figure 7. Calculated potential energy profile of the reaction **B1** (**A7**) + H_2, the addition of the second hydrogen molecule to the $Zr(\mu$-$N_2)Zr$ core of the complex **A1**. Numbers given in parentheses were obtained upon constraining the Zr-P distances to 2.

Quasi-IRC calculations show that TS **B2** connects the molecular complex (H_2)**B1** with the oxidative addition product **B3**, $[p_2n_2]Zr(\mu$-η^2-*cis*-HNNH)(μ-H)Zr(H)[p_2n_2]. Structure **B3** is calculated to be 5.7 kcal/mol lower in the Zr-

P unconstrained, but 7.9 kcal/mol higher in the Zr-P constrained geometry optimization, respectively, than the corresponding reactants. This difference in the energy of **B3** for different optimization schemes reflects added crowding about the Zr^2 atom in **B3**. With the new Zr^2-H^3 bond, Zr^2 becomes roughly seven-coordinated. Thus, the Zr^2-$N^1(H^1)$ bond goes from 2.181Å (**B1**) to 2.209Å (**B2**) to 2.272Å (**B3**). The Zr^2-N^2 bond, across which H^3-H^4 is adding, changes to an even larger extent, 2.122Å (**B1**) → 2.320Å (**B2**) → 2.317Å (**B3**). Again, as noted before in reaction **A1** + H_2, the bridging N-N bond distance changes by only ~0.02Å. This small decrease probably reflects a slight strengthening of the bridging N-N bond due to weakening of the Zr-N bonds, even though a new N^2-H^4 bond is formed eclipsed with the existing N^1-H^1 bond in **B3**. The results obtained by the unconstrained calculations are very close to those presented above for the constrained calculations. Exceptions are the Zr^1-H^2 and Zr^2-H^2 bond distances. In the constrained calculations, the Zr^2-H^2 bond distance expands from 2.212Å in **B1** to 2.305Å in **B3**, while the Zr^1-H^2 bond distance decreases from 2.047Å in **B1** to 2.011Å in **B3**. Thus, in **B3** the Zr "bridging" hydrogen atom becomes an almost localized Zr^1-H^2 bond, balancing the new Zr^2-H^4 bond. In contrast, in the unconstrained calculations, Zr^2-H^2 bond distance decreases from 2.223Å (**B1**) to 2.138Å (**B3**), and the Zr^1-H^2 bond distance does not change at all. The calculated differences between the unconstrained and constrained calculations are the results of the dissociation of one of the unconstrained phosphine ligands at the Zr^2-center to make room for the new Zr^2-H^3 bond.

The conformation of the approach in **B2** naturally produces intermediate **B3** with the new Zr-H and N-H bonds aligned parallel on the outwardly bent face of the Zr_2N_2 core, as in the case of **A3** and the first H_2 addition. However, the energetics of the reactions **A1**+ H_2 → **A3** and **B1** + H_2 → **B3** are slightly different; with the constrained optimization, the former is exothermic by 4.9 kcal/mol, while the latter is endothermic by 7.9 kcal/mol, which, as explained above, is due to the "crowding" around (~7 coordinate) Zr^2 in **B3**.

2.4.2   Reactions of complex B3

From intermediate **B3**, the reaction in general may proceed via three different pathways, **I.a**, **I.b** and **I.c**, which will be discussed separately.

2.4.2.1   Path I.a
Path **I.a** is a reverse reaction leading to reactants **B1** + H_2. It takes place with a 11.6 kcal/mol barrier and is exothermic by 7.9 kcal/mol.

2.4.2.2   Path I.b

The second pathway, Path **I.b**, is also a dihydrogen elimination process which involves, as seen in Figure 7, migration of the bridging H^2-ligand to Zr^2 (and H^3) through transition state **B4** to form **B5**, $[p_2n_2]Zr(\mu-\eta^2\text{-}cis\text{-}HNNH)Zr(\eta^2\text{-}H_2)[p_2n_2]$. At the transition state **B4**, H^2 has already moved to the Zr^2 center from the bridging position between Zr atoms in **B3**, and has almost formed the H^2-H^3 bond at 0.880Å. Quasi-IRC calculations from **B4** confirm that this transition state connects **B3** and **B5**. This reaction occurs with a 21.5 kcal/mol barrier and is about 21 kcal/mol endothermic relative to complex **B3** (see Figure 8). **B5** is a weak H_2 molecular complex with the $Zr(\mu-\eta^2\text{-}HNNH)Zr$ core and has a non-coplanar Zr-Zr-N-N structure, as in **B1**, with the N-N bond preserved. The H_2 molecule can dissociate easily to give complex **A15**, $[p_2n_2]Zr(\mu-\eta^2\text{-}cis\text{-}HNNH)Zr[p_2n_2]$, which has been discussed in detail above.

As shown in Figure 7, from **A15** the reaction splits into two paths. The first, which has been partially already discussed, starts with cleavage of the N-N bond via transition state **A16** and leads to the formation of $[p_2n_2]Zr(\mu-HN)_2Zr[p_2n_2]$, **A17**, with a coplanar $Zr(\mu-NH)_2Zr$ core. The process **A15** → **A17** occurs with about 20 kcal/mol barrier (for unconstrained optimization) and is exothermic by 65.8 kcal/mol. However, complex **A17** is not the energetically most favorable product either of the reaction **B1(A7)** + H_2 nor of the reaction started from complex **B3**. Indeed, complex **A17** may coordinate and activate a second hydrogen molecule (if it is still available) at transition state **B7** and lead to formation of the thermodynamically most favorable product, $[p_2n_2]Zr(\mu-NH)(\mu-NH_2)(\mu-H)Zr[p_2n_2]$, **B8**. Our quasi-IRC calculations confirmed **B7** as TS for H-H addition, which also can be seen by analyzing its geometrical parameters. Indeed, at TS **B7** the H-H bond is elongated to 0.997Å, and the formed Zr^1-H^2 and N^1-H^3 bonds shrink to 2.200Å and 1.375Å, respectively. However, according to the quasi-IRC calculations, transition state **B7** does not connect directly to reactants **A17** + H_2 with complex **B8**; instead, it connects to the weakly bound **A17**(H_2) molecular complex (not presented in Figures 6 and 7) with the complex **B8'** (not presented in Figures 6 and 7) having a $(H)Zr(\mu-NH)(\mu-NH_2)Zr$ core. The **B8'** → **B8** isomerization, corresponding to migration of H atom (atom H^2) from a terminal (coordinated to only one of Zr-centers) to a bridging (coordinated to both Zr-centers) position, is expected to be a kinetically easy process, and has been discussed above.

As seen in Figure 7, complex **B8** with the $Zr(\mu-NH)(\mu-NH_2)(\mu-H)Zr$ core is calculated to be 48.7(45.6), 43.0(53.5) and (65.6) kcal/mol lower in energy than reactants **B1(A7)** + H_2, **B3** and **A15** + H_2, respectively. Process **A17** + H_2 → **B8** is found to be 2.3 kcal/mol endothermic, but 4.4 kcal/mol exothermic in the unconstrained and constrained Zr-P calculations, respectively. Both $Zr-N^1(H)$ distances are 2.119Å, while the $Zr-N^2(H_2)$ distances are 0.23Å larger (see Figure 6), as expected from the weaker bonds to a tetrahedral N^2 atom. The bridging Zr^1-H^2 and Zr^2-H^2 bonds are both 2.082Å. As noted above, the Zr-Zr distance is short, 3.179Å, and the N-N bond is completely broken at an internuclear distance of 2.778Å. The three bridging ligands (N^1H, N^2H_2 and H^2) bring the two Zr atoms closer and the absence of a N-N bond concentrate the bonding between the bridging N atoms and the Zr atoms.

Thus, the process from **B3** with the $Zr(\mu-\eta^2\text{-}cis\text{-HNNH})(\mu\text{-H})Zr$ core + $H_2 \rightarrow$ TS **B4** $\rightarrow$ **B5** $\rightarrow$ ($-H_2$) $\rightarrow$ **A15** $\rightarrow$ TS **A16** $\rightarrow$ **A17** $\rightarrow$ ($+H_2$) $\rightarrow$ TS **B7** $\rightarrow$ **B8** occurs with about 21.5, 18.9 and 25.0 kcal/mol barriers at TS's **B4**, **A16** and **B7**, respectively.

The most important finding in this subsection is that a pathway has been found to convert **B1(A7)**, $[p_2n_2]Zr(\mu-\eta^2\text{-NNH})(\mu\text{-H})Zr[p_2n_2]$, to complex **A17** with the $Zr(\mu\text{-NH})_2Zr$ core, by the aid of the second reacting hydrogen molecule. In our previous studies (see above), where we studied reaction of $[p_2n_2]Zr(\mu-\eta^2\text{-}N_2)Zr[p_2n_2]$, **A1**, with the first H_2, and showed that, while the complex $[p_2n_2]Zr(\mu\text{-NH})_2Zr[p_2n_2]$, **A17**, is thermodynamically the most stable product of the reaction, and even more stable than the experimentally observed product **A7(B1)**, it cannot be generated during the reaction of **A1** + H_2 because of the existence of very high barrier (about 55 kcal/mol) for unimolecular rearrangement of **A7** to **A17**. However, our new results presented here point to both the kinetic and thermodynamic feasibilities of the formation of complex **A17**. Addition of H_2 to $[p_2n_2]Zr(\mu-\eta^2\text{-NNH})Zr[p_2n_2](\mu\text{-H})$, **B1(A7)**, followed by re-elimination of H_2, gives **A15** which can be converted into the thermodynamically most favorable product **A17**, with overall barriers of about 20 kcal/mol at TS **B2** (for the second H_2 addition) and at TS **A16**.

A second process that can start from **A15** is the coordination and activation of a second H_2 molecule leads to complex $Zr(\mu\text{-NH})(\mu\text{-NH}_2)(\mu\text{-H})Zr$ **B8**. This process is exothermic by about 21kcal/mol, and occurs with a 21.0 kcal/mol barrier at TS **B6**. As seen in Figure 6, at the transition state **B6** the broken H^2-H^3 bond is elongated to 0.917Å, while the formed Zr^2-H^2 and N^1-H^3 bonds are 2.312 and 3.122Å, respectively. In other words, the H-H activation takes place mainly on one of Zr-centers, followed by migration of H^3 to N^1. This localization is caused by the existence of the N-N bond in TS **B6**, compared with its absence in TS **B7**. Once again, quasi-IRC calculation shows that TS **B6** connects reactants **A15** + H_2 (more precisely, a weakly bound **A15**(H_2) complex not presented in Figures 6 and 7) with complex **B8'** (also, not presented in Figures 6 and 7) having the $(H)Zr(\mu-\eta^2\text{-NH})(\mu-\eta^2\text{-}NH_2)Zr$ core. As discussed above, **B8'** can rearrange into complex **B8** with a small energetic barrier (not studied here) via migration of the H^2 atom from a terminal to a bridging position.

Although both processes (the N-N bond cleavage leading to **A17** followed by H_2 addition to give **B8**, and the direct H_2 addition leading to **B8**) starting from **A15** occur with moderate energetic barriers and are exothermic, the process via **A17** looks slightly more favorable. Since the calculated rate-determining barriers for both processes starting from **A15** are comparable to that corresponding to addition of the first H_2 molecule to **A1**, which is known to occur under laboratory conditions, one expects that the former processes also will be experimentally feasible under appropriate laboratory conditions.

2.4.2.3   Path I.c.

The third pathway starting from **B3**, path **I.c**, involves migration of H^3 from Zr^2 to either of the N^1 or N^2 centers to form the second N-H bond, and leads to complex **B10**. The first step of this process, migration of the H^3 atom from Zr^2 to N^2 (or N^1), occurs through TS **B9**. As seen in Figure 6, there is no significant change in the Zr^2-H^3 bond length of 1.982Å in **B9** compared to its value of 1.977Å in **B3**. The H^3-N^2 distance in **B9** is 1.577Å. In this process, the bridging Zr-H^2-Zr bond lengths become more equal (2.116Å and 2.026Å, respectively) as Zr^2 begins to lose its localized Zr^2-H^3 bond in the migration process. The migration of H^3 from Zr^2 to N^2 has two other major consequences. Firstly, the Zr^2-N^1 bond is essentially broken, being elongated to 3.44Å, with a consequential shortening of the Zr^1-N^1 bond from 2.260Å in **B3** to 1.965Å in **B9**. This latter distance can be considered to be double bond length. The second structural feature is the breaking of the N-N bond, where its distance is 1.958Å in **B9** compared to the normal 1.483Å in **B3**. The breaking of the N-N bond is an interesting consequence of the increased coordination around N^2. The Zr^1-N^2 bond is also elongated to 2.630Å in **B9** from 2.163Å in **B3**, as expected from the incipiently tetracoordinate N^2 in **B9**. The Zr^2-N^2 distance actually contracts from 2.317Å in **B3** to 2.113Å in **B9** due to the bridging H^3 atom. The breaking of the Zr^2-N^1 bond in **B9**, is temporary, as we will see, and apparently serves to allow the strengthening (shortening) of the Zr^2-N^2 bond needed by the migration process. The calculated barrier for H^3 migration from Zr^2 to N^1 is a substantial 41.1(36.1)kcal/mol relative to the **B3** complex.

The quasi-IRC calculations confirm that TS **B9** connects structures **B3** and **B10**. In complex **B10**, the transfer of H^3 from Zr^2 to N^2 is complete, the Zr^1-N^2 bond is completely broken, and $N^1(H^1)$ reattaches itself to Zr^2. This reformation of the latter bond shows that its breaking in **B9** was a step to facilitate the main process. The detachment of the N^1H_2 group from Zr^2

avoids the formation of a four coordinate N^2 atom. The stability of **B10** is considerable, -54.3(-36.5) kcal/mol relative to reactants (**B1** + H_2).

At the next step, **B10** can be converted to **B8** essentially by the reattachment of N^2 to Zr^1. Additional changes involve the partial rotation of the bridging N^2H_2 group to form a tetrahedral N^2 atom with the two Zr and two H atoms, and the contraction of the Zr-Zr bond to 3.179Å in **B8** from to 3.437Å in **B10**. Complex **B8** is actually nearly exactly symmetric with an approximate reflection plane perpendicular to the Zr-Zr axis on the line of the bridging N^2-N^1-H^2 atoms. All our attempts to locate the transition state connecting **B10** and **B8** failed, suggesting that the energetic barrier separating these two structures is small. This conclusion also has been confirmed by partial geometry optimization of the expected transition state between structures **B10** and **B8**. Indeed, starting with structure **B10**, stepping the N^2-Zr^1-N^2 distance from 4.281Å downward in fixed intervals of 0.25Å, and reoptimization of all other geometry parameters at each fixed Zr^1-N^2 distance gives an energy maximum of ~3 kcal/mol(relative to **B10**) at a Zr^1-N^2 distance of 3.25Å. Equilibrium geometry optimization starting from such a "TS" structure results in complex **B10**. Stepping the Zr^1-N^2 distance down to 2.50Å and releasing the Zr^1-N^2 constraint in an equilibrium geometry optimization gives **B8**. The conclusion to be drawn from these results is that there probably is a small barrier between **B10** and **B8**, but we were not able to find the TS structure.

The results presented above and their comparison with those for the A1 + H_2 reaction indicate that addition of the second hydrogen molecule to B1(A7) should be as easy as addition of the first H_2 molecule to A1, which is known to occur at laboratory conditions. Indeed, the rate-determining barriers of the reaction sequence, **A1** + H_2 → **A3** → **A7** and **B1(A7)** → **B3** are calculated to be 21.2 and 19.8 (19.5) kcal/mol, respectively. However, the first process is exothermic by 13 kcal/mol, while the second process is endothermic by 8 kcal/mol.

From the resultant complex **B3**, the process can proceed via three different paths, **I.a**, **I.b** and **I.c**. Path **I.a**, corresponding to reverse reaction **B1(A7)** ← **B3**, is likely to be more preferable because of the lower (11.6 kcal/mol) barrier. Path **I.c** is not feasible because of high, 41.1(36.1) kcal/mol barrier. However, path **I.b** can be competitive to **I.a**; it occurs with a barrier, 21.5 kcal/mol, and leads to **A15**. From **A15** two new processes, the N-N bond cleavage leading to **A17** followed by H_2 addition to give **B8**, and the direct H_2 addition leading to **B8**, could start. Although both processes occur with moderate barriers and are exothermic, the process proceeding via the N-N bond cleavage leading to **A17** and then **B8** looks slightly more favorable. Either of these processes should be experimentally feasible under appropriate laboratory conditions. *In other words, the direct reaction of* *[p$_2$n$_2$]Zr(μ-η^2-N$_2$)Zr[p$_2$n$_2$], A1, with the first H$_2$ molecule cannot produce the thermodynamically most stable product [p$_2$n$_2$]Zr(μ-NH)$_2$Zr[p$_2$n$_2$], A17, because of the very high barrier (about 55 kcal/mol) for the unimolecular rearrangement of A7 to A17. However, product A17, can eventually be formed by the reaction of the second hydrogen molecule with A7.*

2.4.3 Reaction of complex B11

Another reasonable starting point for addition of the second H$_2$ molecule is [p$_2$n$_2$](H)Zr(μ-η^2-NNH)Zr[p$_2$n$_2$], **A3**, called **B11** below, although this complex has not been observed experimentally. **A3** is expected to quickly convert to **A7**, which has been isolated and characterized. Here, the reaction may proceed via two different paths, **II.a** and **II.b**, which differ by the direction of approach of the attacking dihydrogen (see Figures 8 and 9). The first of them, **II.a**, corresponds to coordination of the H$_2$ molecule from above (where there is no Zr-Zr bond, and the N-N bond is protruded, and consequently is easily accessible), while the second pathway, **II.b**, involves coordination of H$_2$ from below, where there is the Zr-Zr bond and the N-N bond is sheltered inside. Both of these pathways lead to complexes of the (H)Zr(μ-η^2-HNNH)Zr(H) type with two bridging N-H bonds parallel to each other. However, in the first case the N-H bonds are located *cis*, while in the second case they are *trans* to each other. Let us discuss these pathways separately.

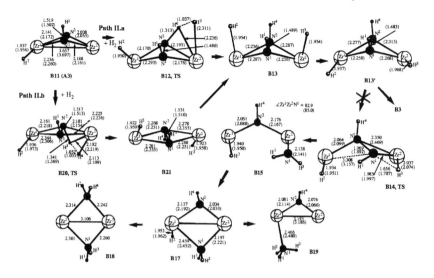

Figure 8. The calculated geometries (distances in Å and angles in deg) of the reactants, intermediates transition states and products of the reaction **B11 (C3)** + H₂, the addition of the second hydrogen molecule to the complex **A1**. For clarity, the ancillary PH₃ and NH₂ ligands are omitted in the illustration. Numbers given in parentheses were obtained upon constraining the Zr-P bond distances to 2.80 Å

2.4.3.1 Path II.a

This path starts by coordination of the H_2 molecule to **B11(A3)** from above, where addition of the H-H bond to the Zr^2 and N^2 atoms occurs. In the four-center transition state, **B12**, for this reaction, as seen in the Figure 8, the H^4-H^3 bond to be broken is elongated from 0.743Å in the free dihydrogen molecule to 1.037Å, while the forming N^2-H^4 and Zr^2-H^3 bonds are calculated to be 1.313Å and 2.311Å, respectively. Approximate IRC calculations starting from **B12** actually lead to **B11''** + H₂ in the reverse (**B12** → **B11**) direction with the constrained geometry optimization. **B11''** (not presented in Figure 8) is an isomer of **B11**, where the PH₃ ligands are located *trans* to each other in both Zr^1 and Zr^2. In the forward direction IRC calculations lead to the product complex **B13**, presented in Figure 8. The barrier height from **B11** + H₂ to **B12** is calculated to be 21.3 kcal/mol, which is only ~2 kcal/mol higher than the **B1→B2** path discussed in the previous section. In the product complex **B13** two N-H bonds are located *cis* to each other. The calculated HN-NH bond distance, 1.489Å, is very close to that in TS **B12**, (1.486Å) and reactant **B11** (1.502Å). The Zr^1-H^2 and Zr^2-H^3 distances, both 1.954Å, are also close to the Zr-terminal hydrogen bond distance, Zr^1-H^2, calculated for the reactant **B11** at 1.956Å, and for TS **B12** to be 1.950Å. The reaction **B11** → **B12** → **B13** is exothermic by 1.6

kcal/mol. Since the calculated barrier at TS **B12** is similar to those for the **A1** → **A2** (see section B.b) and **B1**→ **B2** processes, and the reaction **B11** → **B12** → **B13** is exothermic, one may expect that **II.a** is a feasible reaction path.

In the next step, H^3 migrates from Zr^2 to N^1. This step can start either from **B13** or from its energetically less stable (by 7.2 kcal/mol) isomer **B13'**, where the PH_3 ligands of the Zr^1-center are positioned *cis* (while in **B13** they are *trans*) to each other. Since this step of the reaction is not a crucial one, we studied it only for complex **B13'**.

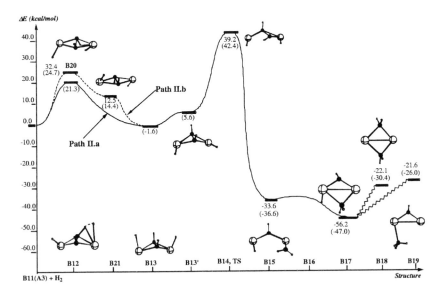

Figure 9. The calculated potential energy profile of the reaction **B11(A3)** + H_2, the addition of the second hydrogen molecule to complex **A1**. For clarity, the ancillary PH_3 and NH_2 ligands are omitted in the illustration. Numbers given in parentheses were obtained upon constraining the Zr-P bond distances to 2.80 Å.

It was found that the migration of H^3 from Zr^2 to N^1 occurs via TS **B14**, where H^3-N^1 is 1.707Å, Zr^2-H^3=2.074Å, and Zr^1-N^1 and N^2-N^1 are stretched to 3.157Å and 1.892Å, respectively. Formation of **B14** is accompanied by a breaking of the Zr^1-N^1 and N^1-N^2 bonds, while the Zr^2-N^1 bond shortens to 1.997Å. The Zr-Zr distance elongates in TS **B14** to 4.155Å, presumably due to the loss of bridging Zr-N^1-Zr bonding and the use of Zr orbitals for the breaking Zr^2-H^3 bond, which still has a bond length of 2.074Å in **B14**. The barrier height for the H^3 migration is calculated to be 36.8 kcal/mol, relative to the **B13'**. Again, as in **A8**, **A14** and **B9**, after the initial H_2 addition and possible bond preserving rearrangements, the subsequent reaction path is effectively blocked by this high barrier.

TS **B14** leads to the complex **B15**. Here all the incipient changes found in TS **B14** have completed; N^1 has two N-H bonds and is essentially localized on Zr^2 and the N-N bond is completely broken. In almost every aspect, **B15** is an analogue to **B10** of the previous reaction path. This is true even energetically; **B15** is 36.6 kcal/mol more stable than reactants ($B1 + H_2$) and 79 kcal/mol below TS **B14** in the Zr-P frozen geometries. For the unconstrained structures, the corresponding stabilities are 33.6 and 72.8 kcal/mol. The similarity in developments along the A (first H_2) and B (second H_2) reaction paths is very clear, both structurally and energetically.

In the final reaction step, the Zr^2-N^1 bond is formed to give **B17**, with N^2-H^4 and N^1-H_2 groups bridging across the two Zr atoms. The localized Zr^1-H^2 bond is preserved but rotated away from Zr^1-N^2 towards the new Zr^1-N^1 bond. The Zr-N^1 bond lengths are longer than Zr-N^2, because of the higher coordination of N^1. The Zr^1-N distances are longer than Zr^2-N because of the Zr^1-H^2 bond. The N^1-N^2 bond is completely broken at 2.864Å, while Zr^1-Zr^2 shortens to 3.364Å, characteristic of Zr-Zr distances in these complexes when the bridging N atoms are not bonded to each other. The exothermicity of **B17** is 47.0 kcal/mol in the frozen Zr-P distance geometry relative to reactants, and 9.4 kcal/mol relative to **B15**. TS **B16**, representing the incipient closing of the Zr^2-N^1 bond and rotation of the Zr^1-H^2 bond has not been found. The barrier for this process is expected to be small. We   also have calculated the products of migration of the terminal H^2-ligand in **B17** from Zr^1 to N^2H and to $N^1H^1H^3$, which are $Zr(\mu\text{-}NH_2)_2Zr$, **B18** and $(NH_3)Zr(\mu\text{-}NH)Zr$, **B19**, respectively. Since these processes are endothermic, by 17 and 21 kcal/mol, respectively, we did not explore the intermediate structures and transition states.

The above presented results show that path **II.a**, **B13** → **B13'** → **B14** → **B15** →…, is kinetically unfavorable and unlikely to proceed further. However, the alternative pathway starting from **B13** or **B13'** and leading to complex **B3**, the direct product of the **B1(A7)** + H_2 reaction, is likely to be feasible. Indeed, **B13** and **B13'** are isomers of the complex **B3**. They lie 1.4 kcal/mol lower and 4.2 kcal/mol higher than **B3**, respectively, and are probably connected by a TS for rotation of Zr^1H^2-group around the Zr^1-N bonds (possibly as seen in **B13'**), or migration of H^2 from the position *cis* to the NH bonds to the *trans* position. Although we did not study this process carefully, one may assume that **B13** can be converted to **B3**, and continue along the allowed path mapped for **B3** in Figures 6 and 7.

2.4.3.2   Path II.b

The second pathway for the reaction **B11(A3)** + H$_2$ leads to complex **B21**, where the N-H bonds are *trans* to each other. It takes the now familiar path; addition of H$_2$ across the Zr^2-N^2 bond, where the original hydrogen atoms are in the N^1-H^1 and Zr^1-H^2 bonds. TS **B20** for this reaction involves four atoms, Zr^2, N^2, H^3 and H^4, in the active regions and has the typical bond distances: N^2-H^4=1.369Å,   H^3-H^4=1.033Å,   H^3-Zr^2=2.219Å   and   Zr^2-N^2=2.154Å. The last bond length is normal, while typical equilibrium bond distances for the others are, N-H=1.035Å, H-H=0.74Å and H-Zr=1.95Å. The barrier height from **B11(A3)** + H$_2$ to **B21** is 32.4(24.7) kcal/mol at the TS **B20**. The ~7 kcal/mol difference in the barrier height is mainly due to **B11(A3)**, which is destabilized more than **B20** under the Zr-P constraint. Of course, this "destabilization" represents the constraints of the [P$_2$N$_2$] ligand and should give more realistic energetics. Thus, 24.7 kcal/mol is the barrier height, which is ~5 and 3.5 kcal/mol higher than the **B1**→**B2** and **B11**→**B12** barriers, respectively. The calculated difference is not sufficiently large to rule out this path as experimentally unfeasible.

TS **B20** leads to complex **B21**. Here, the N-H and Zr-H bonds form a relatively symmetric geometry, moderated by the eclipsed nature of pairs of N-H and Zr-H bonds. Thus, Zr^2-N^2 is 2.355Å and Zr^1-N^2 is 2.251Å. The same difference is found for the Zr^2-N^1 and Zr^1-N^1 bond lengths. It should be noted that the N^1-N^2 bond length is hardly affected in going **B11** → **B20** → **B21**; all three distances being within 0.005Å of each other. This has been found and explained for the previous H-H additions across Zr-N. The two N-H bonds in **B21** are *trans* to each other across the N^1-N^2 bond. This *trans* conformation does not allow hinging at the N^1-N^2 axis that would allow a stronger Zr-Zr bonding interaction. Therefore, the Zr$_2$N$_2$ core in **B21** is essentially planar and the Zr-Zr distance is long at 4.353Å. This distance can be compared with the corresponding value in **A3** (3.697Å) and in **B3** (3.558Å). The latter is the product of **A7** + H$_2$ and is strongly hinged at the N^1-N^2 axis, which is allowed by the two N-H bonds having the *cis* conformation.

Reaction product **B21** is calculated to be 12.5(14.4) kcal/mol above **B11** + H$_2$. This energy difference is higher than **B3** relative to **B1** + H$_2$ and **B13** relative to **B11** + H$_2$ and again indicates that **B11** → **B21** is probably less favorable than **B1** → **B3** and/or **B11** → **B13**. The approximate IRC starting from the TS actually leads to **B11'** (not presented in Figure 8) + H$_2$ in the reverse (**B20** → **B11**) direction. **B11'**(**A3'** in the previous section) closely relates to **B11** and differs from that mainly by having the Zr^1-H^2 bond rotated from being approximately parallel to N^1-H^1 to approximately perpendicular to N^1-H^1 and tilted upward towards the [p$_2$n$_2$] cluster on Zr^1. **B11'** is 1.9(1.3) kcal/mol less stable than **B11**. The structure **B11'**, more likely, is an artifact of the simple [p_2n_2] model for the macrocyclic ligand (for detail see our paper) [13, 14]. In any event, given the similarity in structure and energy of **B11** and **B11'**, we treat them as equivalent, from the point of view of the reaction path. In the forward direction (**B20** → **B21**), the unconstrained geometry optimization leads directly to **B21**, but the Zr-P frozen optimization leads to a form of **B21** (**B21'**, not shown in Figure 8) having one Zr-H bond rotated away from being eclipsed with an N-H bond towards alignment with the N^1-N^2 bond axis. **B21'** is ~6 kcal/mol more stable than **B21**, due to the relief of one eclipsed Zr-H/N-H interaction. However, in **B21'**, the orientation of the [p_2n_2] ligands on the Zr center with the rotated Zr-H bond has the two PH$_3$ groups *cis* to each other, rather than *trans* as in the real [P_2N_2] ligand. The **B21** structure, although more stable than **B21'**, would then seem to be an artificial result of the unconnected [p_2n_2] model in place of the macrocyclic [P_2N_2]. From the reaction path point of view, we treated structures **B21** and **B21'** as the same. Therefore, one concludes that the approximate IRC in both the forward and reverse directions from TS **B20** leads to **B11** (**B11'**) and **B21** (**B21'**), respectively.

The next stage of the reaction mechanism is the motion connecting **B21** with the two bridging N-H bonds *trans* to each other to **B13** with *cis* N-H bonds. **B13** is calculated to be more stable than **B21** by 16.0 kcal/mol in the Zr-P frozen structures. This stabilization can be explained in terms of the stronger Zr...Zr interaction in **B13**, which also resulted in the "bent" eclipsed conformation. Indeed, the Zr...Zr distance is found to be much longer in the staggered conformation than in the eclipsed one. The barrier(s) for **B21** → **B13** can be expected to be small. As pointed out above, in **B13** the Zr$_2$N$_2$ core is hinged at the N^1-N^2 axis. Because of this difference between **B21** and **B13** it is difficult to assess the degree of motion of the [p_2n_2] ligands, if any, in going from **B21** to **B13**. The Zr-N and N-N bond distances do not change much from **B21** to **B13**, as expected from the preservation of all the bond types. The Zr-Zr distance decreases from 4.353Å to 3.970Å (**B13**) due to the hinged bending of the Zr$_2$N$_2$ core. Thus,   the above presented results show that the addition of the second hydrogen molecule to **B11** takes place via path **II.a**, which is kinetically and thermodynamically more favorable than path **II.b**. Indeed, reaction **B11** + H$_2$ → **B13**, path **II.a**, occurs only with the 21.3 kcal/mol barrier and is endothermic by 5.6 kcal/mol. In contrast, reaction **B11** + H$_2$ → **B21**, path **II.b**, requires 24.7 kcal/mol and is endothermic by 14.4 kcal/mol. However, in both cases the reverse reactions, **B13** → **B11** + H$_2$ and **B21** → **B11** + H$_2$, respectively, which occur with only by 15.7 and 10.3 kcal/mol barriers and are exothermic (by 5.9 and 14.4 kcal/mol, respectively) will occur more easily, and the addition of the H$_2$ molecule to complex **B11(A3)** has to compete against this process.

2.5   Addition of the third hydrogen molecule, reaction C.

The complexes that can be starting points for addition of the next molecule of H_2 are **B3** (which we call **C1**), **B13(C4)** and **B21(C7)**. Because of atomic congestion, only Zr-P frozen structures were optimized for the expected TS's and products of these reactions.

2.5.1   Reaction of complex C1

Let us start our discussion from the reaction of **C1** with a dihydrogen molecule. As seen in Figure 10, the addition of H_2 to **C1** occurs across a Zr-N bond. Reactant **C1** has Zr^2-H^3, N^1-H^1, N^2-H^4 and Zr^1-H^2-Zr^2 bonds. The third H_2 molecule can then attach to the hydrogen-bare Zr^2 and one of the bridging nitrogen atoms (N^1 in our case) to form a N^1··H^6··H^5··Zr^1 ring. As seen in the TS, **C2**, active site internuclear distances have their usual values: H^5-H^6 =1.048Å, H^6-N^1=1.294Å and H^5-Zr^1=2.251Å. The last value is somewhat longer than in similar TS's discussed in earlier sections, probably because of the additional general crowding of a larger number of atoms. The Zr^2-N^1 bond is elongated to 2.722Å because of the high coordination around N^1. The Zr^1-N^1 distance is constrained by the active site ring. The Zr-Zr distance is 3.579Å, typical for systems with a bridging Zr-H-Zr bond. The N^1-N^2 bond is preserved at 1.456Å. The product of reaction **C1** + H_2 is **C3**. Each Zr atom has a Zr-H bond, but Zr^1-H^5 is longer because of stronger bonding of Zr^1 to N^1. The Zr^2-N^1 bond is broken at 3.312Å, but the N^1-N^2 bond is maintained at 1.473Å. The Zr-Zr distance is essentially unchanged in the **C1** → **C2** → **C3** transformation. The barrier height at **C2** is calculated to be a relatively high, 28.8 kcal/mol. **C3** is calculated to be 3 kcal/mol above reactants (**C1** + H_2).

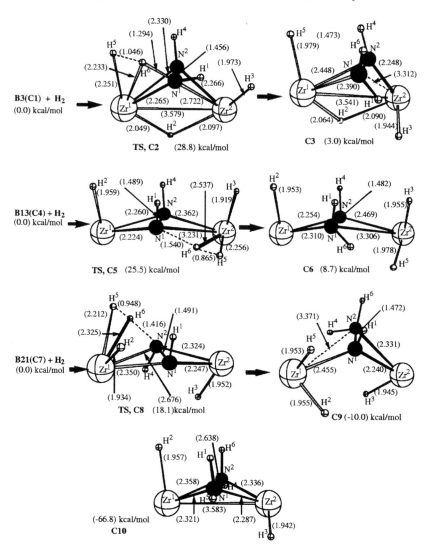

Figure 10. The calculated geometries (distances in Å and angles in deg) and energetics of the transition states and products of the reactions **B3** (**C1**) + H₂, **B13** (**C4**) + H₂ and **B21** (**C7**) + H₂, the addition of the third hydrogen molecule to the complex **A1**. For clarity, the ancillary PH₃ and NH₂ ligands are omitted in the illustration. All these results are obtained upon constraining the Zr-P bond distances to 2.80 Å.

2.5.2   Reaction of complex B13

Starting from **C4**, addition of the third dihydrogen molecule takes place, as expected, across one of the two longer Zr-N bonds. As seen in Figure 7, in TS **C5**, the dihydrogen H^5-H^6 bond being broken is elongated to 0.865Å. Meantime the forming H^6-N^1 and Zr^2-H^5 bonds are found to be 1.540Å and 2.256Å, respectively. These values of the forming and breaking bonds during the reaction clearly indicate an early (reactant-like) character of TS **C5**. At **C5**, the Zr^1-N^1 and Zr^2-N^2 bond distances have changed only slightly, while, as expected, the Zr^2-N^1 and Zr^2-N^2 bond distances changed significantly; the first of them is completely broken, while the other is elongated by 0.13Å. The N^1-N^2 bond is preserved at 1.489Å. The product **C6** has a completely broken Zr^1-N^1 bond, but the N^1-N^2 bond is preserved at 1.482Å. Zr^2 has two Zr-H bonds and Zr^1 has only one. The **C4** → **C6** reaction is calculated to be 8.7 kcal/mol endothermic. The barrier height at **C5** is calculated to be 25.5 kcal/mol.

2.5.3   Reaction of complex B21

As shown above, **C7** has the same type and number of bonds as **C4**, differs only in conformation, and is 16 kcal/mol less stable than **C4(B13)**. One may expect that the products of reaction **C7** and **C4** with another H_2 molecule also will be different from each other by conformational changes. Therefore, we will not discuss the geometries of transition state **C8** and product **C9**. Here, we would only like to point out that the **C7** + H_2 → **C8** → **C9** reaction takes place with an 18.1 kcal/mol barrier and is exothermic by 10.0 kcal/mol.

We did not investigate possible processes starting from **C3**, **C6** and **C9**. However, we have calculated complex **C10** with the (H)Zr(μ-NH$_2$)$_2$Zr(H) core, which can be a result of multiple (or single) rearrangements of complexes **C3**, **C6** and **C9**. As seen in Figure 10, the main structural features of **C10** are two bridging NH$_2$-groups and a terminal hydrogen atom on each Zr-center. The calculated N-N distance, 2.638Å, in **C10** is significantly longer than that for structures **C3**, **C6** and **C9**, and thus the N-N bond is completely broken in **C10**. **C10** is calculated to be significantly lower by energy than **C3**, **C6** and **C9**; for example, it lies about 55.1 kcal/mol lower than **C9**.

In summary, we note that, for **B3(C1)** and **B13(C4)**, the thermodynamically most favorable products of the addition of two H_2 molecule to **A1**, the calculated barrier heights of reactions with the third H_2 molecule are a few kcal/mol larger than those for the first and the second H_2 addition processes we reported above. Therefore, by adjusting the reaction conditions or by optimizing ligands, the reaction of the third H_2 molecule may be possible.

2.6   Summary

From these studies, one may draw the following conclusions:

1. The reaction of the model complex, **A1**, $[p_2n_2]Zr(\mu-\eta^2-N_2)Zr[p_2n_2]$, where $[p_2n_2]$ = $(PH_3)_2(NH_2)_2$ with a hydrogen molecule proceeds via a 21 kcal/mol barrier at the "metathesis-like" transition state, **A2**, for the H-H bond activation, and produces the diazenidohydride complex, **A3**, and diazenido-μ-hydride complex, **A7**. Complex **A7** lies a few kcal/mol lower than **A3**, and is the only observed product of the experimental analog of the calculated **A1** + H_2 reaction. However, the experimentally observed diazenido-μ-hydride complex, **A7**, is not the lowest energy structure on the reaction path. The hydrazono complex **A13** with a bridging NH_2 and the hydrado complex **A17** with two bridging NH units are calculated to be more stable than **A7** by about 40-50 kcal/mol. However, these complexes cannot be generated by the reaction of **A1** + H_2 at ambient conditions because of very high (nearly 55-60 kcal/mol) barriers at **A10** and **A14** separating them from **A7**.

2. The addition of a hydrogen molecule to **B1** (the addition of the second H_2 to **A7**) , the experimentally observed product of the reaction **A1** + H_2, can take place to give product **B3** with a 19.5 kcal/mol barrier, which is 1.2 kcal/mol smaller than that for the **A1** + H_2 → **A7** reaction. Since the addition of the first H_2 molecule to **A1** is known to occur at laboratory conditions, *one predicts that the addition of the second hydrogen molecule to A1 (or the addition of the H_2 molecule to B1(A7)) should also be feasible.*

3. From the product **B3**, which lies by 5.7 kcal/mol lower than reactants **B1** + H_2, the process will most likely proceed via either channel **I.a** (the reverse dihydrogen elimination reaction **B3** → **B1** + H_2) or channel **I.b** (another dihydrogen elimination process, **B3** → **A15** + H_2). Both processes have relatively moderate barriers, 25.5 and 35.1 kcal/mol, respectively. Path **I.c** is not feasible because of the high barrier of 41.1(36.1) kcal/mol. Later in the sequence of reactions, at complex **A15**, channel **I.b**, may split into two new pathways leading to the same product **B8** with the $Zr(\mu-\eta^2-NH)(\mu-H)(\mu-\eta^2-NH_2)Zr$ core. The first pathway, **A15** → TS(**A16**) → **A17** → TS(**B7**) → **B8**, proceeding via N-N bond cleavage leading to **A17** and then H_2 addition to give **B8**, is slightly more favorable than the second pathway of direct H_2 addition to **A15** to give **B8**. *Thus, complex A17, $[p_2n_2]Zr(\mu-NH)_2Zr[p_2n_2]$, the thermodynamically most stable but kinetically not accessible product of the reaction of*

[p₂n₂]Zr(μ-η²-N₂)Zr[p₂n₂], **A1**, *with the first H₂, is now obtained with the aid of the second reacting hydrogen molecule.*

4. Reaction of **B11(A3)** with H₂ occurs with a 21.3 kcal/mol barrier, via path **II.a** and leads to complex **B13** or **B13'**, where the N-H bond are located *cis* to each other. Later, **B13** and **B13'** which are isomers of the complex **B3**, rearranges to complex **B3**, and can follow the path allowed for **B3**.
5. Present preliminary calculations suggest that addition of the third H₂ molecule to **A1** is kinetically less favorable than the first two.

   Thus, the findings presented above and their comparison with those available from experiment indicate that addition of the second (and third) hydrogen molecule to complex *[p₂n₂]Zr(μ-η²-N₂)Zr[p₂n₂]*, **A1** should be feasible under appropriate laboratory conditions, and formation of ammonia from dinitrogen and dihydrogen molecule could be catalytic process (see Figure 11). We encourage experimentalists to check our theoretical prediction.

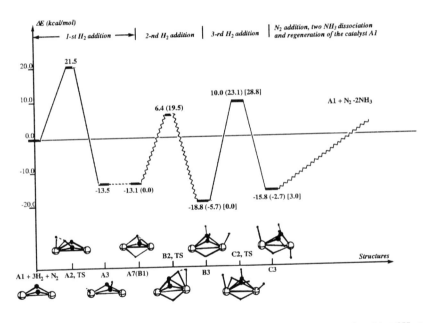

Figure 11. The calculated rate-determining barriers and energies of the reaction **A1** + 3H₂ + N₂ ↔ **A1** + N₂ − 2NH₃. Numbers given without parenthesis, and in (..) and [..] correspond to the first, second and third dihydrogen addition process, respectively.

3.   CONCLUDING REMARK

In spite of the numerous experimental and theoretical recent developments, the catalytic activation and fixation of the dinitrogen molecule still remains a challenge to chemistry. The studies presented above demonstrate that the major problem for dinitrogen activation comes from the high kinetic stability of the N≡N triple bond. While nature has found a unique way (nitrogenase) to perform this difficult task, scientists are still looking for a breakthrough in this field of the chemistry. So far, all the reactions to activate of N≡N triple bond under laboratory conditions require the coordination of a N_2 molecule to the transition metal centers to form a $M_n(N_2)$ complex. Numerous data suggest that the stronger the M_n-N_2 bonds the easier N≡N bond cleavage, which could occur via various ways including protonation, nucleophilic addition, hydrogenation and coordination of another transition metal center. However, the strong M-N_2 bond also makes dinitrogen fixation extremely difficult because of the difficulties in utilization of the formed M≡N bond. Therefore, the search for new and efficient ways for utilization of a strong M≡N bond becomes one of the important tasks. Another important and more effective way could be insertion of the N_2 ligand into a M-L bond, which requires comprehensive studies. In the solution of these problems the close collaboration of experimentalists and theoreticians is absolutely necessary.

For example, the above presented theoretical results and comparison of those with available experiment clearly indicate that addition of the second (and third) hydrogen molecule to complex $[p_2n_2]Zr(\mu-\eta^2-N_2)Zr[p_2n_2]$, **A1** should be feasible under appropriate laboratory conditions, and formation of ammonia from dinitrogen and dihydrogen molecules could be a catalytic process (see Figure 11). This conclusion should be tested by experimentalists.

REFERENCES

1   (a) Howard, J. B.; Rees, D. C. *Chem.Rev.* **1996**, *96* , 2965-2982. (b) Burgess, B. K.; Lowe, D. J. *Chem.Rev.* **1996**, *96*, 2983-3011. (c) Eady, R. R. *Chem.Rev.* **1996**, *96*, 3013-3030. (d) Eady, R. R. *Adv. Inorg. Chem.* **1991**, *36* , 77. (e) Kim, J.; Rees, D. C. *Nature* **1992**, *360*, 553. (f) Kim, J.; Rees, D. C. *Science* **1992**, *257*, 1677. (g) Dees, D. C.; Chan, M. K.; Kim, J. *Adv. Inorg. Chem.* **1993**, *40*, 89.

2   Ertl, G. *in Catalytic Ammonia Synthesis*, Jennings, J.R. Eds., Plenum, New York, 1991.

3   See: (a) Leigh, G.J. *New J. Chem.* **1994**, *18*, 157-161; (b) Leigh, G.J. *Science* **1998**, *279*, 506-508, and references therein

4   Vol'pin, M.E.; Shur, V.B. *Doklady Acad. Nauk SSSR* **1964**, *156*, 1102.

5   Bazhenova, T.A.; Shilov, A.E. *Coord. Chem. Rev.* **1995**, *144*, 69.

6   (a) Laplaza, C. E.; Cummins, C. C. *Science* **1995**, *268*, 861. (b) Laplaza, C. E.; Johnson, M. J. A.; Peters, J. C.; Odom, A. L.; Kim, E.; Cummins, C. C.; George, G. N.; Pickering, I. J. *J. Am. Chem. Soc.* **1996**, *118*, 8623. (c) Laplaza, C. E.; Odom, A. L.; Davis, W. M.; Cummins, C. C. *J. Am. Chem. Soc.* **1995**, *117*, 4999-5000.

7   See: Hidai, M.; Ishii, Y. *Bull. Chem. Soc. Jpn.* **1996**, *69*, 819-831, and references therein.

8   Nishibayashi, Y.; Iwai, S.; Hidai, M. *Science* **1998**, *279*, 540-542.

9   Fryzuk, M. D.; Love, J. B.; Rettig, S. J.; Young, V. G. *Science* **1997**, *275*, 1445-1447.

10 a) Fryzuk, M. D.; Haddad, T. S.; Rettig, S. J. *J. Am. Chem. Soc.* **1990**, *112*, 8185. (b) Fryzuk, M. D.; Haddad, T. S.; Mylvaganam, M.; McConville, D. H.; Rettig, S. J. *J. Am. Chem. Soc.* **1993**, *115*, 2782. (c) Cohen, J. D.; Mylvaganam, M.; Fryzuk, M. D.; Loehr, T. M. *J. Am. Chem. Soc.* **1994**, *116*, 9529. (d) Fryzuk, M. D.; Love, J. B.; Rettig, S. J. *Organometallics* **1998**, *17*, 846.

11 (a) Musaev, D. G. *Russ. J. Inorg. Chem.* **1988**, *33*, 3207. (b) Bauschlicher, C. W., Jr.; Pettersson, L. G. M.; Siegbahn, P. E. M. *J. Chem. Phys.* **1987**, *87*, 2129. (c) Siegbahn, P.E. M. *J. Chem. Phys.* **1991**, *95*, 364. (d) Blomberg, M. R. A.; Siegbahn, P. E. M. *J. Am. Chem. Soc.* **1993**, *115*, 6908.

12 Basch, H.; Musaev, D. G.; Morokuma, K.; Fryzuk, M. D.; Love, J. B.; Seidel, W. W.; Albinati, A.; Koetzle, T. F.; Klooster, W. T.; Mason, S. A.; Eckert, J. *J. Am. Chem. Soc.* **1999**, *121*, 523-528.

13 Basch, H.; Musaev, D. G.; Morokuma, K. *J. Am. Chem. Soc.* **1999**, *121*, 5754-5761.

14 Basch, H.; Musaev, D. G.; Morokuma, K. *Organometallics* **2000**, *19*, 3393-3403.

15 (a) Becke, A. D. *Phys. Rev. A* **1988**, *38*, 3098. (b) Becke, A. D. *J. Chem. Phys.*, **1993**, *98*, 5648. (c) Lee, C.; Yang, W.; Parr, R. G. *Phys. Rev.*, **1988**, *B37*, 785. (d) Stephens, P. J.; Devlin, F. J.; Chabalowski, C. F.; M. J. Frisch, M. J., *J. Phys. Chem.*, **1994**, *98*, 11623.

16 Frisch, M. J. et al. *Gaussian 94*, Gaussian, Inc.: Pittsburgh PA, USA, **1995**.

17 (a) Stevens, W. J.; Basch, H.; Krauss, M. *J. Chem. Phys.* **1984**, *81*, 6026. (b) Stevens, W. J.; Krauss, M.; Basch, H.; Jasien, P. G. *Can. J. Chem.* **1992**, *70*, 612.

18 (a) Serron, S.; Nolan, S. P.; Moloy, K. G. *Organometallics* **1996**, *15*, 4301. (b) Serron, S.; Luo, L.; Stevens, E. D.; Nolan, S. P.; Jones, N. L.; Fagan, P. J. *Organometallics* **1996**, *15*, 5209.

19 Yates, B.; Musaev, D. G.; Basch, H.; Morokuma, K. to be published.

20 Basch, H.; Musaev, D. G.; Morokuma, K.; Fryzuk, M. D.; Love, J. B.; Seidel, W. W.; Albinati, A.; Koetzle, T. F.; Klooster, W. T.; Mason, S. A.; Eckert, J. *J. Am. Chem. Soc.* **1999**, *121*, 523-528.

Index

σ-bond metathesis 97, 197, 204, 341, 355

[2+2] pathway 255, 260
[3+2] pathway 255, 260

1,3 hydrogen shift 145

activation of N-N triple bond 326
agostic complexes 156
agostic interactions 27, 30, 38, 43, 150,
    195, 201
alkene to carbene isomerization 149
alkyne to vinylidene isomerization 141,
    146
alkynyl hydride complex 144
allyl ligand 231, 238, 245
allylic alcohol epoxidation 305
anti-lock-and-key mechanism 110, 121
atactic polymers 47

B3LYP method 9, 30, 34, 92, 96, 98, 102,
    115, 146, 206, 260, 329
B3PW91 method 139, 150
basis set superposition error 34
benzoxantphos ligand 176
bimolecular hydrogen shift 145
bis-imine ligands 60
BisP* ligand 131
bite angle 175, 182

BLYP method 9, 42
Boltzmann distribution 178
BP86 method 9, 30, 35, 38, 40, 45, 167,
    169
BPW91 method 40
branched aldehyde 168, 180
branched polymer 68
bridging hydride ligand 338
Brookhart catalysts 57

carbene ligand 149, 269, 275
Car-Parrinello method 17, 28, 39, 249
CASPT2 method 9, 16
CASSCF method 8
CCSD(T) method 8, 30, 41, 92, 164
cinchona ligands 255
cinnamic acid 260
cobalt catalysts 163
cobalt complexes 145
combinatorial chemistry 18
copper complexes 85
counterion effects 35
coupling of CO_2 with acetylene 84
cycloaddition 270
C-H complexes 147, 151
chain growing 61, 63
chain termination 61, 65
Chalk-Harrod mechanism 224
charge decomposition analysis 129
chelation effect 36
chromium catalysts 265, 275, 295

density functional theory 6, 9, 29, 33, 37,
    163, 199, 257, 275, 293
    See also B3LYP method, B3PW91
        method, BLYP method, BP86
        method, BPW91 method
diazenidohydride complex 336
diboration 206
dihydrogen elimination 345
DiPAMP ligand 109
DIPHOS ligand 168
diphosphine ligands 109, 166, 173, 328
direct attack of peroxo 297, 301
dissociation of CO 273, 275
dissociation of phosphine 102, 195, 206
dissociation of pyrazole 228
Dötz reaction 269
DuPHOS ligand 112, 118, 124

363

Catalysis by Metal Complexes

Series Editors:
   R. Ugo, *University of Milan, Milan, Italy*
   B.R. James, *University of British Colombia, Vancouver, Canada*

1. F.J. McQuillin: *Homogeneous Hydrogenation in Organic Chemistry.* 1976
    ISBN 90-277-0646-8

2. P.M. Henry: *Palladium Catalyzed Oxidation of Hydrocarbons.* 1980
    ISBN 90-277-0986-6

3. R.A. Sheldon: *Chemicals from Synthesis Gas.* Catalytic Reactions of CO and H$_2$.
    1983   ISBN 90-277-1489-4

4. W. Keim (ed.): *Catalysis in C$_1$ Chemistry.* 1983   ISBN 90-277-1527-0

5. A.E. Shilov: *Activation of Saturated Hydrocarbons by Transition Metal Complexes.*
    1984   ISBN 90-277-1628-5

6. F.R. Hartley: *Supported Metal Complexes.* A New Generation of Catalysts. 1985
    ISBN 90-277-1855-5

7. Y. Iwasawa (ed.): *Tailored Metal Catalysts.* 1986   ISBN 90-277-1866-0

8. R.S. Dickson: *Homogeneous Catalysis with Compounds of Rhodium and Iridium.*
    1985   ISBN 90-277-1880-6

9. G. Strukul (ed.): *Catalytic Oxidations with Hydrogen Peroxide as Oxidant.* 1993
    ISBN 0-7923-1771-8

10. A. Mortreux and F. Petit (eds.): *Industrial Applications of Homogeneous Catalysis.*
    1988   ISBN 90-2772-2520-9

11. N. Farrell: *Transition Metal Complexes as Drugs and Chemotherapeutic Agents.*
    1989   ISBN 90-2772-2828-3

12. A.F. Noels, M. Graziani and A.J. Hubert (eds.): *Metal Promoted Selectivity in Organic Synthesis.* 1991   ISBN 0-7923-1184-1

13. L.I. Simándi (ed.): *Catalytic Activation of Dioxygen by Metal Complexes.* 1992
    ISBN 0-7923-1896-X

14. K. Kalyanasundaram and M. Grätzel (eds.): *Photosensitization and Photocatalysis Using Inorganic and Organometalic Compounds.* 1993   ISBN 0-7923-2261-4

15. P.A. Chaloner, M.A. Esteruelas, F. Joó and L.A. Oro: *Homogeneous Hydrogenation.*
    1994   ISBN 0-7923-2474-9

16. G. Braca (ed.): *Oxygenates by Homologation or CO Hydrogenation with Metal Complexes.* 1994   ISBN 0-7923-2628-8

17. F. Montanari and L. Casella (eds.): *Metalloporphyrins Catalyzed Oxidations.* 1994
    ISBN 0-7923-2657-1

18. P.W.N.M. van Leeuwen, K. Morokuma and J.H. van Lenthe (eds.): *Theoretical Aspects of Homogeneous Catalisis.* Applications of *Ab Initio* Molecular Orbital Theory. 1995   ISBN 0-7923-3107-9

19. T. Funabiki (ed.): *Oxygenases and Model Systems.* 1997   ISBN 0-7923-4240-2

20. S. Cenini and F. Ragaini: *Catalytic Reductive Carbonylation of Organic Nitro Compounds.* 1997   ISBN 0-7923-4307-7

21. A.E. Shilov and G.P. Shul'pin: *Activation and Catalytic Reactions of Saturated Hydrocarbons in the Presence of Metal Complexes.* 2000   ISBN 0-7923-6101-6

22. P.W.N.M. van Leeuwen and C. Claver (eds.): *Rhodium Catalyzed Hydroformylation.* 2000   ISBN 0-7923-6551-8

23. F. Joó: *Aqueous Organometallic Catalysis.* 2001   ISBN 1-4020-0195-9

24. R.A. Sánchez-Delgado: *Organometallic Modeling of the Hydrodesulfurization and Hydrodenitrogenation Reactions.* 2002   ISBN 1-4020-0535-0

25. F. Maseras and A. Lledós (eds.): *Computational Modeling of Homogeneous Catalysis.* 2002   ISBN 1-4020-0933-X

KLUWER ACADEMIC PUBLISHERS – BOSTON / DORDRECHT / LONDON

* Volume 1 is previously published under the Series Title:
Homogeneous Catalysis in Organic and iNorganic Chemistry